Applications of Adsorption and Ion Exchange Chromatography in Waste Water Treatment

Edited by
Inamuddin
Amir Al-Ahmed

The ion-exchange process is a natural phenomenon and mankind has been using this technique since the early days of civilisation. With the progress of technologies and concepts, we got a better understanding of this technique and increased its application horizon. Like in other research areas, nanotechnology has also penetrated heavily into this field, and has helped develop smart materials with better properties for application in adsorption and ion-exchange chromatography. A large amount of research was carried out in this field in the last few decades, showing the importance of these materials and technologies.

Water treatment is receiving great attention worldwide, due to the increasing demand of drinking water and hence the need to recycle polluted water sources. Keeping this importance in mind, this book "*Applications of Adsorption and Ion Exchange Chromatography in Waste Water Treatment*" has been edited with contributions from well know experts in the field, who have been working on different ion-exchange materials and technologies for many years.

Applications of Adsorption and Ion Exchange Chromatography in Waste Water Treatment

Edited by

Inamuddin[1] and Amir Al-Ahmed[2]

[1]Department of Applied Chemistry, Aligarh Muslim University, Aligarh, India.

[2]Centre of Research Excellence in Renewable Energy, King Fahd University of Petroleum & Minerals, Dhahran, Saudi Arabia

Published as part of the book series
Materials Research Foundations
Volume 15 (2017)
ISSN 2471-8890 (Print)
ISSN 2471-8904 (Online)

Print ISBN 978-1-945291-32-6
ePDF ISBN 978-1-945291-33-3

Distributed worldwide by

Materials Research Forum LLC
105 Springdale Lane
Millersville, PA 17551
USA
http://www.mrforum.com

Manufactured in the United States of America
10 9 8 7 6 5 4 3 2 1

Table of Contents

Preface

The ion-exchange process is a natural phenomenon and mankind has been using this technique since the early days of civilisation. For example, ancient societies were using some soils, sands, natural zeolites and plants as tools for improving the quality of drinking water by way of desalting or softening. However, they may have not been aware of the actual phenomenon occurring in the process. With the progress of technologies and concepts, we got a better understanding of this technique and increased its application horizon. The development of newer materials also helped this progress. Like in other research areas, nanotechnology has also penetrated heavily into this field, and has helped develop smart materials with better properties for application in adsorption and ion-exchange chromatography. A large amount of research was carried out in this field in the last few decades, sighting importance of these materials and technologies. At a glance, ion-exchange techniques and materials have their applications in various fields, such as water treatment, separation and preconcentration of metal ions, nuclear separations, nuclear medicine, synthesis of organic pharmaceutical compounds, catalysis, redox systems, electro-dialysis, hydrometallurgy, effluent treatment, ion-exchange membranes, chemical and biosensors, ion memory effect, ion-exchange fibers, ion-selective electrodes, proton conductors and so on.

Among all these application *water treatments* is receiving great attention worldwide, due to the increasing demand of drinking water and the need to recycle polluted water sources. Keeping this importance in mind, this book *"Applications of Adsorption and Ion Exchange Chromatography in Waste Water Treatment"* has been edited with contributions from well know experts in the field, who have been working on different ion-exchange materials and technologies for many years. These scientists agreed to share their research expertise as well as visions for future developments. We, the editors, would like to express our gratitude to all contributing authors, publishers, and other research groups for granting us the copyright permissions to use their illustrations.

One of the editors, Dr. Amir Al-Ahmed is thankful to the director of Centre of Research Excellence in Renewable Energy at King Fahd University of Petroleum & Minerals, Saudi Arabia for his continuous support. Finally, we would like to acknowledge the sincere support of Mr. Thomas Wohlbier of Materials Research Forum LLC in evolving this book into its final shape.

Inamuddin
Department of Applied Chemistry
Aligarh Muslim University
Aligarh, India.

Amir Al-Ahmed
Centre of Research Excellence in Renewable Energy
King Fahd University of Petroleum & Minerals
Dhahran, Saudi Arabia.

Chapter 1

Remediation of dyes from industrial wastewater using low-cost adsorbents

Abu Nasar*, Sadia Shakoor

Department of Applied Chemistry, Faculty of Engineering and Technology, Aligarh Muslim University Aligarh – 202 002, India

* abunasaramu@gmail.com

Abstract

Water pollution is one of the leading environmental issue triggering serious problems for living organisms. Various toxic substances are being introduced into the water bodies from various human activities and the removal of such substances from the water and wastewater has become a core interest for scientists and researchers around the globe over the past few decades. Dyes are an important class of water pollutants having various negative impacts on the living organisms. The removal of color from dye-bearing effluents is a major problem due to the difficulty in treating such wastewaters by conventional treatment methods. Although a number of successful physical, chemical, and biological processes have been applied, however, the cost-effective removal of color from effluents remains a challenge for wastewater treatment.

Keywords

Environmental Pollution, Water Pollution, Wastewater Treatment, Dyes, Adsorption, Low-cost Adsorbent

Contents

1.   Introduction

Since the middle of the 19th century, things have started happening in quite disproportions putting the ecological system out of balance. The population explosion, prosperous society with an aspiration for a vast array of products, automobiles, greater energy use leading to increased radiations, increase in food production needs, etc. are some of the potent factors responsible for creating this imbalance. Science and technology brought in this revolutionary change for human life. Modernization made man's life more and more comfortable. Today one can travel faster, speak or send a message to distant places through our modern means of communication. Villages have become growing cities as a result of industrialisation. It was the industrial revolution that gave birth to environmental pollution as we know it today.

One of the greatest problems that the world is facing today is the environmental pollution which is increasing with every passing year and causing severe damage to our earth. Today, environmental pollution is occurring on a vast and unprecedented scale globally. The word "Pollution" is derived from a Latin word "Polluere" which means "to defile" or "to make dirty". Pollution is an undesirable change in the physical, chemical or biological characteristics of air, land and water that makes the environment unhealthy to live and creates potential health hazards to living organisms.

The Royal Commission on Environmental Pollution in U.K. in its third report [1] defined the term "Pollution" as "Introduction by man into the environment of substance or energy liable to cause hazards to human health, harm to living resources and ecological systems, damage to structure or amenity or interference with legitimate uses of the environment."

In fact, environmental pollutants are the executing agents of environmental pollution. A pollutant is a substance or energy introduced into the environment that has undesirable effects or adversely affects the usefulness of a resource. Environmental pollutants can be classified into two types – *Biodegradable pollutants* and *non-biodegradable pollutants*.

Biodegradable Pollutants: Biodegradable pollutants are the ones that can be broken down into simpler, harmless substances in due course of time by the action of microorganisms (like bacteria). For example, domestic wastes, faecal matter, vegetable wast, etc.

Non-biodegradable Pollutants: Non-biodegradable pollutants cannot be broken down into simple and harmless substances by the action of microorganisms. For example, plastics, polythene bags, pesticides, glass, heavy metals, etc.

2. Forms of environmental pollution

The environmental pollution may be classified into different major forms such as air, water, soil, noise pollutions, etc.

2.1 Air pollution

Air pollution is the introduction of particulates, biological fragments and other harmful materials into the earth's atmosphere resulting in diseases, allergies and damage to the natural environment. Sources of air pollution can be natural or anthropogenic. Air pollutants can be classified as *primary* and *secondary.*

Primary air pollutants are introduced into the environment by natural sources and human activities as well. Some common examples of primary pollutants are given below.

Carbon monoxide (CO): It is produced by the incomplete combustion of fuels such as natural gas, wood or coal, vehicular emission, cigarette smoking, etc. It is also generated naturally by the plants during the oxidation of methane (swamps, bogs, rice paddies, etc.) and decomposition of chlorophyll.

Sulphur oxides (SO_x): Sulphur compounds are commonly emitted into the atmosphere during volcanic eruptions. Sulphur oxides are sprayed out as aerosols from the sea during rough weather. These oxides are also produced during the biogenic decomposition of sulphur-containing organic compounds on land and in sea. Anthropogenic sources of SO_x are burning of fossil fuels (coal, petroleum and its products), smelting of sulphide ores, manufacture of sulphuric acid, etc.

Nitrogen oxides (NO_x): They are produced during thunderstorms and by soil microorganisms. These are also formed by fossil-fuel based power plants and vehicles.

Toxic metals: Arsenic, cadmium, lead, mercury, chromium, etc. and their compounds are released into the air by industrial and other activities.

Volatile organic compounds (VOC): Xylene, toluene, methane, 1,3-butadiene, benzene, formaldehyde, etc. are examples of VOC. Sources of VOC include paints, varnishes, air fresheners, fuel oil, cleaning products, etc.

Particulates: These are the tiny particles of solid or liquid suspended in air. They originate from agricultural operations, dust storms, forest and grassland fires, volcanoes, burning of fossil fuels, construction and demolition activities, vehicles and aerosols.

Ammonia: Major source of ammonia in the atmosphere is the decay of organic matter and fertilisers.

Chlorofluorocarbons (CFCs): These are released into the atmosphere from air conditioners, refrigerators, aerosol, etc.

Secondary air pollutants: These pollutants are not emitted directly into the atmosphere. They are formed in the air when primary air pollutants interact with each other.

Ground-level ozone: Ground level ozone is formed due to the reaction of NO_x, carbon monoxide and VOC in the presence of sunlight.

Smog: The term 'smog' is derived from two words, namely, smoke and fog. Two different types of smog are caused by the low-pressure gradient weather patterns depending on the kind of emissions and the intensity of radiation. The **London smog** is formed in winters consisting of a mixture of gaseous and solid aerosols and natural fog. It results from the accumulation of smoke from coal burning having high sulphur content leading to the production of high concentrations of sulphuric acid in fog droplets. On the other hand, the **Los Angeles smog** is formed during sunny summer days in regions where automobile emission is very high. It is formed as a result of interactions among nitrogen oxides, reactive hydrocarbons and sunlight.

2.2 Water pollution

Water pollution is the contamination of water bodies by the release of industrial wastes, domestic sewage, chemical contaminants, urban and agricultural runoff containing fertilisers and pesticides, eutrophication, littering, etc. Water pollution will be discussed in detail in Section 3.

2.3 Soil pollution

Soil pollution is the degradation of land in the presence of xenobiotic chemicals or alteration in the natural environment of the soil. It is typically caused by domestic, agricultural and industrial activities and improper disposal of wastes. Some common and harmful soil pollutants are pesticides (herbicides, fungicides, insecticides, etc.), heavy metals (lead, arsenic, mercury, copper, cadmium, etc.), petroleum hydrocarbons, organic solvents and polynuclear aromatic hydrocarbons (naphthalene, benzopyrene, etc.). Crops and plants are grown on polluted soil act as media for the transportation of the pollutants to living organisms. Long-term exposure of living bodies to such soil can affect their genetic make-up, causing congenital illnesses and chronic health problems that cannot be cured easily. Fungi and bacteria (found in the soil binding it together) begin to decline to create additional problem of soil erosion. The fertility of soil slowly diminishes, making land unsuitable for agriculture and any local vegetation to survive. The death of many soil organisms (e.g. earthworms) in the soil leads to an alteration in soil structure.

2.4 Noise pollution

Noise pollution refers to the presence of such levels of noise or sound in the environment that are disturbing, irritating and annoying to living beings. Noise is measured in decibels (dBA). In daily life, people are generally exposed to noise levels ranging from 30–80 dBA. Exposure to noise level greater than 80 dBA leads to stress [2]. Major sources of noise pollution are industries, transport vehicles, household (televisions, domestic gadgets, air conditioners, vacuum cleaners, etc.), public addressing systems (loudspeakers), agricultural machines (tractors, thrashers, tube wells, etc.), defense equipment (artillery, tanks, explosives, etc.), construction works, etc. Noise pollution causes uneasiness and harm living being's mental and physical health. It is one of the major causes of deafness. Noise pollution also leads to cardiac disturbance, sleeplessness, headache, irregular blood pressure psychological imbalance, etc.

3.   Water pollution

Water is one of the world's most precious resources without which life is not possible on earth. As stated by philosopher Thales from Miletus, "*Hydor (Water) is the beginning of everything*". Thales understood that water is life and living organisms cannot survive without it. Over two-thirds of earth's surface is covered with water and less than one-third is taken up by land. As the earth's population is growing day by day, people are putting ever increasing pressure on the earth's water resources. In a sense, our oceans, rivers and other water resources are being squeezed by human activities and also reducing its quality. Today contamination of freshwater systems with a wide variety of pollutants is a subject of great concern.

Water pollution can be defined as the alteration of the physical, chemical or biological characteristics of water which makes it harmful for living organisms and unsuitable for the desired usage. Water pollution is a major global problem. It may be caused by natural sources or human activities and can have detrimental effects on aquatic ecosystems as well as other living organisms.

4.   Wastewater treatment

Before the industrial revolution, the natural purification phenomenon was sufficient to provide water of high purity. However, excessive human interference with the environment has pushed the natural purification processes beyond their limits. Hence, a series of purification operations are required to restore the natural qualities of water. Various water treatment plants are developed and used to purify the water before it is discharged from the industries into the freshwater systems. Water is purified in following four successive stages: preliminary, primary, secondary and tertiary treatments [3, 4].

4.1 Preliminary treatment

The objective of the preliminary treatment is the removal of suspended coarse solids and other large floating materials often found in wastewater. Removal of these materials is necessary to reduce the maintenance and treatment cost of subsequent treatment units. Suspended matter is removed by *screening* whereas the floating matter is eliminated by *skimming.*

In screening, impure water is allowed to pass through screens made of rows of iron bars with a spacing of 1–2 inches. Materials like rags, sticks, polyethene bags, wood pieces, papers, etc. are held back by the spacing in the iron bars. Impurities which are lighter than water such as oil, grease, etc. rise to the surface of the water and can be removed by mechanical skimming. Skimming technique can also be used to remove grit particles by blowing compressed air in polluted water. The air bubbles that are formed attach themselves to grit particles and lift them to the surface from where they can be skimmed off easily.

4.2 Primary treatment

In this stage, colloidal and suspended matters are allowed to settle down as *sludge*. It involves two methods – *sedimentation* and *flocculation.*

Some solids suspended in water are either too fine to be screened out or too heavy to be skimmed off. Such impurities are removed by allowing them to settle down under the influence of gravity. This technique is called *sedimentation* and is carried out in sedimentation tanks. Polluted water is allowed to remain in tanks for 1–3 hours. The sludge thus formed undergoes putrefaction and hence should not be left in the sedimentation tanks for too long. The tanks have mechanical gears to remove sludge at regular intervals.

To get rid of extremely fine particles, the *flocculation* method is used. Such fine particles take a very long time to settle down and hence cannot be removed by sedimentation. However, their rate of settlement can be enhanced up to a considerable extent by adding some flocculating agents such as potash alum, ferrous sulphate, ferric chloride, etc. These fine suspended particles in polluted water bear either positive or negative charge. When flocculating agents are added, positive charges are neutralised by sulphate or chloride ions whereas negative charges are neutralised by Al^{3+}, Fe^{2+} or Fe^{3+} ions. Once the charges are nullified, the particles come in contact with each other, coalesce and rapidly settle down.

4.3   Secondary treatment

Secondary treatment involves the oxidation of dissolved and colloidal organic compounds in the presence of microorganisms. The organic compounds are biodegraded into simpler and harmless compounds with the consumption of oxygen. The aerated conditions are obtained by *trickling filters*, *activated sludge tanks* or by *oxidation ponds*.

Trickling filters consist of circular or rectangular beds packed with stones, gravel, etc. which serve as a habitat for bacteria, fungi and other microorganisms. A part of the sludge from the primary settling tank is applied to the bed from above and polluted water is allowed to pass over these beds. Gradually with time, a biotic community is established as a gelatinous layer on the surface of the bed. This layer contains bacteria, fungi, algae, etc. When water trickles through this biological layer, the organic impurities are broken down in the presence of dissolved oxygen into simpler compounds. Also, the microorganisms already present in water are retained on the beds and supplement the purification phenomenon.

In *activated sludge method*, water containing organic pollutants is aerated and some amount of sludge settles down. Soon microorganisms start inhabiting the sludge. Next, when a fresh batch of polluted water is aerated over this sludge, it gets purified more efficiently than the previous one. More and more sludge accumulates and a greater density of microorganisms settles on it. Hence, after each purification slot, the sludge becomes more activated. After each slot, the aeration time is reduced because the purification is achieved in shorter duration.

Oxidation ponds are used in warmer climates to purify polluted water through an interaction between bacteria and algae. Polluted water is made to flow through the ponds (shallow lagoons with an average depth of 1 meter) at a slow speed. The bacteria in the pond decompose the biodegradable organic matter with the consumption of dissolved oxygen generating carbon dioxide, nitrates and phosphates. These nitrates and phosphates are consumed by algae and carbon dioxide is utilised in photosynthesis liberating oxygen. Thus, the dissolved oxygen which was consumed by bacteria is restored and a fresh cycle can be again started.

4.4   Tertiary treatment

Tertiary treatment of water removes the impurities that remain after the first three stages of purification. These impurities are mainly the soluble inorganic impurities. Different techniques are used in this stage depending on the nature of pollutants.

Chlorination refers to the addition of gaseous chlorine or compounds containing active chlorine such as bleaching powder. It disinfects the pathogenic bacteria by inactivating the enzymes that are essential for the life processes of bacteria. It controls the growth of undesirable algae in water treatment plants. It also eliminates the odours associated with the anaerobic decomposition of organic matter in water.

The impurities present in water are oxidised to harmless materials using suitable oxidising agents such as hydrogen peroxide. The hydroxyl radical generated by the action of ultraviolet light on H_2O_2 attacks the inorganic and organic pollutants. The sulphur compounds are oxidised to sulphates, phosphorus compounds to phosphates, cyanides to cyanates and halogen compounds to halides. This method is known as *wet oxidation*.

In *reverse osmosis*, impure water is placed above a semi-permeable membrane and is subjected to high pressure. As a result, pure water flows down the semi-permeable membrane and the solute molecules having larger diameters than the pores of the semi-permeable membrane are retained by the membrane.

The *Electrodialysis* technique is used to purify brackish water (water containing a higher concentration of ionic impurities). The equipment consists of an electrolytic cell divided into three compartments by two semi-permeable membranes. The orifices on the membrane near the cathode are coated with negatively charged ions. This membrane repels anions but attracts cations. Similarly, the orifices of the membrane near anode are coated with positively charged ions which repel cations but attract anions. When current is induced, cations present in impure water pass through the cation-permeable membrane and discharge at the cathode. Similarly, anions present in impure water pass through the anion-permeable membrane and discharge at the anode. The central compartment is devoid of any impurity and contains pure water.

During the tertiary stage of water purification, about 1 ppm of fluoride is added, either in the form of sodium fluoride or as sodium hexafluorosilicate (IV). This process is called *fluoridation*. Fluoride is essential to protect teeth against dental decay.

The *ion-exchange technique* is also used for the purification of brackish water. The equipment consists of a cation exchanger coupled with an anion exchanger. Each ion-exchanger contains a resin (high molecular weight polymeric material with ionic functional groups). As water containing ionic impurities is passed over these ion exchangers, the functional groups can exchange ions. Impure water is first passed through cation exchanger. The cations present in impure water are exchanged with the H^+ ions of the resin (cation exchanger contains ionic sulphonic acid group). When water is passed through an anion exchanger (containing quaternary ammonium hydroxide group), the anions present in the impure water are exchanged with the hydroxide ions of the resin. The H^+ ions and OH^- ions combine to form water.

Several compounds which are resistant to biodegradation may persist in water even after primary and secondary treatment. *Adsorption* technique is used for the elimination such impurities (heavy metals and dyes) from water by passing the water over a bed of granulated active charcoal. When the charcoal becomes saturated with impurities, it is heated at about 1800 K in vacuum to remove impurities from the used adsorbent so that it can be used again.

5.   Dyes

A dye is a coloured substance that has an affinity for the substrate to which it is being applied. The phenomenon of absorption of light is important for the sensation of colour. However, it was recognised in 1870 that unsaturation is essential for the light absorption and hence colour sensation. Otto Witt, a German chemist, put forward his theory of colour and constitution and is known as a *chromophore-auxochromes theory*. The main points of this theory are:

1.   The colour of an organic compound is mainly due to the presence of unsaturated groups known as *chromophores* (German: *Chroma* means colour and *protein* means to bear). The compound bearing the chromophoric group is called *chromogen* [5, 6]. For example, azo (–N=N–), carbonyl (C=O), methine (–CH=), and nitro (–NO$_2$) groups, etc.

2.   The greater the number of chromophores, the greater is the intensity of colour.

3.   Certain substituents fail to produce colour by themselves but they deepen the colour due to the chromophoric group already present. Such groups are called *auxochromes* (German: *auxanein* means to increase). Common auxochrome groups include hydroxyl (–OH), amino (–NR$_2$, –NHR, –NH$_2$) groups, halogens, etc.

5.1   Classification of dyes

Dyes can be classified in several ways. Some bases of classification are given below:

- On the basis of source of dyes
- On the basis of nature of chromophore
- On the basis of application of dyes

(a) Classification based on source of dye

On the basis of the source from which the dye is obtained, dyes can be classified as *natural* and *synthetic*.

Natural dyes: These dyes are obtained from natural sources. Most of the natural dyes have plant origin (extracted from leaves, flowers, roots, berries, roots, bark, etc.). Some natural dyes are also obtained from insects and mineral compounds. For example, indigo, madder, saffron, osage orange, safflower, madder, etc. Natural dyes are of two types-*additive* and *substantive* dyes. Additive dyes require a mordant to fix to the fiber (for example, madder). Substantive dye does not require the use of any mordant (for example, cochineal and safflower).

Synthetic dyes: Dyes prepared from organic or inorganic compounds are known as synthetic dyes. While searching for a remedy for malaria, William Henry Perkin, an English chemist, serendipitously discovered the first synthetic dye, Mauveine. Since then thousands of different synthetic dyes were manufactured and used due to their low cost and colour fastness. Based on this classification, textile dyes can be categorised into various groups, namely, direct, vat, sulphur, organic pigments, reactive, dispersed, acidic, azoic, basic, oxidative, developed, mordant, and solvents dyes.

Table 1. Classification of dyes based on chromophoric group

Dye category	Chromophoric group	Representative example	Colour
Acridine dyes		Acridine orange	Orange
Anthraquinone dyes		Alizarin	Crimson
Arylmethane dyes		Auramine O	Yellow

		\newline Crystal violet	\newline Violet
Azo dyes	–N=N–	\newline Aniline yellow	\newline Yellow
Nitroso dyes	–N=O	\newline Naphthol green B	\newline Green
Xanthene dyes		\newline Rhodamin 6G	\newline Yellow fluorescent
Indophenol dyes		\newline Indophenol blue	\newline Blue

(b) Classification based on the nature of chromophore

On the basis of the nature of the chromophoric group present in the dye molecule structure, they can be classified into various categories which are listed in Table 1.

(c) Classification based on the application of dyes

On the basis of application, dyes can be classified into 9 categories:

Reactive dyes: These dyes have reactive groups which form covalent bonds with –OH, – NH_2 or –SH groups on the fiber. Reactive dyes are extensively used in the textile industry because of a wide variety of colour shades and ease of application. The hydrolysis of the reactive groups in the side reaction lowers the degree of fixation. It is estimated 10–50% does not react with fabric and remain hydrolysed in the water phase. The problem of coloured effluents is, therefore, identified to be mainly because of the use of reactive dyes. For example, Reactive yellow HE6G, Reactive red, Reactive violet C2R, Reactive golden yellow MR, Reactive yellow MGR, etc.

Acid dyes: These are anionic compounds and used for dyeing basic group-containing fabrics like wool, polyamide and silk. These are applied to the fabric under acidic conditions which cause protonation of basic groups. The process is reversible and dyes are removed from fabrics during washing. For example, Acid black, Acid blue S5R, Acid green 20, Acid red 119, etc.

Basic dyes: These are cationic compound that are used for dyeing acid group-containing fibers, usually synthetic fibers like modified polyacrylate. They generally give intense shades but have poor light fastness. They are used for dyeing silk and wool directly. For example, Crystal violet, methylene blue, safranin, basic fuschin, etc.

Direct dyes: They dyes have a high affinity for cellulose fibers and bind to them through Vander Waals forces. The common salt or Glauber's salt is often used with direct dye to promote dyeing process because the presence of excess sodium ions favours the establishment of equilibrium with the minimum amount of dye. In this case, the dyeing process is reversible and exhibit poor wash fastness. For example, Direct orange 39, Direct blue 86, Direct red 10, etc.

Mordant dyes: These dyes have poor affinity for the fiber and require pre-treatment of the fiber with the mordant usually metal salts (such as chromium and iron salts). They are used for dyeing wool, leather, silk, paper and modified cellulose fibers. For example, Chrome blue 2K, Alizarin red S, Celestine blue B, Eriochrome cyanine R, etc.

Disperse dyes: They are specifically used to dye synthetic fibers like cellulose acetate, polyester, polyamide, acryl, etc. They are insoluble in water but in the actual fibres themselves. Its diffusion requires swelling of the fiber, either by high temperature (>120°C) or with the help of chemical softener so that the finely ground particles can penetrate. For example, Disperse red, disperse blue, Disperse violet, Disperse yellow, Disperse green, etc.

Vat dyes: They dyes are insoluble in water but their reduced forms are soluble. These dyes are, therefore, applied in their reduced forms (reduced forms are obtained by treating the dye with some reducing agent such as alkaline sodium dithionite). When the reduced dye is adsorbed on the fiber, the original insoluble dye is reformed upon oxidation with air or chemicals. Vat dyes offer excellent fastness but they are quite expensive. For example, Vat blue 1, Vat orange 3, Vat yellow 1, etc.

Sulphur dyes: These are polymeric aromatic compounds containing heterocyclic S-ring. They are mainly used for dyeing cellulose fibers. Dyeing process involves reduction and oxidation of the dye. They become soluble when reduced to sodium sulphide and exhibit affinity for cellulose. But on exposure to air, they get oxidised to insoluble dye inside the fiber. For example, sulphur black 1, Sulphur red 1, Sulphur orange 1, Sulphur brown 21, Sulphur green 12, etc.

Solvent dyes: These dyes are non-ionic compounds soluble in organic solvents. They are used as a solution in an organic solvent. For examples, Solvent red 24, Solvent yellow 124, solvent blue 35, Solvent orange 5, Solvent black 3, etc.

5.2   Dyes as a source of colour contaminant in water

The discovery of the first synthetic dye, *Mauveine*, in 1856 by William Henry Perkin led the way to the synthesis of a wide variety of dyes to be used for various purposes. There are more than 10,000 commercially available dyes with over 7×10^5 tonnes of dyestuff produced annually [7]. Dyes are widely used in a variety of industries such as textiles, rubber, paper, plastics, printing, leather, cosmetics, pharmaceuticals, food, etc., to colour their products. The textile industry is one of the largest sectors globally consuming substantial amounts of water in its manufacturing processes. As a result, a large amount of coloured wastewater is generated which is discharged into the freshwater systems without sufficient treatment. It is estimated that 2% of dyes produced annually is discharged in effluents from associated industries [8]. Discharge of dye-bearing wastewater into natural streams and rivers poses severe problems to the aquatic life, food web and causes damage to the aesthetic nature of the environment.

5.3 Harmful effects of dyes

Colour is the first contaminant to be recognised in wastewater. The presence of an even very small amount of dyes in water is highly visible and hence undesirable. Also, dyes have harmful effects on living organisms. Dyes absorb and reflect sunlight entering the water and so can interfere with the growth of bacteria and hinder photosynthesis in aquatic plants. The problems become grave due to the fact that the complex aromatic structures of the dyes render them ineffective in the presence of heat, light, microbes and even oxidising agents and degradation of the dyes become difficult [7]. Dyes can cause allergic dermatitis, skin irritation, cancer, mutation, etc. Hence, these pose a serious threat to human health and water quality, thereby becoming a matter of vital concern.

Keeping the essentiality of colour removal, concerned industries are required to treat the dye-bearing effluents before dumping it into the water bodies. Thus, the scientific community shoulders the responsibility of contributing to the waste treatment by developing effective dye removal technique.

6. Techniques available for the removal of dyes from wastewater

In this section, the available treatment methods will be discussed with special emphasis on dye removal. These techniques can be broadly classified into three categories: Biological, chemical and physical [9].

6.1 Biological treatment

In this method, microorganisms such as bacteria, yeasts, fungi and algae are used to degrade different pollutants [10]. Biological methods can be broadly classified into two types- *aerobic* and *anaerobic treatment method*. The aerobic method involves the usage of free or dissolved oxygen by microorganisms to decompose the organic matter whereas, in the anaerobic method, decomposition of the organic wastes occurs in the absence of oxygen. However, the application of biological treatment methods is restricted due to methodological limitations such as the requirement of a large land area, sensitivity towards diurnal variations, the toxicity of some chemicals, less flexibility in design and unsatisfactory colour elimination [11].

6.2 Chemical treatment

Chemical methods include coagulation, flocculation, precipitation–flocculation with $Fe(II)/Ca(OH)_2$, electroflotation, electrokinetic coagulation, conventional oxidation methods by oxidising agents (ozone), electrochemical processes, advanced oxidative processes, etc. The chemical techniques are often expensive, create disposal problems due to the accumulation of concentrated sludge and require a high input of electrical energy [12,13].

6.3   Physical treatment

Different physical methods such as membrane filtration (nanofiltration, reverse osmosis, electrodialysis, etc.) and adsorption techniques are widely used for the removal of dyes from wastewater. The main disadvantage associated with membrane filtration is the limited lifetime of the membrane. Fouling of the membrane occurs after a certain period of time and requires periodic replacement and thereby, reduces the economic viability of the process [14]. To overcome these limitations the adsorption method has been very commonly used. In fact, it is one of the preferred methods used for the removal of dyes from wastewater because of the various advantages such as ease of operation, the simplicity of design, avoidance of secondary pollution, insensitivity to toxic pollutants, low cost and, therefore, economic feasibility. [15, 16].

7.   Adsorption

Adsorption is a surface phenomenon and may be defined as the "phenomenon of attracting and retaining the molecules of a substance on the surface of a liquid or a solid resulting into a higher concentration of the molecules on the surface". The adsorption process involves two components adsorbent and adsorbate. The adsorbent is the substance on the surface of which adsorption takes place while adsorbate is the substance which is being adsorbed on the surface of the adsorbent.

The process of adsorption arises due to the presence of unbalanced or residual forces at the surface of the liquid or solid phase. In a bulk of the material, every molecule is equally attracted from all sides and hence the net force experienced by each molecule in the bulk is zero. However, molecules present on the surface are not wholly surrounded by other molecules and therefore experience a net inward force towards the bulk.

7.1   Types of adsorption

There are two types of adsorption processes, namely, *physisorption* and *chemisorption*.

(a) Physisorption

It involves the attraction between the adsorbate molecules and the adsorbent surface via weak van der Waals forces. Physical adsorption occurs with the formation of a multilayer of adsorbate on adsorbent [17, 18].

Characteristics of physisorption

- It occurs at very low temperatures and its magnitude decreases with the rise in temperature
- Heat evolved in physisorption is low, varying between 4–40 kJ/mol
- In case of physisorption of gases over solids, the extent of adsorption increases with increase in pressure
- It is reversible
- It is not specific with respect to adsorbent
- The extent of adsorption increases with the increase in surface area of the adsorbent

(b) Chemisorption

Chemisorption is said to have taken place when the affinity between adsorbate and adsorbent is a chemical force or chemical bond. Chemisorption occurs with the formation of a monolayer of adsorbte on adsorbent [19].

Characteristics of chemisorption

- Chemisorption occurs at low as well as high temperatures and its magnitude increases with the rise in temperature
- Heat of adsorption is very high, varying between 40–400 kJ/mol
- The chemisorption is not appreciably affected by small changes in pressure
- Chemisorption is irreversible in nature
- It is highly specific and occurs only if there is some possibility of chemical bonding between adsorbent and adsorbate
- Like physisorption, chemisorption also increases with increase of surface area of the adsorbent

7.2   Factors affecting adsorption

Adsorption on a solid surface depends on a number of factors such as [20]:

- Surface area
- Nature of the adsorbate
- pH of the solution
- Temperature
- Nature of adsorbate

Surface Area: Adsorption is a surface phenomenon and the extent of adsorption is proportional to a specific surface area. The specific surface area can be defined as that portion of the total surface area that is available for adsorption. Thus, if the solid is more finely divided and more porous, the extent of adsorption accomplished per unit weight of solid adsorbent is greater.

Nature of adsorbate: The adsorption is also influenced by the solubility of the adsorbate in the solvent. Generally, the extent of adsorption increases with a decrease in solubility due to respective increase in solute-solvent interaction.

pH of the solution: The pH of a solution from which adsorption occurs influences the extent of adsorption. Since in most cases, the adsorption occurs via the association of hydronium or hydroxide ions, the process is greatly affected by pH of the medium. Alkaline medium offers favourable condition for the adsorption of cationic dye while anionic dye can be best adsorbed in acidic medium.

Temperature: Since adsorption is accompanied by evolution or absorption of heat depending upon the nature of the adsorbent-dye interaction, the magnitude of adsorption is also dependent on temperature.

Nature of adsorbent: The physiochemical nature of the adsorbent imparts profound effects on both rate and capacity for adsorption. Adsorption is affected by the presence of functional groups and other structural characteristics. Different adsorbents adsorb a dye in a different way because of their different characteristic intrinsic nature.

8.   Selection of adsorbent

Activated carbon has been used as a most conventional adsorbent for the treatment of wastewater because of its excellent adsorption ability [21]. However, widespread use of activated carbon is restricted due to its high cost, difficulties in its regeneration and loss of adsorbent during regeneration. Ultimately these factors reduce its economic feasibility [22]. Literature survey reveals that a number of low-cost substitutes have successfully been developed and used for the treatment of dyes from aqueous solutions. Materials like chitin, chitosan, modified cotton, keratin, fly ash, oil cakes, fruit peels, barks, seeds, fruit wastes and much more have been used extensively as adsorbents for the removal of dyes from wastewater. Some typical examples [23-106] of such materials are presented in Table 2.

Table 2. Summary of the adsorbents used for the removal of dyes from wastewater.

Adsorbent	Dye	Reference
Acid-treated pine cone powder	Congo red	[23]
Aleurites Moluccana seeds	Rhodamine B	[24]
Aleurites Moluccana	Methylene blue	[24]
Avocado seed powder	Crystal violet	[25]
Bamboo leaves	Methylene blue	[26]
Banana fibre	Methylene blue	[27]
Biochar-palm bark	Methylene blue	[28]
Biochar-eucalyptus	Methylene blue	[28]
Blast furnace slag	Acid red 138	[29]
Blast furnace slag	Acid green 27	[29]
Bottom ash	Carmoisine A	[30]
	Fast Green FCF	[31]
Breadnut peel	Malachite green	[32]
Canola residues	Methylene blue	[33]
Carbon slurry	Methylene blue	[34]
	Crystal violet	[34]
	Meldola blue	[34]
	Chrysoidine G	[34]
	Ethyl orange	[34]
Carbon slurry	Metanil yellow	[34]
	Acid blue 113	[34]

Cerastoderma lamarcki shell	Malachite green	[35]
Citrus limetta peel	Eriochrome black T	[36]
Clay	Reactive Red 120	[37]
Coconut fibre	Methylene blue	[27]
Coconut husk-based activated carbon	2,4,6-trichlorophenol	[38]
Cotton stalk	Methylene blue	[39]
Curcuma angustifolia scales	Basic violet 14	[40]
De-oiled soya	Eosin yellow	[41]
	Carmoisine A	[30]
De-oiled soya	Fast Green FCF	[31]
Dicentrarchus labrax scales	Acid Blue 121	[42]
Eucalyptus sawdust	Brilliant green	[43]
Eucalyptus sheathiana bark	Methylene blue	[44]
Fly ash	Direct black	[45]
Giombo persimmon seed	Toluidine Blue	[46]
Hectorite	Congo red	[47]
Hen feather	Amido black 10B	[48]
	Congo red	[49]
Hen feather	Brilliant yellow	[50]
Hevea brasiliensis seed shell	Crystal violet	[51]
Illitic clay	Methylene blue	[52]
Jatropha curcas pods	Remazol brilliant blue R	[53]
Jujuba seeds	Congo red	[54]

Kaolin-Bentonite mixture	Congo red	[55]
Kaolin	Congo red	[55]
Kenaf fibre	Methylene blue	[56]
Lemna minor biomass	Acid blue 113	[57]
Longan shell	Methylene blue	[58]
Lotus leaf	Methylene blue	[59]
Melaleuca diosmifolia	Methylene blue	[60]
Melaleuca diosmifolia	Acridine orange	[60]
	Malachite green	[60]
Melon peel	Methylene blue	[61]
Mineral waste from coal mining	Aztrazon blue	[62]
Modified natural bentonite	Congo red	[63]
Modified sphagnum peat moss	Malachite green	[64]
Montmorillonite	Crystal violet	[65]
	Basic red 18	[66]
Moroccan clay	Malachite green	[67]
	Methylene blue	[67]
	Basic red 46	[67]
	Methylene blue	[68]
	Malachite green	[68]
	Methyl orange	[68]
NaOH-modified rice husk	Crystal violet	[69]
Natural Clay	Methylene blue	[70]

Neem sawdust	Malachite green	[71]
Olive pomace	Basic green 4	[72]
Orange peel	Methylene blue	[73]
Palm kernel fibre	Methylene blue	[74]
	Crystal violet	[74]
Peanut husk	Methylene blue	[75]
Perlite	Maxilon Blue 5G	[76]
Pine cone	Acid Black 26	[77]
	Acid Green 25	[77]
	Acid Blue 7	[77]
	Congo red	[23]
Pine tree leaves	Basic red 46	[78]
Pinus radiate	Methylene blue	[79]
Pistachio hull waste	Methylene blue	[80]
Red mud	Acid Blue 15	[81]
Rejected tea	Methylene blue	[82]
Rice husk	Direct Red 31	[83]
	Direct Orange 26	[83]
River sand	Methylene blue	[84]
Rosa canina galls	Methylene blue	[85]
	Crystal violet	[85]
Saklıkent mud	Brilliant green	[86]
Sawdust	Methylene blue	[27]

Sepiolite	Maxilon red GRL	[87]
Sesame hull	Methylene blue	[88]
Simarouba glauca seed shell	Malachite green	[89]
Soybean hull	Safranin	[90]
Spent tea leaves	Crystal violet	[91]
Tea dust	Crystal violet	[92]
Water chestnut	Rhodamine B	[93]
Water hyacinth	Methylene blue	[94]
Water hyacinth root	Dark blue-GL	[95]
Wheat bran	Reactive red 180	[96]
	Reactive orange 16	[96]
	Reactive black 5	[96]
	Direct red 80	[96]
	Acid red 42	[96]
Wheat bran	Acid yellow 199	[96]
Zeolite	Reactive Red 239	[97]
	Reactive Blue 250	[97]

9.   Conclusion

Water pollution has become a plague in our modern society. Rapid increase in industrialisation has resulted in the generation of a variety of toxic pollutants such as detergents, acids, agro-chemicals, heavy metals, dyes, etc. Dyes are an important class of pollutants which are extensively used in various industries like textiles, leather, paper and pulp, pharmaceuticals, paint, cosmetics, plastic, etc. The presence of dyes in water even in trace amounts can be easily recognised and is aesthetically undesirable. Dyes largely affect the photosynthetic activities of aquatic plants by reducing the transmission of sunlight into the water bodies. Moreover, many dyes are toxic and even carcinogenic thus affecting the aquatic biota and human health. Various methods such as adsorption, coagulation, advanced oxidation, ozonation, Fenton's process, reverse osmosis and biological processes have been used for the removal of dyes from wastewater. Amongst them, adsorption is one of the most effective processes because of the various advantages associated with it such as ease of operation, simple design, avoidance of secondary pollution, low cost, economic feasibility, etc. In recent years, there has been an increasing trend in using low-cost adsorbents based on agricultural, industrial and domestic wastes. Inexpensive locally available and effective raw and treated materials could be used in place of expensive commercial activated carbon in the removal of dyes from its aqueous solution.

References

[1]   E. Ashby, The Royal Commission on Environmental Pollution, UK. Third report: Pollution in some British Estuaries and coastal waters. Published in London, H.M.S.O., 1972, 126–128.

[2]   M. Bond, Plagued by noise, New Scientist, November 2016, 14–15.

[3]   J. Cross, Water purification – surviving the changes. Chem. Brit. 26 (1990) 579–580.

[4]   K. Imhoff and C.M. Fair, Sewage treatment. Second edition. John Wiley & Sons, Inc., New York, 1929, pp. 162–213.

[5]   R.M. Christie, Colour chemistry. The Royal Society of Chemistry, Cambridge, UK, 2001, pp. 1–200. https://doi.org/10.1039/9781847550590-00001

[6]   C. Raghavacharya, Color removal from industrial effluents – a comparative review of available technologies, Chem. Eng. World. 32 (1997) 53–54.

[7]   C.I. Pearce, J.R. Lloyd and J.T. Guthrie, The removal of colour from textile wastewater using whole bacterial cells: a review, Dyes Pigments. 58 (2003) 179–196. https://doi.org/10.1016/S0143-7208(03)00064-0

[8]   S.J. Allen, B. Koumanova, Decolourisation of water/wastewater using adsorption, J. Univ. Chem. Technol. Metall. 40 (2003) 175–192.

[9]   T. Robinson, G. McMullan, R. Marchant, P. Nigam, Remediation of dyes in textile effluent: a critical review on current treatment technologies with a proposed

alternative, Bioresource Technol. 77 (2001) 247–255.
https://doi.org/10.1016/S0960-8524(00)00080-8

[10] G. McMullan, C. Meehan, A. Conneely, N. Kirby, T. Robinson, P. Nigam, I.M. Banat, R. Marchant, W.F. Smyth, Microbial decolourisation and degradation of textile dyes. Appl. Microbiol. Biotechnol. 56 (2001) 81–87.
https://doi.org/10.1007/s002530000587

[11] G. Crini, Non-conventional low-cost adsorbents for dye removal: a review, Bioresource Technol. 97 (2006) 1061–1085.
https://doi.org/10.1016/j.biortech.2005.05.001

[12] J.S. Kace, H.B. Linford, Reduced cost flocculation of a textile dyeing wastewater, J. Water Pollut. Control Fed. 47 (1975) 1971–1980.

[13] J.W. Lee, S.P. Choi, R. Thiruvenkatachari, W.G. Shim, H. Moon, Evaluation of the performance of adsorption and coagulation processes for the maximum removal of reactive dyes, Dyes Pigments. 69 (2006) 196–203.
https://doi.org/10.1016/j.dyepig.2005.03.008

[14] A. Maartens, A. Swart, E. Jacobs, Feed-water pretreatment: methods to reduce membrane fouling by natural organic matter, J. Membr. Sci. 163 (1999) 51–62.
https://doi.org/10.1016/S0376-7388(99)00155-6

[15] A. Dabrowski, Adsorption, from theory to practice. Adv. Colloid Interf. Sci. 93 (2001) 135–224. https://doi.org/10.1016/S0001-8686(00)00082-8

[16] Z. Abbasi, M. Alikarami, Kinetics and thermodynamics studies of acetic acid adsorption from aqueous solution by peels of banana, Biochem. Bioinfo. 1(1) (2012) 1–7.

[17] C.T. Rettner, D.J. Auerbach, Chemical Dynamics at the Gas-Surface Interface, J. Phys. Chem. 1996, 100(31) (1996), 13021–13033.

[18] E.R. Alley, Water quality control handbook. McGraw-Hill Inc, New York. 2000.

[19] W.J. Weber, Physiochemical processes for water quality control. Wiley, New York, 1972, 199–259.

[20] B.R. Puri, L.R. Sharma and M.S. Pathania, Principles of Physical Chemistry. 45th Edition. Vishal Publishing Co. India. 1179–1186.

[21] T. Robinson, B. Chandran, P. Nigam, Removal of dyes from a synthetic textile dye effluent by bio-sorption on apple pomace and wheat straw, Water Res. 36 (2002) 2824–2830. https://doi.org/10.1016/S0043-1354(01)00521-8

[22] Y.S. Al-Degs, M.A.M. Khraisheh, S.J. Allen, M.N. Ahmad, Adsorption characteristics of reactive dyes in columns of activated carbon, J. Hazard. Mater. 165 (2009) 944–949. https://doi.org/10.1016/j.jhazmat.2008.10.081

[23] S. Dawood, T.K. Sen, Removal of anionic dye Congo red from aqueous solution by raw pine and acid-treated pine cone powder as adsorbent: Equilibrium, thermodynamic, kinetics, mechanism and process design, Water Res. 46 (6) (2012) 1933–1946. https://doi.org/10.1016/j.watres.2012.01.009

[24] D.L. Postai, C.A. Demarchi, F. Zanatta, D.C.C. Melo, C.A. Rodrigues, Adsorption of rhodamine B and methylene blue dyes using waste of seeds of Aleurites Moluccana, a low cost adsorbent, Alexandria Eng. J. 55 (2016) 1713–1723. https://doi.org/10.1016/j.aej.2016.03.017

[25] A. Bazzo, M.A. Adebayo, S.L.P. Dias, E.C. Lima, J.C.P. Vaghetti, E.R. Oliveira, A.J.B. Leite, F.A. Pavan, Avocado seed powder: characterization and its application for crystal violet dye removal from aqueous solutions, Desalin. Water Treat. 57 (2016) 15873–15888. https://doi.org/10.1080/19443994.2015.1074621

[26] L. Zhu, Y. Wang, T. He, L. You, X. Shen, Assessment of Potential Capability of Water Bamboo Leaves on the Adsorption Removal Efficiency of Cationic Dye from Aqueous Solutions, J. Polym. Environ. 24 (2016) 148–158. https://doi.org/10.1007/s10924-016-0757-8

[27] R. Karthik, R. Muthezhilan, A.J. Hussain, K. Ramalingam, V. Rekha, Effective removal of Methylene Blue dye from water using three different low-cost adsorbents, Desalin. Water Treat. 57 (2016) 10626–10631. https://doi.org/10.1080/19443994.2015.1039598

[28] L. Sun, S. Wan, W. Luo. Biochars prepared from anaerobic digestion residue, palm bark, and eucalyptus for adsorption of cationic methylene blue dye: Characterization, equilibrium, and kinetic studies, Bioresource Technol. 140 (2013) 406–413. https://doi.org/10.1016/j.biortech.2013.04.116

[29] D. Zhao, Q. Qiu, Y. Wang, M. Huang, Y. Wu, X. Liu, T. Jiang, Efficient removal of acid dye from aqueous solutions via adsorption using low-cost blast-furnace slag, Desalin. Water Treat. (2016) 1–10.

[30] V.K. Gupta, A. Mittal, A. Malviya, J. Mittal, Adsorption of carmoisine A from wastewater using waste materials—Bottom ash and deoiled soya, J. Colloid Interf. Sci. 335 (2009) 24–33. https://doi.org/10.1016/j.jcis.2009.03.056

[31] A. Mittal, D. Kaur, J. Mittal, Batch and bulk removal of a triarylmethane dye, Fast Green FCF, from wastewater by adsorption over waste materials. J. Hazard. Mater. 163(2–3) (2009) 568–577. https://doi.org/10.1016/j.jhazmat.2008.07.005

[32] H.I. Chieng, L.B.L. Lim, N. Priyantha, Enhancing adsorption capacity of toxic malachite green dye through chemically modified breadnut peel: equilibrium, thermodynamics, kinetics and regeneration studies, Environ. Technol. 36(1) (2015) 86–97. https://doi.org/10.1080/09593330.2014.938124

[33] D. Balarak, J. Jaafari, G. Hassani, Y. Mahdavi, I. Tyagi, S. Agarwal, V.K. Gupta, The use of low-cost adsorbent (Canola residues) for the adsorption of methylene blue from aqueous solution: Isotherm, kinetic and thermodynamic studies, Colloid Interf. Sci. Comm. 7 (2015) 16–19. https://doi.org/10.1016/j.colcom.2015.11.004

[34] V.K. Gupta, Suhas, I. Tyagi, S. Agarwal, R. Singh, M. Chaudhary, A. Harit, S. Kushwaha, Column operation studies for the removal of dyes and phenols using a low cost adsorbent, Global J. Environ. Sci. Manage. 2(1) (2016) 1–10.

[35] S.Y. Kazemi, P. Biparva, E. Ashtiani, Cerastoderma lamarcki shell as a natural, low cost and new adsorbent to removal of dye pollutant from aqueous solutions: Equilibrium andkinetic studies, Ecol. Eng. 88 (2016) 82–89. https://doi.org/10.1016/j.ecoleng.2015.12.020

[36] U.V. Ladhe, S.K. Wankhede, V.T. Patil, P.R. Patil, Adsorption of Eriochrome black T from aqueous solutions on activated carbon prepared from mosambi peel, J. Appl. Sci. Environ. Sanit. 6(2) (2011) 149–154.

[37] E. Errais, J. Duplay, F. Darragi, I.M. Rabet, A. Aubert, F. Huber, G. Morvan, Efficient anionic dye adsorption on natural untreated clay: Kinetic study and thermodynamic parameters. Desalination. 275(1–3) (2011) 74–81. https://doi.org/10.1016/j.desal.2011.02.031

[38] B.H. Hameed, I.A.W. Tan, A.L. Ahmad, Adsorption isotherm, kinetic modeling and mechanism of 2,4,6 trichlorophenol on coconut husk-based activated carbon, Chem. Eng. J. 144 (2) (2008) 235–244. https://doi.org/10.1016/j.cej.2008.01.028

[39] H. Deng, J. Lu, G. Li, G. Zhang, X. Wan, Adsorption of methylene blue on adsorbent materials produced from cotton stalk, Chem. Eng. J. 172(1) (2011) 326–334. https://doi.org/10.1016/j.cej.2011.06.013

[40] S. Suresh, R.W. Sugumar, T. Maiyalagan, A low cost adsorbent prepared from Curcuma angustifolia scales for removal of Basic violet 14 from aqueous solution. Indian J. Chem. Technol. 21(56) (2014) 368–378.

[41]   A. Mittal, D. Jhare, J. Mittal, Adsorption of hazardous dye Eosin Yellow from aqueous solution onto waste material De-oiled Soya: Isotherm, kinetics and bulk removal, J. Mol. Liq. 179 (2013) 133–140. https://doi.org/10.1016/j.molliq.2012.11.032

[42]   D. Uzunoğlu, A. Ozer, Adsorption of Acid Blue 121 dye on fish (Dicentrarchus labrax) scales, the extracted from fish scales and commercial hydroxyapatite: equilibrium, kinetic, thermodynamic, and characterization studies, Desalin. Water Treat. 57 (2016) 14109–14131 https://doi.org/10.1080/19443994.2015.1063091

[43]   V.S. Mane, P.V.V. Babu, Studies on the adsorption of Brilliant Green dye from aqueous solution onto low-cost NaOH treated sawdust. Desalination 273 (2–3) (2011) 321–329. https://doi.org/10.1016/j.desal.2011.01.049

[44]   S. Afroze, T.K. Sen, T.M. Ang, Adsorption performance of continuous fixed bed column for the removal of methylene blue (MB) dye using Eucalyptus sheathiana bark biomass, Res. Chem. Intermed. 42(3) (2016) 2343–2364. https://doi.org/10.1007/s11164-015-2153-8

[45]   Saakshy, K. Singh, A.B. Gupta, A.K. Sharma, Fly ash as low cost adsorbent for treatment of effluent of handmade paper industry-Kinetic and modelling studies for direct black dye, J. Cleaner. Prod. 112(1) (2016) 1227–1240. https://doi.org/10.1016/j.jclepro.2015.09.058

[46]   M.S. Bretanha, G.L. Dotto, J.C.P. Vaghetti, S.L.P. Dias, E.C. Lima and F.A. Pavan, Giombo persimmon seed (GPS) an alternative adsorbent for the removal Toluidine Blue dye from aqueous solutions, Desalin. Water Treat., 2016, 1–12.

[47]   C. Xia, Y. Jing, Y. Jia, D. Yue, J. Ma, X. Yin, Adsorption properties of congo red from aqueous solution on modified hectorite: Kinetic and thermodynamic studies, Desalination 265(1–3) (2011) 81–87. https://doi.org/10.1016/j.desal.2010.07.035

[48]   A. Mittal, V. Thakur, V. Gajbe, Adsorptive removal of toxic azo dye Amido Black 10B by hen feather, Environ. Sci. Pollut. Res. 20 (2013) 260–269. https://doi.org/10.1007/s11356-012-0843-y

[49]   A. Mittal, V. Thakur, J. Mittal, H. Vardhan, Process development for the removal of hazardous anionic azo dye Congo red from wastewater by using hen feather as potential adsorbent, Desalin. Water Treat. 52 (2014) 227–237. https://doi.org/10.1080/19443994.2013.785030

[50]   A. Mittal, V. Thakur, V. Gajbe, Evaluation of adsorption characteristics of an anionic azo dye Brilliant Yellow onto hen feathers in aqueous solutions, Environ. Sci. Pollut. Res. 19 (2012) 2438–2447. https://doi.org/10.1007/s11356-012-0756-9

[51]   I.A.W. Tan, B.H. Hameed, Removal of crystal violet dye from aqueous solutions using rubber (hevea brasillensis) seed shell-based biosorbent, Desalin. Water Treat. 48 (2012) 174–181. https://doi.org/10.1080/19443994.2012.698810

[52]   D. Ozdes, C. Duran, H.B. Senturk, H. Avan, B. Bicer, Kinetics, thermodynamics, and equilibrium evaluation of adsorptive removal of methylene blue onto natural illitic clay mineral, Desalin. Water Treat. 52 (2014) 208–218. https://doi.org/10.1080/19443994.2013.787554

[53]   P. Sathishkumar, M. Arulkumar, T. Palvannan, Utilization of agro-industrial waste Jatropha curcas pods as an activated carbon for the adsorption of reactive dye Remazol Brilliant Blue R (RBBR), J. Cleaner Prod., 22(1) (2012) 67–75. https://doi.org/10.1016/j.jclepro.2011.09.017

[54]   M.C.S. Reddy, L. Sivaramakrishna, A.V. Reddy, The use of an agricultural waste material, Jujuba seeds for the removal of anionic dye (Congo red) from aqueous medium, J. Hazard. Mater. 203–204 (2012) 118–127. https://doi.org/10.1016/j.jhazmat.2011.11.083

[55]   O.T. Ogunmodedel, A.A. Ojo, E. Adewole, O.L. Adebayo, Adsorptive removal of anionic dye from aqueous solutions by mixture of Kaolin and Bentonite clay: Characteristics, isotherm, kinetic and thermodynamic studies, Iranica J. Energy Environ. 6(2) (2015) 147–153.

[56]   D.K. Mahmoud, M.A.M. Salleh, W.A.W.A. Karim, A. Idris, Z.Z. Abidin, Batch adsorption of basic dye using acid treated kenaf fibre char: Equilibrium, kinetic and thermodynamic studies, Chem. Eng. J. 181–182 (2012) 449–457. https://doi.org/10.1016/j.cej.2011.11.116

[57]   D. Balarak, Y. Mahdavi, F. Ghorzin, S. Sadeghi, Biosorption of acid blue 113 dyes using dried Lemna minor biomass, Sci. J. Env. Sci. 5(1) (2016) 152 –158.

[58]   Y. Wang, L. Zhu, H. Jiang, F. Hu, X. Shen, Application of longan shell as non-conventional low-cost adsorbent for the removal of cationic dye from aqueous solution, Spectrochim. Acta A. 159 (2016) 254–261. https://doi.org/10.1016/j.saa.2016.01.042

[59] X. Han, W. Wang, X. Ma, Adsorption characteristics of methylene blue onto low cost biomass material lotus leaf, Chem. Eng. J. 171(1) (2011) 1–8. https://doi.org/10.1016/j.cej.2011.02.067

[60] S. Kuppusamy, P. Thavamani, M. Megharaj, K. Venkateswarlu, Y.B. Lee, R. Naidu, Potential of Melaleuca diosmifolia as a novel, nonconventional and low-cost coagulating adsorbent for removing both cationic and anionic dyes, J. Ind. Eng. Chem. 37 (2016) 198–207. https://doi.org/10.1016/j.jiec.2016.03.021

[61] C. Djelloul, O. Hamdaoui, Dynamic adsorption of methylene blue by melon peel in fixed-bed columns, Desalin. Water Treat. 56 (2015) 2966–2975.

[62] A. dos Santos, M.F. Viante, P.P. Anjos, N. Naidek, M.P. Moises, E.G. Castro, A.J. Downs, C.A.P. Almeida, Removal of astrazon blue dye from aqueous media by a low-cost adsorbent from coal mining, Desalin. Water Treat. 57 (2016) 27213–27225. https://doi.org/10.1080/19443994.2016.1167627

[63] M. Toor, B. Jin, Adsorption characteristics, isotherm, kinetics, and diffusion of modified natural bentonite for removing diazo dye, Chem. Eng. J. 187(1) (2012) 79–88. https://doi.org/10.1016/j.cej.2012.01.089

[64] F. Hemmati, R. Norouzbeigi, F. Sarbisheh, H. Shayesteh, Malachite green removal using modified sphagnum peat moss as a low-cost biosorbent: Kinetic, equilibrium and thermodynamic studies, J. Taiwan. Inst. Chem. Eng. 58 (2016) 482–489. https://doi.org/10.1016/j.jtice.2015.07.004

[65] G.K. Sarma, S.S. Gupta, K.G. Bhattacharyya, Adsorption of Crystal violet on raw and acid-treated montmorillonite, K10, in aqueous suspension, J. Env. Manage., 171 (2016) 1–10. https://doi.org/10.1016/j.jenvman.2016.01.038

[66] B.A. Fil, K.Z. Karakas, R. Boncukcuoglu, A.E. Yilmaz, Removal of cationic dye (basic red 18) from aqueous solution using natural Turkish clay, Global NEST J. 15(4) (2013) 529–541.

[67] K.A. Bennani, B. Mounir, M. Hachkar, M. Bakasse, A. Yaacoubi, Adsorption of cationic dyes onto Moroccan clay: Application for industrial wastewater treatment, J. Mater. Environ. Sci. 6(9) (2015) 2483–2500.

[68] R. Elmoubarki, F.Z. Mahjoubi, H. Tounsadi, J. Moustadraf, M. Abdennouri, A. Zouhri, A. ElAlbani, N. Barka, Adsorption of textile dyes on raw and decanted Moroccan clays: Kinetics, equilibrium and thermodynamics, Water Res. Ind. 9 (2015) 16–29. https://doi.org/10.1016/j.wri.2014.11.001

[69]   S. Chakraborty, S. Chowdhury, P.D. Saha, Adsorption of Crystal Violet from aqueous solution onto NaOH-modified rice husk, Carbohyd. Polymer. 86 (2011) 1533– 1541. https://doi.org/10.1016/j.carbpol.2011.06.058

[70]   H. Ouasif, S. Yousfi, M.L. Bouamrani, M. El Kouali, S. Benmokhtar, M. Talbi, Removal of a cationic dye from wastewater by adsorption onto natural adsorbents, J. Mater. Environ. Sci. 4(1) (2013) 1–10.

[71]   S.D. Khattri, M.K. Singh, Removal of malachite green from dye wastewater using neem sawdust by adsorption, J. Hazard. Mater. 167(1–3) (2009) 1089–1094. https://doi.org/10.1016/j.jhazmat.2009.01.101

[72]   O. Kocer, B. Acemioglu, Adsorption of Basic green 4 from aqueous solution by olive pomace and commercial activated carbon: process design, isotherm, kinetic and thermodynamic studies, Desalin. Water Treat. 57 (2016) 16653–16669. https://doi.org/10.1080/19443994.2015.1080194

[73]   G.O. El-Sayed, Removal of methylene blue and crystal violet from aqueous solutions by palm kernel fiber, Desalination, 272(1–3) (2011) 225–232. https://doi.org/10.1016/j.desal.2011.01.025

[74]   P. Velmurugan, V. Rathinakumar, G. Dhinakaran, Dye removal from aqueous solution using low-cost adsorbent, Int. J. Environ. Sci. 1(7) (2011) 1492–1503.

[75]   J. Song, W. Zou, Y. Bian, F. Su, R. Han, Adsorption characteristics of methylene blue by peanut husk in batch and column modes, Desalination, 265(1–3) (2011) 119–125. https://doi.org/10.1016/j.desal.2010.07.041

[76]   O. Demirbas, M. Alkan, Adsorption kinetics of a cationic dye from wastewater, Desalin. Water Treat. 53 (2015) 3623–3631. https://doi.org/10.1080/19443994.2013.874705

[77]   N.M. Mahmoodi, B. Hayati, M. Arami, C. Lan, Adsorption of textile dyes on Pine Cone from colored wastewater: Kinetic, equilibrium and thermodynamic studies, Desalination, 268(1–3) (2011) 117–125. https://doi.org/10.1016/j.desal.2010.10.007

[78]   F. Deniz, S. Karaman, Removal of Basic Red 46 dye from aqueous solution by pine tree leaves, Chem. Eng. J. 170(1) (2011) 67–74. https://doi.org/10.1016/j.cej.2011.03.029

[79]   T.K. Sen, S. Afroze, H.M. Ang, Equilibrium, Kinetics and Mechanism of Removal of Methylene Blue from Aqueous Solution by Adsorption onto Pine Cone Biomass

of Pinus radiate, Water Air Soil Pollut. 218 (2011) 499–515.
https://doi.org/10.1007/s11270-010-0663-y

[80]   G. Moussavi, R. Khosravi, The removal of cationic dyes from aqueous solutions
        by adsorption onto pistachio hull waste, Chem. Eng. Res. Des. 89 (2011) 2182–
        2189. https://doi.org/10.1016/j.cherd.2010.11.024

[81]   D. Balarak, Y. Mahdavi, S. Sadeghi, Adsorptive removal of acid blue 15 dye
        (AB15) from aqueous solutions by red mud: characteristics, isotherm and kinetic
        studies, Sci. J. Environ. Sci. 4(5) (2015) 102–112.

[82]   N. Nasuha, B.H. Hameed, Adsorption of methylene blue from aqueous solution
        onto NaOH-modified rejected tea, Chem. Eng. J. 166(2) (2011) 783–786.
        https://doi.org/10.1016/j.cej.2010.11.012

[83]   Y. Safa, H.N. Bhatti, Kinetic and thermodynamic modeling for the removal of
        Direct Red31 and Direct Orange26 dyes from aqueous solutions by rice husk,
        Desalination, 272(1–3) (2011) 313–322.
        https://doi.org/10.1016/j.desal.2011.01.040

[84]   A.B. Halim, K.K. Han, M.M. Hanafiah, Removal of Methylene Blue from Dye
        Wastewater Using River Sand by Adsorption, Nature Environ. Pollut. Technol.
        14(1) (2015) 89–94.

[85]   E. Bagda, The feasibility of using Rosa canina galls as an effective new biosorbent
        for removal of methylene blue and crystal violet, Desalin. Water Treat. 43 (2012)
        63–75. https://doi.org/10.1080/19443994.2012.672203

[86]   Y. Kismir, A.Z. Aroguz, Adsorption characteristics of the hazardous dye Brilliant
        Green on Saklıkent mud, Chem. Eng. J. 172(1) (2011) 199–206.
        https://doi.org/10.1016/j.cej.2011.05.090

[87]   O. Demirbas, Y. Turhan, M. Alkan, Thermodynamics and kinetics of adsorption of
        a cationic dye onto sepiolite, Desalin. Water Treat. 54 (2015) 707–714.
        https://doi.org/10.1080/19443994.2014.886299

[88]   Y. Feng, F. Yang, Y. Wang, L. Ma, Y. Wu, P.G. Kerr and L. Yang, Basic dye
        adsorption onto an agro-based waste material – Sesame hull (Sesamum indicum
        L.), Bioresource Technol. 102 (2011) 10280–10285.
        https://doi.org/10.1016/j.biortech.2011.08.090

[89]   B. Jeyagowri, R.T. Yamuna, Potential efficacy of a mesoporous biosorbent
        Simarouba glauca seed shell powder for the removal of malachite green from

aqueous solutions, Desalin. Water Treat. 57 (2016) 11326–11336. https://doi.org/10.1080/19443994.2015.1042060

[90]   V. Chandane, V.K. Singh, Adsorption of safranin dye from aqueous solutions using a low-cost agro-waste material soybean hull, Desalin. Water Treat. 57 (2016) 4122–4134. https://doi.org/10.1080/19443994.2014.991758

[91]   S.K. Bajpai, A. Jain, Equilibrium and Thermodynamic Studies for Adsorption of Crystal Violet onto Spent Tea Leaves (STL), Water. 4 (2012) 52–71.

[92]   M.M.R. Khan, M.W. Rahman, H.R. Ong, A.B. Ismail, C.K. Cheng, Tea dust as a potential low-cost adsorbent for the removal of crystal violet from aqueous solution, Desalin. Water Treat. 57 (2016) 14728–14738. https://doi.org/10.1080/19443994.2015.1066272

[93]   T.A. Khan, M. Nazir, E.A. Khan, Adsorptive removal of rhodamine B from textile wastewater using water chestnut (Trapa natans L.) peel: adsorption dynamics and kinetic studies, Toxicol. Environ. Chem. 95(6) (2013) 919–931. https://doi.org/10.1080/02772248.2013.840369

[94]   K. Murali, R.N. Uma, Removal of basic dye (methylene blue) using low cost biosorbent: water hyacinth, Int. J. Adv. Eng. Technol. 7(2) (2016) 386–391.

[95]   M.J. Alam, B.C. Das, M.W. Rahman, B.K. Biswas, M.M.R. Khan, Removal of dark blue-GL from wastewater using water hyacinth: a study of equilibrium adsorption isotherm, Desalin. Water. Treat. 56 (2015) 1520–1525. https://doi.org/10.1080/19443994.2014.950996

[96]   M.T. Sulak, H.C. Yatmaz, Removal of textile dyes from aqueous solutions with eco-friendly biosorbent, Desalin. Water Treat. 37 (2012) 169–177. https://doi.org/10.1080/19443994.2012.661269

[97]   E. Alver, A.U. Metin, Anionic dye removal from aqueous solutions using modified zeolite: Adsorption kinetics and isotherm studies, Chem. Eng. J. 200–202 (2012) 59–67. https://doi.org/10.1016/j.cej.2012.06.038

Chapter 2

Ion-exchange kinetics of alkaline metals on the surface of carboxymethyl cellulose Sn(IV) phosphate composite cation exchanger

Ali Mohammad[1,]* Mohd Imran Ahamed[2], Arshi Amin[1], Inamuddin[1],

[1]Department of Applied Chemistry, Faculty of Engineering and Technology, Aligarh Muslim University (AMU), Aligarh, 202002, India

[2]Department of Chemistry, Faculty of Science, Aligarh Muslim University, Aligarh–202002, India

*alimohammad08@gmail.com

Abstract

Approximated Nernst-Plank equation was used to study the kinetics and mechanism for the ion-exchange processes like Mg^{2+}-H^+, Ca^{2+}-H^+, Sr^{2+}-H^+ and Ba^{2+}-H^+ on the surface of carboxymethyl cellulose Sn(1V) phosphate composite nano-rod like cation-exchanger. The kinetics of exchange was favoured under the particle diffusion controlled phenomenon. Some physical parameters i.e. fractional attainment of equilibrium $U(\tau)$, self diffusion coefficients (D_o), energy of activation (E_a) and entropy of activation (ΔS^*) have been estimated.

Keywords

Organic–Inorganic Composite Materials, Cation-Exchanger, Ion-Exchange Kinetics, Carboxymethyl Cellulose Sn(IV) Phosphate

Contents

1.   Introduction

Applications of organic-inorganic composite materials have been explored in the fields of heterogeneous catalysis [1,2] protective coatings [3] solid polymer electrolyte membrane fuel cells [4,5] ion-selective membrane electrodes [6,7] gas perm-selectivity [8], [9] ion transport [10] and ion exchange [11]. In most of these fields, information related to the ion exchange kinetics and the mobility of counter ions in the lattice structure is needed. It is obvious that the kinetics studies envisage the three aspects of ion exchange process viz the mechanism of ion exchange, the rate determining step and the rate laws obeyed by the ion exchange system. Hence, in this study carboxymethyl cellulose Sn(IV) phosphate nanorod like composite cation-exchanger was selected to evaluate the ion-exchange mechanism occurring over the surface of the cation-exchanger.

2.   Experimental

2.1   Reagents and instruments

The main reagents viz stannic chloride, $SnCl_4 5H_2O$ (95%), carboxymethyl cellulose sodium salt, *tri*-sodium orthophosphate dodecahydrate, $Na_3PO_4.12H_2O$ (98%) and N-Cetyl-N,N,N-trimethyl ammonium bromide, $C_{19}H_{42}BrN$ (CTAB) (99%) used for the synthesis of the composite material were obtained from Central Drug House (CDH) Pvt. Ltd., India. Solutions for kinetic measurement were made using analytical reagent grade nitrate salts of Mg, Ca, Sr and Ba (99%) obtained from Central Drug House Pvt. Ltd. India. A digital pH meter (Elico LI-10, India) to adjust the pH and a water bath incubator shaker for all equilibrium studies having a temperature variation of ± 0.5 °C (MSW-275, India) were used.

2.2 Preparation of organic-inorganic composite cation-exchange material

Organic-inorganic composite cation exchanger carboxymethyl cellulose Sn(IV) phosphate composite cation exchange material was prepared as reported by Ali Mohammad et al. [12].

2.3 Kinetic measurements

Composite cation exchanger particles of mean radii ~125 μm (50-70 mesh) in H^+ form were used to evaluate various kinetic parameters. The rate of exchange was determined by limited batch technique as follows: A total of twenty milliliter fractions of the 0.03 M metal ion solutions (Mg, Ca, Sr and Ba) were shaken with 200 mg of the cation-exchanger in H^+-form in several stoppered conical flasks at desired temperatures [25, 35, 50 and 65 ($\pm$ 0.5) °C] for different time intervals (1.0, 2.0, 3.0, 4.0 and 25 min). The supernatant liquid was removed immediately and determinations were made as usual by ethylene diamine tetra acetic acid (EDTA) titrations[12]. Each set was repeated four times and the mean values were taken for the calculation.

3. Results and discussions

In this study, carboxymethyl cellulose Sn(IV) phosphate composite nanorod-like cation exchange material was prepared by the sol-gel method as reported by Mohammad et al. [13]. Equilibrium studies showed that 20 min. were required for the establishment of equilibrium at 35 °C for Mg^{2+}-H^+ Ca^{2+}-H^+, Sr^{2+}-H^+ and Ba^{2+}-H^+ exchanges. Therefore, 20 min. was assumed to be the infinite time of exchange for all exchange systems. Thus, kinetic measurements were made under conditions favoring a particle diffusion-controlled ion-exchange phenomenon for all metal ions because a study of the concentration effect on the rate of exchange at 35 °C showed that the initial rate of exchange was proportional to the metal ion concentration and τ versus time (t) (t in min) plots are also straight lines passing through the origin at and above 0.03 M of metal ion concentration (data not shown). The kinetic results are expressed in terms of the fractional attainment of equilibrium, $U(\tau)$ with time according to the equation:

$$U(\tau) = \frac{\text{the amount of exchange at time } 't'}{\text{the amount of exchange at infinite time}} \tag{1}$$

Plots of $U(\tau)$ versus time (t) (t in min.), for Mg^{2+}-H^+, Ca^{2+}-H^+, Sr^{2+}-H^+ and Ba^{2+}-H^+ (Fig. 1) respectively. Each value of $U(\tau)$ will have a corresponding value of τ, a dimensionless time parameter. On the basis of the Nernst-Planck equation, the numerical results can be expressed by explicit approximation [14–16]:

$$U(\tau) = \{ 1 - \exp [\pi^2 (f_1(\alpha)\tau + f_2(\alpha)\tau^2 + f_3(\alpha)\tau^3)]\}^{1/2} \tag{2}$$

where τ is the half time of exchange $= \overline{D}_{H^+}t / r_o^2$, α is the mobility ratio $= \overline{D}_{H^+} / \overline{D}_{M^{2+}}$, r_o is the particle radius, $\overline{D}_{H^+}$ and $\overline{D}_{M^{2+}}$ are the interdiffusion coefficients of counter ions H^+ and M^{2+} respectively, in the exchanger phase. The three functions $f_1(\alpha)$, $f_2(\alpha)$ and $f_3(\alpha)$ depend upon the mobility ratio (α) and the charge ratio $(Z_{H^+} / Z_{M^{2+}})$ of the exchanging ions. Thus, they have different expressions as given below. When the exchanger is taken in the H^+-form and the exchanging ion is M^{2+}, for $1 \leq \alpha \leq 20$, as in the present case, the three functions have the values-

$$f_1(\alpha) = - \frac{1}{0.64 + 0.36\ \alpha^{0.668}}$$

$$f_2(\alpha) = - \frac{1}{0.96 - 2.0\ \alpha^{0.4635}}$$

$$f_3(\alpha) = - \frac{1}{0.27 + 0.09\ \alpha^{1.140}}$$

The value of τ was obtained by solving equation (2) using a computer. The plots of τ versus time (t) at the four temperatures for Mg^{2+}-H^+, Ca^{2+}-H^+, Sr^{2+}-H^+ and Ba^{2+}-H^+ exchanges as shown in Fig. 2 are straight lines passing through the origin, confirming the particle diffusion control phenomenon for M^{2+}–H^+ exchanges at a metal ion concentration of 0.03 M. The slopes (S values) of various τ versus time (t) plots are given in Table 1. The S values are related to $\overline{D}_{H^+}$ as follows:

$$S = \overline{D}_{H^+} / r_o^2 \tag{3}$$

The values of $- \log \overline{D}_{H^+}$ obtained by using equation (3) plotted against $1/T$ are straight lines as shown in Fig. 3, thus verifying the validity of the Arrhenius relation:

$$\overline{D}_{H^+} = D_o \exp(- E_a / RT) \tag{4}$$

D_0 is obtained by extrapolating these lines and using the intercepts at the origin. The activation energy (E_a) is then calculated with the help of the equation (4), putting the value of $\overline{D}_{H^+}$ at 273 K. The entropy of activation (ΔS^*) was then calculated by substituting D_o in equation (5).

$$D_0 = 2.72 d^2 (kT / h) \exp(\Delta S^* / R) \tag{5}$$

where d is the ionic jump distance taken as 5×10^{-10} m, k is the Boltzmann constant, R is the gas constant, h is Plank's constant and T is taken as 273 K. The values of the diffusion coefficient (D_o), energy of activation (E_a) and entropy of activation (ΔS^*), thus obtained are summarized in Table 3. The positive values of activation energy indicated that the minimum energy is required to facilitate the forward (M^{2+}–H^+) ion-exchange process. Negative values of the entropy of activation (ΔS^*) suggest a greater degree of order achieved during the forward ion-exchange (M^{2+}–H^+) process.

4. Conclusion

In this study, carboxymethyl cellulose Sn(IV) phosphate composite cation exchanger was prepared and used to study the kinetic and mechanism for the ion-exchange processes like Mg^{2+}-H^+, Ca^{2+}-H^+, Sr^{2+}-H^+ and Ba^{2+}-H^+. The kinetic exchange in the forward direction (M^{2+}-H^+) on the surface of this composite cation exchanger is being governed by the particle diffusion controlled phenomenon. Some physical parameters have also been determined which showed that the ion exchange process is spontaneous and feasible.

Acknowledgements

The authors are thankful to Department of Applied Chemistry, Z. H. College of Engineering and Technology, A.M.U. (Aligarh) for providing research facilities.

References

[1]   H. Li, Z. Zheng, M. Cao, R. Cao, Stable gold nanoparticle encapsulated in silica-dendrimers organic–inorganic hybrid composite as recyclable catalyst for oxidation of alcohol, Microporous Mesoporous Mater. 136 (2010) 42–49.

[2]   K. Dallmann, R. Buffon, Sol–gel derived hybrid materials as heterogeneous catalysts for the epoxidation of olefins, Catal. Commun. 1 (2000) 9–13.

[3]   S. Chaudhari, P.P. Patil, Corrosion protective poly(o-ethoxyaniline) coatings on copper, Electrochim. Acta. 53 (2007) 927–933.

[4]   Y. Zhang, H. Zhang, C. Bi, X. Zhu, An inorganic/organic self-humidifying composite membranes for proton exchange membrane fuel cell application, Electrochim. Acta. 53 (2008) 4096–4103.

[5]   J.L. Malers, M.-A. Sweikart, J.L. Horan, J.A. Turner, A.M. Herring, Studies of heteropoly acid/polyvinylidenedifluoride–hexafluoroproylene composite membranes and implication for the use of heteropoly acids as the proton

conducting component in a fuel cell membrane, J. Power Sources. 172 (2007) 83–88.

[6]   A.A. Khan, Inamuddin, Applications of Hg(II) sensitive polyaniline Sn(IV) phosphate composite cation-exchange material in determination of Hg^{2+} from aqueous solutions and in making ion-selective membrane electrode, Sensors Actuators B Chem. 120 (2006) 10–18.

[7]   I. Nabi, S.A. Alam, Z., A Cadmium Ion-selective Membrane Electrode Based on Strong Acidic Organic-inorganic Composite Cation-exchanger: Polyaniline Ce(IV) Molybdate, Sens. Transd. J. (S T E-Digest). 2008 (92AD) 87.

[8]   C. Guizard, A. Bac, M. Barboiu, N. Hovnanian, Hybrid organic-inorganic membranes with specific transport properties, Sep. Purif. Technol. 25 (2001) 167–180.

[9]   Z. Alam, Inamuddin, S.A. Nabi, Synthesis and characterization of a thermally stable strongly acidic Cd(II) ion selective composite cation-exchanger: Polyaniline Ce(IV) molybdate, Desalination. 250 (2010) 515–522.

[10]   P. Lacan, C. Guizard, P. Le Gall, D. Wettling, L. Cot, Facilitated transport of ions through fixed-site carrier membranes derived from hybrid organic-inorganic materials, J. Memb. Sci. 100 (1995) 99–109..

[11]   A. Nilchi, A. Khanchi, H. Atashi, A. Bagheri, L. Nematollahi, The application and properties of composite sorbents of inorganic ion exchangers and polyacrylonitrile binding matrix, J. Hazard. Mater. 137 (2006) 1271–1276.

[12]   A. Mohammad, Inamuddin, A. Amin, Surfactant assisted preparation and characterization of carboxymethyl cellulose Sn(IV) phosphate composite nano-rod like cation exchanger, J. Therm. Anal. Calorim. 107 (2012) 127–134.

[13]   K.G. Varshney, A.A. Khan, S. Rani, Forward and reverse ion-exchange kinetics for Na^+-H^+ and K^+-H^+ exchanges on crystalline antimony (V) silicate, Colloids and Surfaces. 25 (1987) 131–137.

[14]   S. Kodama, K. Fukui, A. Mazume, Relation of Space Velocity and Space Time Yield, Ind. Eng. Chem. 45 (1953) 1644–1648.

[15]   F. Helfferich, M.S. Plesset, Ion Exchange Kinetics. A Nonlinear Diffusion Problem, J. Chem. Phys. 28 (1958) 418–424.

[16]   M.S. Plesset, F. Helfferich, J.N. Franklin, Ion Exchange Kinetics. A Nonlinear Diffusion Problem. II. Particle Diffusion Controlled Exchange of Univalent and Bivalent Ions, J. Chem. Phys. 29 (1958) 1064–1069.

Chapter 3

Removal of nitrogen containing compounds by adsorption: a review

Ali Mohammad[1,]* Mohd Imran Ahamed[2], Arshi Amin[1], Inamuddin[1]

[1]Department of Applied Chemistry, Faculty of Engineering and Technology, Aligarh Muslim University (AMU), Aligarh, 202002, India

[2]Department of Chemistry, Faculty of Science, Aligarh Muslim University, Aligarh–202002, India

* alimohammad08@gmail.com

Abstract

Nitrogen occurs in all living organism. It is a key constituent in a number of compounds which are chemically and biologically of great importance. In this review, chemically and biologically active nitrogen containing compounds specially pyridine, nicotinic acid and aniline are discussed. These nitrogen containing compounds play an important role for the survival of human beings, animals as well as plants. Therefore, their importance and the threats they impose on living beings and on the environment are described briefly. Therefore, these compounds are being removed from the environment by various methods. However, in this review the adsorption method for the removal of pyridine, nicotinic acid and aniline is being discussed.

Keywords

Pyridine, Nicotinic Acid, Aniline, Adsorption Materials

Contents

1. Nitrogen and its compounds

Nitrogen occurs in all living organism. It is a key constituent in a number of compounds which are chemically and biologically of great importance. Nitrogen is an important constituent of amino acids, proteins and nucleic acids (DNA and RNA). As part of the symbiotic relationship, the plant converts the 'fixed' ammonium ion to nitrogen oxides and amino acids to form proteins and other molecules, (e.g., alkaloids). Nitrogen compounds are basic building blocks in animal biology as well. Nitrogen is a constituent of molecules in every major drug class in pharmacology and medicine. Nevertheless, some of the most toxic and carcinogenic of known chemicals are nitrogen containing compounds. In this review, chemically and biologically active nitrogen containing compounds especially pyridine, nicotinic acid and aniline are discussed. A literature survey for the removal of pyridine, nicotinic acid and aniline by adsorption is presented.

1.1 Nicotinic acid

Also known as **vitamin B$_3$, niacin** and **vitamin PP** is an organic nitrogen compound one of the forty to eighty essential human nutrients. **Nicotinic acid** is a precursor to NAD+/NADH and NADP+/NADPH, which play essential metabolic roles in living cells [1]. Niacin is involved in both DNA repair and the production of steroid hormones in the adrenal gland. Niacin blocks the breakdown of fats in adipose tissue. Pharmacological doses of niacin (1.5 - 6 g per day) occasionally lead to side effects such as skin flushing, itching, dry skin, eczema, nausea, liver toxicity etc. Side effects of liver damage, hyperglycemia, cardiac arrhythmias and birth defects have also been reported [2]. Niacin at extremely high doses can cause life-threatening acute toxic reactions [3]. Tertiary nitrogen-containing compounds like nicotinic acid are suspected to have carcinogenic behaviour, and hence their systematic study might reveal some interesting results. Several important studies have been carried out and are listed as shown in Table 1.

Table 1 List of adsorption studies performed on nicotinic acid.

S. No.	Adsorbent	Remarks	Ref.
1.	Bentonite and dodecyl ammonium-bentonite	Dodecylammonium bentonite (DAB) and bentonite (B) were used as sorbents for nicotine (N), nicotinic acid (NA), iso-nicotinic acid (INA), nicotinic acid hydrazide (NH), and isonicotinic acid hydrazide (INH).	[4]
2.	Monolayer of mercapto acetic acid-coated gold electrode	A novel and sensitive method for the determination of nicotinic acid was proposed and based on the voltammetric behaviour of nicotinic acid on a self-assembled monolayer of the mercapto acetic acid-coated gold electrode.	[5]
3.	Imprinted polymer stationary phase	The adsorption isotherms of nicotine amide and nicotinic acid and the competitive adsorption isotherms of nicotine amide and nicotinic acid on the imprinted stationary phase are determined using rectangular pulse frontal analysis and static method.	[6]
4.	–	The separations of nicotinic acid and its derivatives analysed by adsorption and reversed phase TLC and HPLC were compared.	[7]
5.	–	Separations of nicotinic acid and its derivatives on different stationary phases (silica gel, a mixture of silica gel and kieselguhr, polyamide 11, and RP-18) have been compared.	[8]
6.	Poly(acrylamide/ maleic acid) hydrogels (AAm/MA)	AAm/MA hydrogel sorbed only nicotine and did not sorb nicotinamide, nikethamide and nicotinic acid in the binding experiments.	[9]
7.	Polycrystalline gold electrode	Adsorption behaviour of three isomeric pyridine carboxylic acids (picolinic acid: 2-PCA, nicotinic acid: 3-PCA, and isonicotinic acid: 4-PCA) in 0.1 M (M = mol dm^{-3}) $HClO_4$ solution on a smooth polycrystalline gold electrode surface was investigated by in situ infrared reflection absorption spectroscopy (IRAS).	[10]

8.	(110) Face of silver	The adsorption of nicotinic acid, nicotinamide and nipecot amide at Ag(110) single crystal electrode has been studied from aqueous solutions of 0.1 KF.	[11]
9.	Silver sol surface	Adsorption of picolinic and nicotinic acids on a silver sol surface has been investigated over a wide range of solution pH by surface-enhanced Raman scattering.	[12]
10.	Activated charcoal	The effects of various factors on the adsorption of nicotinic acid onto and desorption from activated charcoal was investigated in vitro.	[13]
11.	Mercury electrode	Differential capacity measurements are reported for the adsorption of nicotinic acid (Nc) at the mercury electrode in water using buffered solutions at various pH and aqueous solutions with KF as a supporting electrolyte.	[14]
12.	Activated carbon	Activated carbon has been used for the adsorption of nicotinic acid.	[15]
13.	Platonized platinum electrode	The electrosorption of C-14 labelled nicotinic acid was studied at a Platonized platinum electrode in both acid (1 mol dm^{-3} H_2SO_4) and alkaline (0.1 mol dm^{-3} NaOH) media.	[16]
14.	Hydroxyapatite	Adsorption of nicotinic and isonicotinic acid derivatives by hydroxyapatite from aqueous solutions.	[17]
15.	Monodispersed sols of α-Fe_2O_3 and $Cr(OH)_3$	The adsorption of nicotinic, picolinic, and di-picolinic acids on α-Fe_2O_3 and $Cr(OH)_3$ sols consisting of spherical particles of narrow size distributions was measured at solute concentrations up to 4×10^{-4} mole dm^{-3} and over a range of pH values at 25°C.	[18]

1.2 Pyridine

Pyridine is used as an important precursor in manufacturing of vitamins, agrochemicals and pharmaceuticals as well as solvents and reagents for dyes and rubber [19,20]. However, increased demand for pyridine leads to its disposal in water bodies'directly as industrial residue or indirectly as breakdown products and consequently harms both animals and plants in aquatic systems [21,22]. Pyridine is harmful if inhaled, swallowed or absorbed through the skin. Effects of acute pyridine intoxication include dizziness, headache, nausea, salivation, loss of appetite and may progress into abdominal pain, liver damage, pulmonary congestion, unconsciousness and even death [23-26]. Various methods such as biodegradation, photodegradation, catalytic oxidation, liquid membrane separation and adsorption have been developed to remove pyridine from wastewater [27–29]. Some of the effective adsorption techniques for the removal of pyridine are listed in Table 2.

Table 2 Adsorption techniques used for removal of pyridine.

S. No.	Adsorbent	Remarks	Ref.
01.	Zeo adsorbents	Zeo adsorbents on the basis of copper forms of synthetic zeolite ZSM5 and natural zeolite of the clinoptilolite type (CT) have been studied taking into account their environmental application in removing harmful pyridine (py) from liquid and gas phase.	[30]
02.	Fenton oxidation	Isolated Pseudomonas pseudo alcaligenes-KPN in batch culture experiments, wherein the residual pyridine and 3-cyanopyridine removal efficiency was observed to be 84% and >99%, respectively.	[31]
03.	Scandium oxide	This investigation showed that scandium oxide, which is a recycled catalyst, is capable of removing organic nitrogen compounds from fuels.	[32]
04.	Natural phosphate rock and two synthetic mesoporous hydroxyapatites	Experiments performed by the batch method showed that the sorption process occurs by a first order reaction for both pyridine and phenol. In contrast, the Freundlich model was able to describe sorption isotherms for phenol but not for pyridine.	[33]
05.	Bio-zeolite	A specific bio-zeolite composed of mixed bacteria (a pyridine-degrading bacterium and a quinoline – degrading bacterium) and modified zeolite was used for biodegradation and adsorption in two types of wastewater: sterile synthetic and coking wastewater.	[34]
06	Paracoccussp	The effect of different co-substrates including glucose, ammonium chloride and trace elements on biodegradation and removal of pyridine by Paracoccussp. KT-5 was investigated.	[35]

08.	Microporous and mesoporous materials	The adsorption process primarily occurs in the pores of Ti-HMS, which is confirmed by comparing the adsorptive denitrogenation performance of Ti-HMS with uncalcined Ti-HMS.	[36]
09.	Rice husk ash (RHA) and granular activated carbon (GAC)	The simultaneous removal of Pyridine (Py), α-picoline (αPi), and γ-picoline (γPi) from aqueous solutions by sorption onto rice husk ash (RHA) and granular activated carbon (GAC).	[37]
10.	Shewanella putrefaciens	The corn-cob packed bio-trickling filter inoculated by S. putrefaciens should provide excellent performance in the removal of gaseous pyridine.	[38]
11.	Bagasse flyash (BFA)	The simultaneous removal of Pyridine (Py) and its derivatives 2-picoline (2Pi) and 4-picoline (4Pi) by adsorption from aqueous solutions using bagasse fly ash (BFA) as an adsorbent.	[39]
12.	Rice husk ash (RHA) and granular activated carbon (GAC).	The adsorption of pyridine (Py) from synthetic aqueous solutions by rice husk ash (RHA) and commercial grade granular activated carbon (GAC) and reports on the kinetic, equilibrium and thermodynamic aspects of Py sorption.	[40]
13.	Carbon nanotubes	The original CNTs and four treated CNTs were first used as adsorbents to remove pyridine from water and the adsorption isotherms of pyridine on CNTs were studied. At the same time, the effect of pH, temperature, and the adsorption kinetics of the adsorption of pyridine were also evaluated.	[29]
14.	Bagasse fly ash (BFA)	The present study examines the adsorption of pyridine (Py) from aqueous solutions, using bagasse fly ash (BFA), which is a solid waste that is generated from bagasse-fired boilers, as an adsorbent.	[41]
15.	π-Complexation sorbents	The sulphur and nitrogen compounds in transportation fuels (gasoline, diesel and jet fuels) can be selectively removed by a new class of adsorbents, referred to as π-complexation sorbents.	[42]
16.	Activated carbons	Investigates the ability of activated carbons developed from coconut shell to adsorb α-picoline, β-picoline, and γ-picolin from aqueous solution.	[43]

17.	Low cost activated carbons	Activated carbons developed from agricultural waste materials were characterised and utilised for the removal of pyridine from wastewater. The results indicate that the Langmuir adsorption isotherm model fits the data better as compared to the Freundlich adsorption isotherm model.	[44]
18.	Polymeric ligand exchangers (PLE)	The chelating resins with nitrogen donor atoms served as excellent metal hosting polymers and pyridine nitrogen atoms in polymer phase bind with metal ions more effectively than amine nitrogen atoms.	[45]
19.	High-area C-cloth electrodes	The use of high-area carbon cloth (C-cloth) electrodes as quasi-3-dimensional interfaces, coupled with in situ UV-Vis spectrophotometric techniques and scanning kinetics for quantitatively monitoring the adsorption/desorption processes and simultaneously obtaining kinetic parameters are described.	[46]
20.	Rh Odo coccus sp. KCTC3218	Pyridine, a representative nitrogen compound in heavy oil - was degraded by Rh Odo coccus sp. KCTC3218 in a water-heavy oil two-phase system. This microorganism formed floes which could be a barrier to mass transfer between the cells in floes and the pyridine dissolved in water.	[47]
21.	Combined ozonation/fluidised bed biofilm treatment	3-Methylpyridine (MP) and 5-ethyl-2-methylpyridine (EMP) were quantitatively removed in batch ozonation. In continuous combined experiments, wastewater was fed to a fluidised bed biofilm reactor with a mixed culture. The liquid was circulated through an ozonation bubble column. Ozone supply was controlled to keep the dissolved ozone concentration at a low level in the oxidation reactor.	[48]
22.	Hydroxyls in NaHZSM-5zeolite	The effect of removal of some hydroxyls by dehydroxylation and of pyridine sorption on the acid strength of OH groups remaining in the zeolite and contributing to the 3609 cm^{-1} band was studied.	[49]
23.	Spent rundle oil shale	Adsorption of pyridine from aqueous solutions onto the solids generated by the processing of Rundle oil shale shows that the adsorption isotherm is of Langmuir type (L-4) with two plateaux.	[50]
24.	Coal, coal extracts and coal residues	Techniques for the removal of pyridine and quinoline from coal, coal extracts and coal residues have been developed.	[51]

1.3   Aniline ($C_6H_5NH_2$)

Aniline also known as phenylamine, aminobenzene or benzylamine is a nitrogen-containing organic compound, widely used as raw materials and intermediate chemicals in the manufacturing of pesticides, herbicides, rubber, dyestuff, varnishes, organic paints and pigments, azo dyes, pharmaceuticals, petrochemicals and other industries [52,53]. These are considered to be very toxic water pollutants even in very low concentration to aquatic life and human beings [54–58]. Due to negative impact of aniline on environment and its derivatives in wastewater, various methods such as biological degradation [59–63], catalytic oxidation process [64–67], electrochemical techniques [68] , irradiation treatment [69], adsorption (Han et al., 2006; Xie et al., 2007) and other methods (Datta, 2003; Devulapalli and Jones, 1999) have been proposed for the successful removal of aniline compounds from wastewater. The adsorption of aniline on various adsorbents has been listed in Table 3.

Table 3 List of adsorbents used for removal of aniline, its isomers and aniline derivatives by adsorption techniques.

S. No.	Adsorbent	Remarks	Ref.
16.	Hyper-cross-linked Macronet resin (MN200)	This work was conducted to evaluate the sorption performance of hyper-cross-linked Macronet resin (MN200) compared to the granular activated carbon in order to remove phenol and aniline from aqueous solution in both single and binary solutions.	[76]
17.	poly(ethylene terephthalate) (PET) fibers	Experiments were conducted in aniline monomer and hydrochloric acid solution with the variables such as contact time, initial concentration, and temperature, which can enhance the equilibrium adsorption capacity to aniline of poly(ethylene terephthalate) (PET) fibres.	[77]
18.	Nanostructured Co_3O_4/CeO_2 catalyst	Kinetic modelling of catalytic wet air oxidation (CWAO) of aniline was investigated in a bubble reactor over a nanostructured Co_3O_4 (10 wt %)/CeO_2 catalyst.	[78]

19.	Hyper cross-linked polymeric resin (AH-1)	In the present work, a modified hyper-cross-linked polymeric resin (AH-1) with tertiary amino groups was adopted for adsorbing aniline from aqueous solution, and a traditional hyper crosslinked polymeric adsorbent NDA-100 was selected for comparison.	[78]
20.	Au(111)electrode	*Insitu* scanning tunnelling microscopy (STM) was used to study the adsorption and polymerization of aniline on Au(111) single-crystal electrode in 0.1 M perchloric acid and 0.1 M benzene sulfonic acids (BSA) containing 30 mM aniline, respectively.	[79]
21.	Polymeric adsorbent by multiple phenolic hydroxyl groups	A hyper-cross-linked polymeric adsorbent functionalized with multiple phenolic hydroxyl groups HJ-03 was prepared in this study and its adsorptive characteristics for p-nitro aniline from aqueous solution were studied as compared with Amberlite XAD-4.	[80]
22.	Activated carbon from the shell of caryacathayensis.	The conditions for preparation of activated carbon from walnut (Caryacathayensis S.) shell were studied with phosphate method and optimised with orthogonal experiments taking phosphate concentration, carbonising temperature and time as influential factors, methylene blue adsorption, iodine value and yield as indexes, the adsorption process was analysed in thermodynamics.	[81]
23.	Polyvinylchloride (PVC) and polystyrene (PS)	The fibres had bigger hydrophilicity, the better ability of acid, and alkali corrosion resistance, so they had better practical application value. This type of ion exchange fibres had faster absorption property and better working stability to aniline and could be used repeatedly, so they were applied for treatment of waste water containing aniline with a promising prospect.	[82]

24.	Polyaniline (PANI) multiwalled carbon nanotubes (MWCNTs)	The polyaniline multiwalled carbon nanotube magnetic composite was prepared by plasma-induced graft technique and its application for removal of aniline and phenol.	[83]
25.	α-zirconium phosphate (α-Zr P)	Molecular dynamics simulation of adsorption of aniline by α-zirconium phosphate.	[84]
26.	Wetland soil	Batch equilibrium experiments were performed to assess adsorption and desorption characteristics of nitrobenzene and aniline by wetland soil.	[85]
27.	Polymethacrylic acid silicon dioxide. (PMAA/SiO$_2$)	The adsorption properties and mechanism of PMAA/SiO$_2$ towards aniline were researched through batch and column adsorption methods.	[86]
28.	Activated carbon fibres (ACFs)	Activated carbon fibres (ACFs) were prepared for the removal of p-nitroaniline (PNA) from the cotton stalk by chemical activation with NH$_4$H$_2$PO$_4$ and subsequent physical activation with steam.	[87]
29.	Phenolic resin	Dumwald-Wagner model can illuminate the adsorption process of aniline onto LM-4 much better due to the higher correlation coefficient. Kannan-Sundaram model indicates that the adsorption of aniline onto LM-4 is a favourable adsorption.	[88]
30.	Activated carbon and hyper-cross-linked polymeric resin (MN200)	Kinetic adsorption of phenol and aniline from aqueous solution onto activated carbon and hyper cross-linked polymeric resin MN200 were evaluated in the single and binary system.	[89]
31.	β-cyclodextrin polymer (β-CDP)	Removal of aniline from aqueous solutions by the β-cyclodextrin polymer (β-CDP) using batch adsorption experiments.	[90]
32.	Cr-bentonite	The sorption characteristics of aniline on Cr-bentonite prepared using synthetic wastewater containing chromium was investigated in a batch system at 30 °C.	[91]

33.	Rice bran	The effects of particle size, bio sorbent dosage, pH and temperature on biosorption capacity were studied with static experiments, and the adsorption process was analysed in thermodynamics and kinetics, and the adsorption mechanism analysed by infrared spectroscopy.	[92]
34.	Carboxylated diaminoethane sporopollenin (CDAE-S)	A dynamic method called stepwise frontal analysis (SFA) was used to derive equilibrium sorption data of copper and aniline on a sporopollenin (CDAE-S) solid phase.	[93]
35.	Polymeric adsorbent and XAD-4 (highly adsorbent resin)	Phenolichydroxylgroupsmodifiedhyper-cross-linked polymeric adsorbent HJ-02 was prepared and it was applied to remove p-nitroaniline in aqueous solution as compared with Amberlite XAD-4.	[94]
36.	Cross-linked starch sulphate	A new environment-friendly adsorbent, cross-linked starch sulphate (CSS), was prepared and used to adsorb aniline from aqueous solution.	[52]
37.	Phenolic resin	Using liquid paraffin as disperse phase, span80 as a dispersant, ethylene glycol as a porogen, adsorbents of LM-3 and LM-4 were prepared by inverse suspension polymerization, and their performance was compared with LM-1 and LM-2 which prepared by solution polycondensation.	[95]
38.	Polyacrylamide silicon dioxide (PAM/SiO$_2$)	In this paper, functional monomer acrylamide (AM) was grafted step by step onto the surface of silica gel particles using 3-methacryloxypropyl trimethoxysilane (MPS) as a coupling agent and the grafted particle PAM/SiO$_2$ was prepared.	[96]
39.	Activated carbon fibre	Activated carbon fibre prepared from cotton stalk was used as an adsorbent for the removal of p-nitroaniline (PNA) from aqueous solutions.	[97]

40.	Activated carbon	Activated carbon prepared from cotton stalk fibre has been utilised as an adsorbent for the removal of 2-nitroaniline from aqueous solutions.	[98]
41.	Si (5 5 12) -2×1surface	A scanning tunnelling microscopy and first-principles calculations study of the adsorption structures of aniline on a Si (5 5 12) -2×1surface.	[99]
42.	CuO doped activated carbon	Adsorption of aniline, benzene and pyridine from water on a copper oxide doped activated carbon (CuO/AC) at 30 °C and oxidation behaviour of the adsorbed pollutants over CuO/AC in a temperature range up to 500 °C are investigated in TG and tubular reactor/MS systems.	[100]
43.	Anatase TiO$_2$ (100)surface	A computational technique based on semi-empirical SCF-MO method MSINDO has been used for the investigation of adsorption and initial oxidation step for the photodegradation of aniline on anatase TiO$_2$ (100) surface.	[101]
44.	Modified montmorillonite	Adsorption of phenol and aniline onto original and with quaternary ammonium salts (QASs)-modified montmorillonite was described by sorption isotherms of type III and II respectively	[102]
45.	Multi-walled carbon nanotubes	Aqueous adsorption of a series of phenols and anilines by a multiwalled carbon nanotube material (MWCNT15), which depends strongly on the solution pH and the number and types of solute groups was investigated in this study.	[58]
46.	XDA-1 resin	A combined physical-biological method consisted of the physical adsorption of aniline from hypersaline effluents by resins and the biodegradation of the adsorbate and the regeneration of the adsorbent in the subsequent stage.	[103]

47.	A solid catalytic adsorbent	The focus of the work described was the development of a solid catalytic adsorbent material capable of being regenerated, with the ability to adsorb aniline from aqueous solution and the subsequent catalytic oxidation of the adsorbed aniline.	[104]
48.	Cu-Beta adsorbent materials	The aim of this research was to develop a solid regenerable catalytic adsorbent capable of removing aniline from aqueous solutions.	[66]
49.	Multiwalled carbon nanotubes	In this work, oxidised multiwalled carbon nanotubes (MWCNTs-COOH) coated on the outer surface of the fused-silica tube and inserted in the polyether ether ketone (PEEK) tubing, which was fixed directly on the six-port injection valve to substitute for the sample loop.	[105]
50.	Silicate monoliths (HOM-2)	Hexagonally highly ordered monolithic mesoporous silica and aluminosilicate composites (HOM-2 and Al/HOM-2, respectively) were used as adsorbents for removal of aniline in aqueous solution.	[106]
51.	Dimethyl di tallowy ammonium montmorillonite	The expansion behaviour of an organically modified montmorillonite during the adsorption of increasing amounts of an organic pollutant: 2-chloroaniline (2-CA).	[107]
52.	Sodium tetraborate-modified Kaolinite clay adsorbent	Raw Kaolinite clay obtained from Ubulu-Ukwu, Delta State of Nigeria and its sodium tetraborate (NTB)-modified analogue was used to adsorb aniline blue dye.	[108]
53.	Fuel oil fly ash	Fuel oil fly ash has been tested as a low-cost carbon-based adsorbent of 2-chlorophenol (CP), 2-chloroaniline (CA) and methylene blue (MB) from aqueous solutions.	[109]
54.	Macroporous kaolin	Adsorption kinetics of small molecular mass anilines (<205 g/mol) over macroporous kaolin was studied.	[110]

55.	Mesoporous materials, Na-AlMCM-41 and Na-AlSBA-15catalysts	In this work, the water adsorption and adsorbed aniline interaction onto the mesoporous materialsNa-AlMCM-41 was studied and compared with aniline adsorption onto Na-AlSBA-15.	[111]
56.	Imprinted polymer(MIP)	The synthesised MIP was then tested by the equilibrium adsorption method, and the MIP demonstrated high removal efficiency to the aniline.	[112]
57.	Polyurethane foams, diffuse reflectance spectroscopy	The adsorption of aniline as its 4-nitrophenylazo derivative was studied on polyurethane foams depending on phase contact time, reagent concentration, and type of adsorbent.	[113]
58.	Zeolite ZSM-5	Zeolite ZSM-5 was investigated with a view to examining its potential to function as both adsorbent and catalyst in the removal of aniline from aqueous solutions.	[114]
59.	Multi-walled carbon nanotubes	The thermodynamic behaviours of aniline adsorption on the surface of chemical modified carbon nanotubes (CNTs) are studied, which indicate to understand their surface chemical characteristics.	[72]
60.	Organo-clays	The aims of this study were to make use of organo-clays (i.e., Cloisite-10A, Cloisite-15A, Cloisite-30B and Cloisite-93A), to remove p-nitrophenol, phenol and aniline of organic pollutants.	[115]
61.	Carboxylated polymeric adsorbent ZK-1	In this study, a carboxylated polymeric adsorbent ZK-1 was synthesised for enhanced removal of p-nitroaniline (PNA) from aqueous solution.	[116]
62.	Basic activated carbon	The interaction of phenol and aniline with the surface of highly micro porous ash-free carbon was systematically studied in aqueous solutions in the pH range 3-11 under oxic conditions.	[55]
63.	Activated carbon fibre	Electrosorption-enhanced solid-phase microextraction (EE-SPME) based on activated carbon fibre (ACF) was developed for the determination of aniline in aqueous solution.	[117]

64.	Activated carbons(ACs)	The heterogeneity of activated carbons (ACs) prepared from different precursors is investigated on the basis of adsorption isotherms of aniline from dilute aqueous solutions at various pH values.	[118]
65.	Carboxylated polymeric sorbent	In this study, a carboxylated styrene-divinylbenzene (St-DVB) polymeric sorbent (CSPS) was prepared for enhanced removal of p-chloro aniline from aqueous solution.	[119]
66.	Polymer adsorbents(NDA-100, XAD-4, NDA-16 and NDA-1800)	The adsorption equilibria of phenol and aniline on nonpolar polymer adsorbents (NDA-100, XAD-4, NDA-16 and NDA-1800) were investigated in single- and binary-solute adsorption systems at 313 K.	[56]
67.	Non-polar adsorbents	Adsorption equilibria of phenol and aniline onto nonpolar macro-reticular adsorbents were investigated in single and binary-solute aqueous systems at 293 K and 313 K.	[120]
68.	Gold nanoparticles	Surface-enhanced Raman scattering (SERS) spectra of p-, m-, and o-nitroaniline (PNA, MNA, and ONA) adsorbed on gold nanoparticles were studied, respectively, in a gold colloidal solution and on dried gold-coated filter paper.	[121]
69.	Montmorillonite and kaolinite	The sorption-desorption behaviour of the environmentally hazardous industrial pollutants and certain pesticides degradation products, 3-chloroaniline, 3,4-dichloroaniline, 2,4,6-trichloroaniline, 4-chlorophenol, 2,4-dichlorophenol and 2,4,6-trichlorophenol on the reference clay kaolinite KGa-1 and Na-montmorillonite SWy-l.	[122]
70.	Activated carbon fibres	This study describes the anodic polarisation of activated carbon fibres (ACFs), which can enhance the adsorption rate and capacity of aniline.	[71]

71.	Carbon nanotubes (MWNT)	Aniline-modified multi-walled carbon nanotubes (MWNT) have been obtained through interaction between carboxylated MWNT and aniline in aqueous solution.	[123]
72.	Multi-walled carbon nanotubes(designated as c-MWNTs and a-MWNTs)	Based on the π-π* electron interaction between aniline monomers and functionalized MWNT and hydrogen bonding interaction between the amino groups of aniline monomers and the carboxylic acid/acylchloride groups of functionalized MWNT, aniline molecules were adsorbed and polymerised on the surface of MWNTs.	[124]
73.	Hyper cross-linked polymeric adsorbents	Competitive and cooperative simultaneous adsorptions of phenol and aniline from aqueous solutions by hyper-cross-linked polymeric adsorbents (NDA103, NDA101, NDA100) were investigated.	[125]
74.	Non-polar resin adsorbents	A new model was developed to describe the synergistic adsorption in the binary system (phenol/aniline in aqueous solution) onto non-polar resin adsorbent Amberlite XAD-4 and NDA-100.	[126]
75.	Modified bentonite	The adsorption capacities of clay-organic complexes (bentonite-EDTA and bentonite-HDTMA) are higher than those of bentonite-HNO_3 and pure bentonite.	[127]
76.	Montmorillonite and kaolinite	This study reports the sorption-desorption behaviour of the environmentally hazardous industrial pollutants and certain pesticides degradation products, 3-chloroaniline, 3,4-dichloroaniline, 2,4,6-trichloroaniline, 4-chlorophenol, 2,4-dichlorophenol and 2,4,6-trichlorophenol on the reference clay kaolinite KGa-1 and Na-montmorillonite SWy-1.	[122]

77.	Non-polar macro-reticular adsorbents	The adsorption behaviours of phenol and aniline on non-polar macro-reticular adsorbents (NDA100 and Amberlite XAD4) were investigated in single or binary batch system at 293K and 313K respectively in this study. The results indicated that the adsorption isotherms of phenol and aniline on both adsorbents in both systems fitted well Langmuir equation, which indicated a favourable and exothermic process.	[128]
78.	Bifunctional polymeric resin (LS-2) with sulfonic groups	A hydrophilic bifunctional polymeric resin (LS-2) with sulfonic groups was synthesised, and the adsorption performance of three aniline compounds, aniline, 4-methylaniline, and 4-nitroaniline onto LS-2 was compared with that on the commercial Amberlite XAD-4.	[129]
79.	Bifunctional polymeric adsorbent modified by sulfonic groups	A new bifunctional polymeric resin (LS-2) was synthesised by introducing sulfonic groups onto the surface of the resin during the post-crossing of chloromethyl low cross-linking macroporous polystyrene resin, and the comparison of the adsorption properties of LS-2 with Amberlite XAD-4 toward aniline and 4-methylaniline in aqueous solutions was made.	[53]
80.	Ru/SiO$_2$ catalyst	(Ru/SiO$_2$catalyst) the heterogeneous catalyst was found to be very effective in the complete degradation of aniline and also to be active in the conversion of the -NH$_2$ group in aniline into N$_2$ gas.	[130]
81.	Activated carbons	Microporous carbons of the similar surface area (1200-1500 m^2/g) and porosity but different surface composition were prepared from poly(ethylene terephthalate) (PET) based activated carbon by chemical (cc HNO$_3$) and thermal (700°C) post-treatment. The waste removal capacity was studied by adsorption from buffered aqueous phenol and aniline solutions.	[131]

82.	Silver, Ag(111)	The adsorption of sub-monolayer aniline on the Ag(1 1 1) surface has been characterised using temperature programmed desorption (TPD) and electron energy loss spectroscopy (EELS).	[132]
83.	H-beta zeolites and copper-exchanged beta zeolites	Zeolite beta, a large-pore zeolite, was investigated in this study with a view to examining it as a potential adsorbent for the removal of aniline from aqueous solutions.	[133]
84.	Aminated-macron porous hyper-cross-linked resins	The sorption of phenol and aniline on the synthesised resins in aqueous and non-aqueous solutions was investigated.	[134]
85.	QSAR model by using artificial neural networks (ANNs).	Based on descriptors of n-octanol/water partition coefficients (log K_{ow}), molecular connectivity indices, and quantum chemical parameters, several QSAR models were built to estimate the soil sorption coefficients (log K_{oc}) of substituted anilines and phenols.	[135]
86.	Cobalt(II)-poly(vinyl chloride)-carboxylated diaminoethane (PVC-CDAE) resin	The adsorption of aniline from aqueous solutions onto cobalt(II)-poly(vinyl chloride)-carboxylated diaminoethane (PVC-CDAE) resin has been studied using a mini-column apparatus at $25 \pm 0.1°C$.	[136]
87.	High-area carbon-cloth	The adsorption of aniline compounds; aniline, p-toluidine, 1-napthylamine and sodium salt of diphenylamine-4-sulfonic acid from solutions in H_2O, in 1 M H_2SO_4 or in 0.1 M NaOH onto activated carbon cloth was studied by in situ UV spectroscopy.	[137]
88.	Nanofiltration membranes (polyamide and cellulose acetate)	This study evaluates the performance of two nanofiltration membranes in removing a herbicide: dichloroaniline. The membranes, one polyamide and one cellulose acetate, have a cut-off in the range 150-300 g/mol.	[138]

89.	Kaolinite and montmorillonite	Batch experiments have been performed in order to evaluate the ability of the two reference clays kaolinite (KGa-1) and Na-montmorillonite (SWy-1) to retain three representative chloro anilines: 3-chloroaniline, 3,4-dichloroaniline and 2,4,6-trichloroaniline.	[139]
90.	Commercial activated carbon	Studies on the removal of aniline blue (AB) and acid violet (AV) by commercial activated carbon (CAC), have been made at 30°C.	[140]
91.	Sediment	In natural water systems, sorption of an organic pollutant to soil/sediment is often influenced by coexisting organic compounds.	[141]
92.	Copper-exchangedZSM-5 and unmodifiedH-ZSM-5	Three medium-pore aluminosilicates were investigated with a view to examining their potential as adsorbents for the removal of aniline from aqueous solutions. H-ZSM-5 was exchanged with copper to prepare two different metal-loaded zeolites.	[142]
93.	Alkyl-graftedMCM-41	Molecular selective adsorption of alkyl phenols and alkyl anilines onto n-alkyl graftedMCM-41 with different alkyl chain lengths and Al contents was studied. Octyl groups gave better adsorbent performance than pentyl and dodecyl groups.	[143]
94.	Organo-zeolite	In batch tests, the adsorption capacity of aniline and nitrobenzene onto natural zeolite surface is very low or almost nil but becomes significant upon modifying the zeolite surface by hexadecyltrimethylammonium (HDTMA).	[144]

95.	Particulate soil organic matter	The distribution of TNT* (the sum of TNT and its degradation products), aniline, and nitrobenzene between particulate organic matter (POM), dissolved soil organic matter (DOM), and free compound was studied in controlled kinetic (with and without irradiation) and equilibrium experiments with mixtures of POM and DOM reflecting natural situations in organic-rich soils.	[145]
96.	Microwave-assisted headspace solid-phase microextraction and gas chromatography	Determination of aniline in wastewater was investigated by microwave-assisted headspace solid-phase microextraction (MA-HS-SPME), for one-step in-situ sample preparation, and gas chromatography.	[146]
97.	Carbon materials: a mesoporous high surface area graphite (HSAG) and a microporous activated carbon(AC).	The adsorption behaviour of phenol, aniline and phenol-aniline mixtures in water over carbonaceous material surfaces has been studied.	[147]
98.	Silver and gold colloid	Raman and surface-enhanced Raman scattering (SERS) spectra of o-, m-, and p-nitroaniline in silver and gold colloidal solutions were measured under off-resonance conditions.	[148]
99.	MCM-41mesoporousmaterial	The adsorption of aniline on Na-AlMCM-41 synthesised by us has been characterised by infrared spectroscopy, temperature programmed desorption (TPD), and differential thermal analysis methods.	[149]
100.	Fe-bound and Zn-bound SDS micelles	The co-adsorption of aniline by complex formation with Zn-bound and Fe-bound micelles of anionic surfactant SDS (sodium lauryl sulphate) causes redox reactions.	[150]
101.	Carbon cloth (C-cloth)	Removal of aniline, 2,2'-bipyridyl,4,4'-bipyridyl and their protonated cations from dilute aqueous solutions, simulating polluted wastewaters, by adsorption and electrosorption at high area carbon cloth (C-cloth) electrodes is described.	[151]

102.	Chromium ferrocyanide	The interaction of aniline and p-anisidine with chromium ferrocyanide has been studied.	[152]
103.	Resin-bound cobaltion	Studies have been made of the sorption equilibria of chlorinated anilines in aqueous solution on ligand exchange resin. The chlorinated anilines used included 2-chloroaniline, 3-chloroaniline, 4-chloroaniline and 2,5-dichloroaniline.	[153]
104.	Quasi-3-dimensionalporouselectrodes	Applications are made to the removal of pollutants such as salts of S-containing anions, K-ethylxanthate, phenol, aniline and its sulphate, and choline hydroxide from aqueous solutions at low concentrations.	[154]
105.	Ion-exchanged montmorillonites	The adsorption of 2-(trifluoromethyl)aniline (2TFMA) on the interlamellar surfaces of KSF, K-10, Cu^{2+}-, Ni^{2+}-, Fe^{3+}- and Al^{3+}-montmorillonites has been investigated.	[155]
106.	Activated carbon.	Pilot plant study was made of the extraction of aniline from the soil. It is a two-step integrated plant comprising CO_2supercriticalextraction and pollutant adsorption onto activated carbon.	[156]
107.	Carbon-paste electrodes	The relationship between the energies of adsorption of molecules on active carbons and the peak current constants was determined.	[157]
108.	Cu(II)-montmorillonite	Batch adsorption experiments in the presence of oxygen were performed to study the interlayer reactions of aniline on Cu(II)-montmorillonite in aqueous solutions.	[158]
109.	Dual-cation organo bentonites	The influential factors, mechanisms and characteristics of polar and ionizable organic contaminants, such as p-nitrophenol, phenol, and aniline, and sorption to dual cation organo bentonites from water are investigated systematically and described quantitatively.	[159]

110.	Zeolites	The adsorption sites inside the zeolite channels and the diffusion characteristics of acylated products of 4-aminophenol are analysed in detail.	[160]
111.	Si(100)(2 × 1)	The chemisorption of aniline ($C_6H_5NH_2$) on Si(100)(2 × 1) at room temperature has been studied for the first time with scanning tunnelling microscopy (STM) and spectroscopy (STS).	[161]
112.	Solid adsorbent	Determinations of the adsorption of aniline and pyridine at solid-liquid interfaces from n-hexane solutions have been performed.	[162]
113.	Co-precipitated Co/Al_2O_3 and Ni/Al_2O_3 catalysts	The adsorption and catalytic properties of Co/Al_2O_3 were compared with those of Ni/Al_2O_3 catalysts containing the same metal content and prepared under similar conditions.	[163]
114.	Si(100)-2 × 1	The chemisorption of benzoic acid (C_6H_5-COOH) and aniline (C_6H_5-NH_2) on Si(100)-2 × 1 at room temperature has been studied with high-resolution electron energy loss spectroscopy (HREELS) and low electron energy diffraction (LEED)	[164]
115.	Zeolites	Separation of N-alkyl-substituted anilines by selective sorption on zeolites is investigated	[165]
116.	Smooth polycrystalline platinum electrode	Adsorption of aniline on smooth polycrystal line platinum was studied in 0.5 M H_2SO_4 aqueous medium as a function of the aniline concentration (10^{-8} -10^{-3}M) and electrode potential (-670 to + 1000 mV (MSE)) by potential programmed voltammetry.	[166]
117.	Ni(100)	The bonding and reactions of adsorbed aniline have been characterised on the Ni(100) surface both in hydrogen and in a vacuum with a combination of surface spectroscopies.	[167]
118.	Bentonite	Bentonite was exposed to aniline through batch experiments and flexible wall conductivity tests.	[168]

119.	Activated carbon	This paper examines the influence of molecular oxygen and pH on the adsorption of aniline to F-300 Calgon Carbon.	[169]
120.	Polycrystalline Pt, Rh and Pd electrodes	The different metals show remarkable differences concerning the stability of the adsorbed layers. The anodic desorption products of pre-adsorbed aniline are CO_2 (on Pd) or CO_2, HCN and NO (on Pt and Rh).	[170]
121.	Layer silicate lays and an organic soil	The characteristics of isotherms for aniline adsorption on Ca-saturated kaolinite, montmorillonite, vermiculite, and an organic soil were determined for a wide range of aqueous concentrations at acid and neutral pH.	[171]
122.	Organo-clay	In this study, the time-dependent adsorption and desorption of phenol and aniline on hexadecyl trimethyl-ammonium-montmorillonite (HDTMA-montmorillonite) from aqueous solutions were determined in a stirred-flow chamber.	[172]
123.	Zeolites	The formation of the aniline radical cation was observed on H-mordenite and H-faujasite in which the distribution of aluminium among different framework and non-frame work coordination sites has been characterised by high-resolution solid-state NMR.	[173]
124.	Nonlinear optical	We have measured the optical second harmonic generation (SHG) at a mercury electrode. The influence of the applied potential and of the adsorption of aniline on the nonlinear optical response was studied by simultaneously measured differential capacitance.	[174]
125.	Granular activated carbon samples	Adsorption equilibrium studies of some aromatic organic pollutants in water with some commercially available standard grades of granular activated carbons have been carried out at 35°C.	[175]

126.	Cadmium selenide	Adsorption of ring-substituted aniline derivatives, presumably through the amino group, onto the (0001) face of single-crystal n-CdSe or n-CdS [CdS(e)] profoundly affects the semiconductor's photoluminescence (PL) by effecting charge transfer between surface states and the bulk semiconductor.	[176]
127.	Polycrystalline gold electrode	Combined capacity and electro reflectance measurements have been employed to investigate the adsorption of aniline on a polycrystalline gold electrode in neutral aqueous solution.	[177]
128.	Gold	The electrosorption of aniline on a polycrystalline gold electrode from acidic and neutral electrolyte solutions has been studied using surface enhanced Raman spectroscopy and tensiometry.	[178]
129.	Ag electrode	The adsorption of aniline on an Ag electrode in acidic aqueous solutions has been studied by surface-enhanced Raman scattering.	[179]
130.	Carbon paste electrode	The voltammetric behaviour of aniline and some of its N-alkylated derivatives at a carbon paste (Nujol/graphite) electrode is examined.	[180]
131.	A gold electrode	The adsorption and polymerization of aniline in an acidic electrolyte solution on a gold electrode were investigated using Surface Raman Spectroscopy.	[178]
132.	Silver	The potential- and pH-dependent adsorption of aniline on silver have been studied using Surface Enhanced Raman Spectroscopy.	[181]
133.	Modified silica surfaces	Adsorption of benzene and aniline on C_{18} modified silica columns is investigated as a function of solute concentration and solvent composition.	[182]
134.	Lead	Effects in electro reflectance (ER) at the metal/electrolyte interface are discussed which are due to charge transfer between the adsorbed molecules and the electrode and to intramolecular optical transitions.	[183]

135.	Granular activated carbon.	A linear correlation between adsorptive affinity and acidic properties of adsorbates is evidenced when phenol. aniline and their o-, m-, and p-nitro-derivatives are adsorbed on granular activated carbon.	[184]
136.	Powdered carbon	Adsorption of some phenyl carbamates, phenyl ureas and anilide pesticides on powdered activated carbon was investigated.	[185]
137.	Clay minerals	Adsorption of IPC, CIPC, Linuron, Neburon and Vitavax on bentonite clays (H, Fe and Ca forms) was investigated.	[186]
138.	HYzeolites	The position of the N_{1s} level of aniline adsorbed on HYzeolite undergoes a shift of approximately-2 eV when the calcination temperature exceeds 400°C which is the dehydroxylation temperature for the zeolite.	[187]
139.	Mercury	A preliminary study has been made of the amplitude modulation signal connected with adsorption-desorption of aniline on mercury using a 2 MHz modulation polarograph.	[188]
140.	Montmorillonite and hectorite	Quantitative measurements are made of the adsorption of benzidine and aniline from aqueous hydrochloride solutions by Na-, Li-, and Ca-montmorillonite and of the displaced inorganic cations.	[189]
141.	Alumina and HCl-treated alumina	Infrared spectra of adsorbed aniline on alumina showed bands at 1605, 1575, 1495 and 1470 cm^{-1} in the region of ring stretching vibration.	[190]
142.	Alumina	Linear isotherm free energies of adsorption from n-pentane onto 3.6% $H2O-Al_2O_3$ are reported for 66 nitrogen compounds related to pyridine, aniline or pyrrole.	[191]

Conclusion

Nitrogen is ubiquitous in all living organism in different forms and is also considered key constituent in a number of compounds which are chemically and biologically of great importance. In this review, some of the nitrogen containing compounds useful to the human beings as well as plants with their importance and the threats they impose on living beings and the environment are discussed. They imposed some threats to the human beings and environments. Therefore, the adsorption method for their removal from the environment by using various types of adsorbents is discussed.

References

[1]   R.D.R. M. Cox, Lehninger, L. Albert, Nelson, Lehninger principles of biochemistry, New York, 2000.

[2]   K. Parker, L. Brunton, Goodman, L. Sanford, Lazo, S. John, Gilman, Alfred. Goodman, Gilman's the pharmacological basis of therapeutics, New York,

[3]   M.K. Mittal, T. Florin, J. Perrone, J.H. Delgado, K.C. Osterhoudt, Toxicity from the use of niacin to beat urine drug screening, Ann. Emerg. Med. 50 (2007) 587–590. https://doi.org/10.1016/j.annemergmed.2007.01.014

[4]   G. Akçay, K. Yurdakoç, Removal of nicotine and its pharmaceutical derivatives from aqueous solution by raw bentonite and dodecyl ammonium-bentonite, J. Sci. Ind. Res. (India). 67 (2008) 451–454.

[5]   N. Yang, X. Wang, Thin self-assembled monolayer for voltammetrically monitoring nicotinic acid in food, Colloids Surfaces B Biointerfaces. 61 (2008) 277–281. https://doi.org/10.1016/j.colsurfb.2007.09.002

[6]   Z. Li, G. Yang, S. Liu, Y. Chen, Adsorption isotherms on nicotinamide-imprinted polymer stationary phase, J. Chromatogr. Sci. 43 (2005) 362–366. https://doi.org/10.1093/chromsci/43.7.362

[7]   A. Pyka, J. Sliwiok, A. Niestrój, A comparison of chromatographic separation of selected nicotinic acid derivatives by TLC and HPLC techniques, Acta Pol. Pharm. 60 (2003) 327—333.

[8]   A. Pyka, A. Niestroj, A. Szarkowicz, J. Sliwiok, Use of TLC and RPTLC for separation of nicotinic acid derivatives, J. Planar Chromatogr. – Mod. TLC. 15 (2002) 410–413. https://doi.org/10.1556/JPC.15.2002.6.3

[9]   E. Karadağ, D. Saraydin, Swelling of superabsorbent acrylamide/sodium acrylate hydrogels prepared using multifunctional crosslinkers, Turkish J. Chem. 26 (2002) 863–875.

[10]   K.T. N.Nanbu, F. Kitamura, T. Ohsaka, Adsorption behavior of pyridine carboxylic acids in acidic solution on a polycrystalline gold electrode surface studied by infrared reflection absorption spectroscopy Electrochem. 67 (1999) 1165.

[11]   M.J.-H. M. Miłkowska, The adsorption of nicotinic acid, nicotinamide and nipecotamide from aqueous solutions on the (110) face of silver, Pol. J. Chem. Tech. 70 (1996) 783.

[12]   S.M. Park, K. Kim, M.S. Kim, Adsorption of picolinic and nicotinic acids on a silver sol surface investigated by Raman spectroscopy, J. Mol. Struct. 344 (1995) 195–203. https://doi.org/10.1016/0022-2860(94)08439-O

[13]   L. Roivas, P.J. Neuvonen, Reversible adsorption of nicotinic acid onto charcoal In vitro, J. Pharm. Sci. 81 (1992) 917–919. https://doi.org/10.1002/jps.2600810916

[14]   M. Jurkiewicz-Herbich, Adsorption of nicotinic acid at the mercury—solution interface, J. Electroanal. Chem. 332 (1992) 265–278. doi:10.1016/0022-0728(92)80355-8. https://doi.org/10.1016/0022-0728(92)80355-8

[15]   A.A. M. Qureshi, K. G. Varshney, K. Z. Alam, Adsorption of tertiary nitrogen-containing compounds on activated carbon. I. equilibrium studies of nicotinic acid in aqueous systems, Coll. and Surf. 50 (1990) 7. https://doi.org/10.1016/0166-6622(90)80249-4

[16]   G. Horányi, A radiotracer study of the adsorption and electrocatalytic reduction of nicotinic acid at a platinized platinum electrode, J. Electroanal. Chem. Interfacial Electrochem. 284 (1990) 481–489. https://doi.org/10.1016/0022-0728(90)85052-7

[17]   N. Nambu, T. Nagai, Adsorption of nicotinic and isonicotinic acid derivatives by hydroxyapatite from aqueous solutions., Chem. Pharm. Bull. (Tokyo). 29 (1981) 2093–2096. https://doi.org/10.1248/cpb.29.2093

[18]   C.G. Pope, E. Matijević, R.C. Patel, Adsorption of nicotinic, picolinic and di-picolinic acids on monodispersed sols of α-Fe2O3 and Cr(OH)3, J. Colloid Interface Sci. 80 (1981) 74–83. https://doi.org/10.1016/0021-9797(81)90161-2

[19]   R. L. Ramos, R. O. Perez, O. L. T. Rivera, M. S. B. Mendoza, Gilman's the pharmacological basis of therapeutics., J. Basic Princ. Diffus. Theory, Exp. Appl. 11 (2009) 1.

[20] C. E. Terry, R. P. Ryan, S. S. Leffingwell, Toxicology desk reference: The toxic exposure & medical monitoring index: the toxic exposure and medical monitoring index, 5th ed.

[21] G.D. Henry, De novo synthesis of substituted pyridines, Tetrahedron. 60 (2004) 6043–6061. https://doi.org/10.1016/j.tet.2004.04.043

[22] Z. Yanli, L. Dongguang, Adsorption of pyridine on post-crosslinked fiber, J. Sci. Ind. Res. (India). 69 (2010) 73–76.

[23] G. Aylward, SI Chemical Data, 6th ed., 2008.

[24] IARC Monographs, OSHA, IARC, Washington D.C., 1985.

[25] N. Bonnard, M. T. Brondeau, S. Miraval, F. Pillière, J. C. Protois, O. Schneider. Schneider, Pyridine, Fiche Toxicologique, in: INRS (in French).

[26] Pyridine Summary & Evaluation. IARC Summaries & Evaluations. IPCS INCHEM, in: Int. Agency Res. Cancer (2000).

[27] J. Herrmann, Heterogeneous photocatalysis: fundamentals and applications to the removal of various types of aqueous pollutants, Catal. Today. 53 (1999) 115–129. https://doi.org/10.1016/S0920-5861(99)00107-8

[28] D. Drijvers, H. Van Langenhove, M. Beckers, Decomposition of phenol and trichloroethylene by the ultrasound/H2O2/CuO process, Water Res. 33 (1999) 1187–1194. https://doi.org/10.1016/S0043-1354(98)00308-X

[29] B. Zhao, H. Liang, D. Han, D. Qiu, S. Chen, Adsorption of pyridine from aqueous solution by surface treated carbon nanotubes, Sep. Sci. Technol. 42 (2007) 3419–3427. https://doi.org/10.1080/01496390701511689

[30] M. Reháková, Ľ. Fortunová, Z. Bastl, S. Nagyová, S. Dolinská, V. Jorík, E. Jóna, Removal of pyridine from liquid and gas phase by copper forms of natural and synthetic zeolites, J. Hazard. Mater. 186 (2011) 699–706. https://doi.org/10.1016/j.jhazmat.2010.11.051

[31] K.V. Padoley, S.N. Mudliar, S.K. Banerjee, S.C. Deshmukh, R.A. Pandey, Fenton oxidation: A pretreatment option for improved biological treatment of pyridine and 3-cyanopyridine plant wastewater, Chem. Eng. J. 166 (2011) 1–9. https://doi.org/10.1016/j.cej.2010.06.041

[32] J.W. Bauserman, G.W. Mushrush, H. Willauer, J.H. Wynne, J.P. Phillips, J.L. Buckley, F.W. Williams, Removing organic nitrogen compounds from middle

distillate fuels with a catalyst used as a filtering media, Pet. Sci. Technol. 28 (2010) 1761–1769. https://doi.org/10.1080/10916460903261731

[33] H.Bouyarmane, S.E. Asri, A. Rami, C. Roux, M.A. Mahly, A. Saoiabi, T. Coradin, A. Laghzizil, Pyridine and phenol removal using natural and synthetic apatites as low cost sorbents: Influence of porosity and surface interactions, J. Hazard. Mater. 181 (2010) 736–741. https://doi.org/10.1016/j.jhazmat.2010.05.074

[34] Y. Bai, Q. Sun, R. Xing, D. Wen, X. Tang, Removal of pyridine and quinoline by bio-zeolite composed of mixed degrading bacteria and modified zeolite, J. Hazard. Mater. 181 (2010) 916–922. https://doi.org/10.1016/j.jhazmat.2010.05.099

[35] L. Qiao, J. Wang, Microbial degradation of pyridine by Paracoccus sp. isolated from contaminated soil, J. Hazard. Mater. 176 (2010) 220–225. https://doi.org/10.1016/j.jhazmat.2009.11.016

[36] H. Zhang, G. Li, Y. Jia, H. Liu, Adsorptive removal of nitrogen-containing compounds from fuel, J. Chem. Eng. Data. 55 (2010) 173–177. https://doi.org/10.1021/je9003004

[37] D.H. Lataye, I.M. Mishra, I.D. Mall, Multicomponent sorption of pyridine and its derivatives from aqueous solution onto rice husk ash and granular activated carbon, Pract. Period. Hazardous, Toxic, Radioact. Waste Manag. 13 (2009) 218–228. https://doi.org/10.1061/(ASCE)HZ.1944-8376.0000013

[38] A.K. Mathur, C.B. Majumder, Biofiltration of Pyridine by Shewanella putrefaciens in a corn-cob packed biotrickling filter, CLEAN – Soil, Air, Water. 36 (2008) 180–186. https://doi.org/10.1002/clen.200700090

[39] D.H. Lataye, I.M. Mishra, I.D. Mall, Multicomponent sorptive removal of toxics pyridine, 2-picoline, and 4-picoline from aqueous solution by bagasse fly ash: optimization of process parameters, Ind. Eng. Chem. Res. 47 (2008) 5629–5635. https://doi.org/10.1021/ie0716161

[40] D.H. Lataye, I.M. Mishra, I.D. Mall, Pyridine sorption from aqueous solution by rice husk ash (RHA) and granular activated carbon (GAC): Parametric, kinetic, equilibrium and thermodynamic aspects, J. Hazard. Mater. 154 (2008) 858–870. https://doi.org/10.1016/j.jhazmat.2007.10.111

[41] D.H. Lataye, I.M. Mishra, I.D. Mall, Removal of pyridine from aqueous solution by adsorption on bagasse fly ash, Ind. Eng. Chem. Res. 45 (2006) 3934–3943. https://doi.org/10.1021/ie051315w

[42] R.T.Y. A. J. Hernández-Maldonado, F. H. Yang, G. Qi, Desulfurization of transportation fuels by complexation sorbents: Cu(I)-, Ni(II)-, and Zn(II)-zeolites, J. Chinese Inst. Chem. Eng. 37 (2006) 9.

[43] D. Mohan, K.P. Singh, S. Sinha, D. Gosh, Removal of pyridine derivatives from aqueous solution by activated carbons developed from agricultural waste materials, Carbon N. Y. 43 (2005) 1680–1693. https://doi.org/10.1016/j.carbon.2005.02.017

[44] D. Mohan, K.P. Singh, S. Sinha, D. Gosh, Removal of pyridine from aqueous solution using low cost activated carbons derived from agricultural waste materials, Carbon N. Y. 42 (2004) 2409–2421. https://doi.org/10.1016/j.carbon.2004.04.026

[45] W.D. Henry, D. Zhao, A.K. SenGupta, C. Lange, Preparation and characterization of a new class of polymeric ligand exchangers for selective removal of trace contaminants from water, React. Funct. Polym. 60 (2004) 109–120. https://doi.org/10.1016/j.reactfunctpolym.2004.02.016

[46] J. Niu, B.E. Conway, Development of techniques for purification of waste waters: removal of pyridine from aqueous solution by adsorption at high-area C-cloth electrodes using in situ optical spectrometry, J. Electroanal. Chem. 521 (2002) 16–28. https://doi.org/10.1016/S0022-0728(02)00660-5

[47] J.H. Do, W.G. Lee, K. Theodore, H.N. Chang, Biological removal of pyridine in heavy oil by Rhodococcus sp. KCTC 3218, Biotechnol. Bioprocess Eng. 4 (1999) 205–209. https://doi.org/10.1007/BF02931930

[48] M. Stem, E. Heinzle, G.M. Kut, K. Hunge rbühler, Removal of substituted pyridines by combined ozonatio fluidized bed biofilm treatment, Water Sci. Technol. 35 (1997) 329–335. https://doi.org/10.1016/S0273-1223(97)00042-5

[49] J. Datka, M. Boczar, The effect of partial dehydroxylation and pyridine sorption on the strength of OH groups in NaHZSM-5 zeolite, Zeolites. 11 (1991) 397–400. https://doi.org/10.1016/0144-2449(91)80309-N

[50] S. Zhu, P.R.F. Bell, P.F. Greenfield, Adsorption of pyridine onto spent Rundle oil shale in dilute aqueous solution, Water Res. 22 (1988) 1331–1337. https://doi.org/10.1016/0043-1354(88)90122-4

[51] N.E. Cooke, R.P. Gaikwad, Removal of pyridine and quinoline from coal and coal extracts, Fuel. 63 (1984) 1468–1470. https://doi.org/10.1016/0016-2361(84)90360-0

[52] L. Guo, G. Li, J. Liu, P. Yin, Q. Li, Adsorption of aniline on cross-linked starch sulfate from aqueous solution, Ind. Eng. Chem. Res. 48 (2009) 10657–10663. https://doi.org/10.1021/ie9010782

[53] C. Jianguo, L. Aimin, S. Hongyan, F. Zhenghao, L. Chao, Z. Quanxing, Adsorption characteristics of aniline and 4-methylaniline onto bifunctional polymeric adsorbent modified by sulfonic groups, J. Hazard. Mater. 124 (2005) 173–180. https://doi.org/10.1016/j.jhazmat.2005.05.001

[54] F.S.H. Abram, I.R. Sims, The toxicity of aniline to rainbow trout, Water Res. 16 (1982) 1309–1312. https://doi.org/10.1016/0043-1354(82)90208-1

[55] K. László, E. Tombácz, C. Novák, pH-dependent adsorption and desorption of phenol and aniline on basic activated carbon, Colloids Surfaces A Physicochem. Eng. Asp. 306 (2007) 95–101. https://doi.org/10.1016/j.colsurfa.2007.03.057

[56] W.M. Zhang, Q.J. Zhang, B.C. Pan, L. Lv, B.J. Pan, Z.W. Xu, Q.X. Zhang, X.S. Zhao, W. Du, Q.R. Zhang, Modeling synergistic adsorption of phenol/aniline mixtures in the aqueous phase onto porous polymer adsorbents, J. Colloid Interface Sci. 306 (2007) 216–221. https://doi.org/10.1016/j.jcis.2006.10.056

[57] C. Moreno-Castilla, Adsorption of organic molecules from aqueous solutions on carbon materials, Carbon 42 (2004) 83–94. https://doi.org/10.1016/j.carbon.2003.09.022

[58] K. Yang, W. Wu, Q. Jing, L. Zhu, Aqueous adsorption of aniline, phenol, and their substitutes by multi-walled carbon nanotubes, Environ. Sci. Technol. 42 (2008) 7931–7936. https://doi.org/10.1021/es801463v

[59] S. H. Gheewala, A. P. Annachhatre, Biodegradation of aniline, Water Sci. Technol. 36 (1997) 53–63. https://doi.org/10.1016/S0273-1223(97)00642-2

[60] F. O'Neill, Bacterial growth on aniline: implications for the biotreatment of industrial wastewater, Water Res. 34 (2000) 4397–4409. https://doi.org/10.1016/S0043-1354(00)00215-3

[61] F. Orshansky, N. Narkis, Characteristics of organics removal by PACT simultaneous adsorption and biodegradation, Water Res. 31 (1997) 391–398. https://doi.org/10.1016/S0043-1354(96)00227-8

[62] L. Wang, S. Barrington, J.-W. Kim, Biodegradation of pentyl amine and aniline from petrochemical wastewater, J. Environ. Manage. 83 (2007) 191–197. https://doi.org/10.1016/j.jenvman.2006.02.009

[63]   X. Y. Liu, B. J. Wang, C. Y. Jiang, K. X. Zhao, L. D. Harold, S. J. Liu, Simultaneous biodegradation of nitrogen-containing aromatic compounds in a sequencing batch bioreactor, J. Environ. Sci. 19 (2007) 530–535. https://doi.org/10.1016/S1001-0742(07)60088-6

[64]   G. Deiber, J.N. Foussard, H. Debellefontaine, Removal of nitrogenous compounds by catalytic wet air oxidation. Kinetic study, Environ. Pollut. 96 (1997) 311–319. https://doi.org/10.1016/S0269-7491(97)00047-X

[65]   X. Qi, Decomposition of aniline in supercritical water, J. Hazard. Mater. 90 (2002) 51–62. https://doi.org/10.1016/S0304-3894(01)00330-2

[66]   J. O'Brien, T.F. O'Dwyer, T. Curtin, A novel process for the removal of aniline from wastewaters, J. Hazard. Mater. 159 (2008) 476–482. https://doi.org/10.1016/j.jhazmat.2008.02.064

[67]   J. Garcia, H.T. Gomes, P. Serp, P. Kalck, J.L. Figueiredo, J.L. Faria, Carbon nanotube supported ruthenium catalysts for the treatment of high strength wastewater with aniline using wet air oxidation, Carbon N. Y. 44 (2006) 2384–2391. https://doi.org/10.1016/j.carbon.2006.05.035

[68]   J. Anotai, M.-C. Lu, P. Chewpreecha, Kinetics of aniline degradation by Fenton and electro-Fenton processes, Water Res. 40 (2006) 1841–1847. https://doi.org/10.1016/j.watres.2006.02.033

[69]   L. Wojnárovits, E. Takács, Irradiation treatment of azo dye containing wastewater: An overview, Radiat. Phys. Chem. 77 (2008) 225–244. https://doi.org/10.1016/j.radphyschem.2007.05.003

[70]   F. M. T. Luna, A. A. Pontes-Filho, E. D. Trindade, I. J. Silva, Jr., C. L. Cavalcante, D. C. S. Azevedo, Ind. Eng. Chem. Res. 47 (2008) 3207. https://doi.org/10.1021/ie071476v

[71]   Y. Han, X. Quan, S. Chen, H. Zhao, C. Cui, Y. Zhao, Electrochemically enhanced adsorption of aniline on activated carbon fibers, Sep. Purif. Technol. 50 (2006) 365–372. https://doi.org/10.1016/j.seppur.2005.12.011

[72]   X. Xie, L. Gao, J. Sun, Thermodynamic study on aniline adsorption on chemical modified multi-walled carbon nanotubes, Colloids Surfaces A Physicochem. Eng. Asp. 308 (2007) 54–59. https://doi.org/10.1016/j.colsurfa.2007.05.028

[73]   H. L. Du, L. L. Yao, J. Environ. Prot. Sci. 29 (2003) 23.

[74]   S. Datta, Removal of aniline from aqueous solution in a mixed flow reactor using emulsion liquid membrane, J. Memb. Sci. 226 (2003) 185–201. https://doi.org/10.1016/j.memsci.2003.09.003

[75]   R. Devulapalli, F. Jones, Separation of aniline from aqueous solutions using emulsion liquid membranes, J. Hazard. Mater. 70 (1999) 157–170. doi:10.1016/S0304-3894(99)00134-X. https://doi.org/10.1016/S0304-3894(99)00134-X

[76]   C. Valderrama, J.I. Barios, A. Farran, J.L. Cortina, Evaluation of phenol aniline (Single and binary) removal from aqueous solutions onto hyper-cross-linked polymeric resin (Macronet MN200) and granular activated carbon in fixed-bed column, Water, Air, Soil Pollut. 215 (2011) 285–297. https://doi.org/10.1007/s11270-010-0478-x

[77]   Y. Zhao, Z. Cai, Z. Zhou, X. Fu, Adsorption behavior of monomers and formation of conducting polymers on polyester fibers, J. Appl. Polym. Sci. 119 (2011) 662–669. https://doi.org/10.1002/app.32743

[78]   G. Ersöz, S. Atalay, Kinetic modeling of the removal of aniline by low-pressure catalytic wet air oxidation over a nanostructured CoO4 /CeO2 catalyst, Ind. Eng. Chem. Res. 50 (2011) 310–315. https://doi.org/10.1021/ie1016706

[79]   Y. Lee, S. Chen, H. Tu, S. Yau, L. Fan, Y. Yang, W.-P. Dow, In situ STM revelation of the adsorption and polymerization of aniline on Au(111) electrode in perchloric acid and benzenesulfonic acid, Langmuir. 26 (2010) 5576–5582. https://doi.org/10.1021/la903857x

[80]   C. He, K. Huang, J. Huang, Surface modification on a hyper-cross-linked polymeric adsorbent by multiple phenolic hydroxyl groups to be used as a specific adsorbent for adsorptive removal of p-nitroaniline from aqueous solution, J. Colloid Interface Sci. 342 (2010) 462–466. https://doi.org/10.1016/j.jcis.2009.10.026

[81]   X. J. Yu, C. S. Zhou, Y. X. Wang, L. J. Pang, Preparation of activated carbon from shell of carya cathayensis S. and Its adsorption behavior of aniline, Chinese J. Process Eng. 1 (2010) 65.

[82]   Z.-J. Ding, L. Qi, J.-Z. Ye, Preparation and characterization of PVC/PS composite cation exchange fibers and their adsorption properties on aniline, J. Appl. Polym. Sci. 117 (2010) 1914–1923. https://doi.org/10.1002/app.32114

[83]   D. Shao, J. Hu, C. Chen, G. Sheng, X. Ren, X. Wang, Polyaniline multiwalled
        carbon nanotube magnetic composite prepared by plasma-induced graft technique
        and tts application for removal of aniline and phenol, J. Phys. Chem. C. 114
        (2010) 21524–21530. https://doi.org/10.1021/jp107492g

[84]   R. Y. Chen, J. Zhong, C. R. Gu, C. L. Chen, Molecular dynamics simulation of
        adsorption of aniline by α-zirconium phosphate, J. Theor. Comput. Chem. 9
        (2010) 861–873. https://doi.org/10.1142/S0219633610006043

[85]   C. Song, Y. Song, Adsorption and desorption behaviors of nitrobenzene and
        aniline by wetland soil, in: 2010 4th Int. Conf. Bioinforma. Biomed. Eng., IEEE,
        2010: pp. 1–5. https://doi.org/10.1109/icbbe.2010.5516254

[86]   F. An, X. Feng, B. Gao, Adsorption property and mechanism of composite
        adsorbent PMAA/SiO2 for aniline, J. Hazard. Mater. 178 (2010) 499–504.
        https://doi.org/10.1016/j.jhazmat.2010.01.109

[87]   K. Li, Y. Li, Z. Zheng, Kinetics and mechanism studies of p-nitroaniline
        adsorption on activated carbon fibers prepared from cotton stalk by NH4H2PO4
        activation and subsequent gasification with steam, J. Hazard. Mater. 178 (2010)
        553–559. https://doi.org/10.1016/j.jhazmat.2010.01.120

[88]   S.S.C. X. H. Yuan, Z. R. Han, L. M. Lui, J. Hu, Adsorption of aniline and benzene
        on phenolic resin prepared by inverse suspension polymerization, Gaofenzi Cailiao
        Kexue Yu Gongcheng/Polym. Mater. Sci. Eng. 26 (2010) 36.

[89]   C. Valderrama, J.I. Barios, M. Caetano, A. Farran, J.L. Cortina, Kinetic evaluation
        of phenol/aniline mixtures adsorption from aqueous solutions onto activated
        carbon and hypercrosslinked polymeric resin (MN200), React. Funct. Polym. 70
        (2010) 142–150. https://doi.org/10.1016/j.reactfunctpolym.2009.11.003

[90]   N. Li, X.-L. Xiong, R.-Q. Wang, Adsorption properties of aniline onto beta-
        cyclodextrin epichlorohydrin copolymer, in: 2009 3rd Int. Conf. Bioinforma.
        Biomed. Eng., IEEE, 2009: pp. 1–4. https://doi.org/10.1109/icbbe.2009.5162288

[91]   H. Zheng, D. Liu, Y. Zheng, S. Liang, Z. Liu, Sorption isotherm and kinetic
        modeling of aniline on Cr-bentonite, J. Hazard. Mater. 167 (2009) 141–147.
        https://doi.org/10.1016/j.jhazmat.2008.12.093

[92]   Y.Z. LI Fang-chen, DAI You-zhi, LUO Yue-ping, Adsorption behavior and
        mechanism of aniline on rice bran, Chinese J. Process Eng. (2009) 274.

[93]   O. Gezici, A. Ayar, Stepwise frontal analysis to derive equilibrium sorption data for copper and aniline on functionalized sporopollenin, CLEAN - Soil, Air, Water. 37 (2009) 349–354. https://doi.org/10.1002/clen.200900001

[94]   J. Huang, X. Wang, K. Huang, Adsorption of p-nitroaniline by phenolic hydroxyl groups modified hyper-cross-linked polymeric adsorbent and XAD-4: A comparative study, Chem. Eng. J. 155 (2009) 722–727. https://doi.org/10.1016/j.cej.2009.09.012

[95]   W.C. X. H. Yuan, W. Song, L. M. Lui, J. Hu, Adsorption of aniline and benzene on phenolic resin prepared by inverse suspension polymerization, Sheng, S. S. Cao,Gaofenzi Cailiao Kexue Yu Gongcheng/Polym. Mater. Sci. Eng. (2009).

[96]   F. An, X. Feng, B. Gao, Adsorption of aniline from aqueous solution using novel adsorbent PAM/SiO2, Chem. Eng. J. 151 (2009) 183–187. https://doi.org/10.1016/j.cej.2009.02.011

[97]   K. Li, Z. Zheng, J. Feng, J. Zhang, X. Luo, G. Zhao, X. Huang, Adsorption of p-nitroaniline from aqueous solutions onto activated carbon fiber prepared from cotton stalk, J. Hazard. Mater. 166 (2009) 1180–1185. https://doi.org/10.1016/j.jhazmat.2008.12.035

[98]   K. Li, Z. Zheng, X. Huang, G. Zhao, J. Feng, J. Zhang, Equilibrium, kinetic and thermodynamic studies on the adsorption of 2-nitroaniline onto activated carbon prepared from cotton stalk fibre, J. Hazard. Mater. 166 (2009) 213–220. https://doi.org/10.1016/j.jhazmat.2008.11.007

[99]   S.H. Jang, S. Jeong, J.R. Hahn, The preserved aromaticity of aniline molecules adsorbed on a Si(5 5 12)−2×1 surface, J. Chem. Phys. 130 (2009) 234703. https://doi.org/10.1063/1.3153920

[100] B. Li, Z. Liu, Z. Lei, Z. Huang, Catalytic dry oxidation of aniline, benzene, and pyridine adsorbed on a CuO doped activated carbon, Korean J. Chem. Eng. 26 (2009) 913–918. https://doi.org/10.1007/s11814-009-0153-3

[101] H.S. Wahab, A.D. Koutselos, A computational study on the adsorption and OH initiated photochemical and photocatalytic primary oxidation of aniline, Chem. Phys. 358 (2009) 171–176. https://doi.org/10.1016/j.chemphys.2009.01.013

[102] H. Kostelníkova, P. Praus, M. Turicová, Adsorption of phenol and aniline by original and quaternary ammonium salts-modified montmorillonite, Acta Geodyn. Geomater. 5 (2008) 83–88.

[103] X. Gu, J. Zhou, A. Zhang, P. Wang, M. Xiao, G. Liu, Feasibility study of the treatment of aniline hypersaline wastewater with a combined adsorption/bio-regeneration system, Desalination. 227 (2008) 139–149. https://doi.org/10.1016/j.desal.2007.06.021

[104] O'Brien, Curtin, O'Dwyer, Removal of aniline from waste streams using a combined adsorption and catalytic oxidation approach, Adsorpt. Sci. Technol. 26 (2008) 311–321. https://doi.org/10.1260/026361708787548800

[105] X.-Y. Liu, Y.-S. Ji, H.-X. Zhang, M.-C. Liu, Highly sensitive analysis of substituted aniline compounds in water samples by using oxidized multiwalled carbon nanotubes as an in-tube solid-phase microextraction medium, J. Chromatogr. A. 1212 (2008) 10–15. https://doi.org/10.1016/j.chroma.2008.10.034

[106] T.H. S. A. El-Safty, F. Mizukami, Cationic surfactant templates for newly developed cubic Fd3m silica mesocage structures, Int. J. Environ. Pollut. 34 (2008) 97. https://doi.org/10.1504/IJEP.2008.020785

[107] L. Zampori, P. Gallo Stampino, G. Dotelli, D. Botta, I. Natali Sora, M. Setti, Interlayer expansion of dimethyl ditallowylammonium montmorillonite as a function of 2-chloroaniline adsorption, Appl. Clay Sci. 41 (2008) 149–157. https://doi.org/10.1016/j.clay.2007.10.003

[108] E.I. Unuabonah, K.O. Adebowale, F.A. Dawodu, Equilibrium, kinetic and sorber design studies on the adsorption of aniline blue dye by sodium tetraborate-modified Kaolinite clay adsorbent, J. Hazard. Mater. 157 (2008) 397–409. https://doi.org/10.1016/j.jhazmat.2008.01.047

[109] S. Andini, R. Cioffi, F. Colangelo, F. Montagnaro, L. Santoro, Adsorption of chlorophenol, chloroaniline and methylene blue on fuel oil fly ash, J. Hazard. Mater. 157 (2008) 599–604. https://doi.org/10.1016/j.jhazmat.2008.01.025

[110] F. López-Linares, L. Carbognani, C.S. Stull, P. Pereira-Almao, Adsorption kinetics of anilines on macroporous kaolin, Energy & Fuels. 22 (2008) 2188–2194. https://doi.org/10.1021/ef800084s

[111] O.A. Anunziata, M.B. Gómez Costa, M.L. Martínez, Interaction of water and aniline adsorbed onto Na-AlMCM-41 and Na-AlSBA-15 catalysts as hosts materials, Catal. Today. 133 (2008) 897–905. https://doi.org/10.1016/j.cattod.2007.12.073

[112] J. Yao, X. Li, W. Qin, Computational design and synthesis of molecular imprinted polymers with high selectivity for removal of aniline from contaminated water, Anal. Chim. Acta. 610 (2008) 282–288. https://doi.org/10.1016/j.aca.2008.01.042

[113] E. V. Kuz'mina, L.N. Khatuntseva, S.G. Dmitrienko, Determination of aniline and phenol in waters using polyurethane foams and diffuse reflectance spectroscopy, J. Anal. Chem. 63 (2008) 34–40. https://doi.org/10.1134/S1061934808010073

[114] T.F.O. J. O'Brien, T. Curtin, Removal of organic compounds from waste streams: a combined approach, WIT Trans. Ecol. Environ. 103 (2007) 447. https://doi.org/10.2495/wrm070421

[115] C.H. Ko, C. Fan, P.N. Chiang, M.K. Wang, K.C. Lin, p-Nitrophenol, phenol and aniline sorption by organo-clays, J. Hazard. Mater. 149 (2007) 275–282. https://doi.org/10.1016/j.jhazmat.2007.03.075

[116] K. Zheng, B. Pan, Q. Zhang, W. Zhang, B. Pan, Y. Han, Q. Zhang, D. Wei, Z. Xu, Q. Zhang, Enhanced adsorption of p-nitroaniline from water by a carboxylated polymeric adsorbent, Sep. Purif. Technol. 57 (2007) 250–256. https://doi.org/10.1016/j.seppur.2007.04.017

[117] X. Chai, Y. He, D. Ying, J. Jia, T. Sun, Electrosorption-enhanced solid-phase microextraction using activated carbon fiber for determination of aniline in water, J. Chromatogr. A. 1165 (2007) 26–31. https://doi.org/10.1016/j.chroma.2007.07.048

[118] P. Podkościelny, K. László, Heterogeneity of activated carbons in adsorption of aniline from aqueous solutions, Appl. Surf. Sci. 253 (2007) 8762–8771. https://doi.org/10.1016/j.apsusc.2007.04.057

[119] K. Zheng, B. Pan, Q. Zhang, Y. Han, W. Zhang, B. Pan, Z. Xu, Q. Zhang, W. Du, Q. Zhang, Enhanced removal of p-chloroaniline from aqueous solution by a carboxylated polymeric sorbent, J. Hazard. Mater. 143 (2007) 462–468. https://doi.org/10.1016/j.jhazmat.2006.09.052

[120] W. Zhang, Z. Xu, B. Pan, Q. Zhang, W. Du, Q. Zhang, K. Zheng, Q. Zhang, J. Chen, Adsorption enhancement of laterally interacting phenol/aniline mixtures onto nonpolar adsorbents, Chemosphere. 66 (2007) 2044–2049. https://doi.org/10.1016/j.chemosphere.2006.09.082

[121] W. Ma, Y. Fang, Experimental (SERS) and theoretical (DFT) studies on the adsorption of p-, m-, and o-nitroaniline on gold nanoparticles, J. Colloid Interface Sci. 303 (2006) 1–8. https://doi.org/10.1016/j.jcis.2006.05.001

[122] M.C.G. S. Polati, F. Gosetti, V. Gianotti, Sorption and desorption behavior of chloroanilines and chlorophenols on montmorillonite and kaolinite, J. Environ. Sci. Heal. - Part B Pestic. Food Contam. Agric. Wastes. 41 (2006) 765. https://doi.org/10.1080/03601230600805774

[123] X. Xie, L. Gao, Adsorption modification of carboxylated carbon nanotubes with aniline in aqueous solution, Chem. Lett. 35 (2006) 624–625. https://doi.org/10.1246/cl.2006.624

[124] T.-M. Wu, Y.-W. Lin, Doped polyaniline/multi-walled carbon nanotube composites: Preparation, characterization and properties, Polymer (Guildf). 47 (2006) 3576–3582. https://doi.org/10.1016/j.polymer.2006.03.060

[125] B.P. W. Zhang, J. Chen, Q. Zhang, Competitive and cooperative effect on simultaneous adsorption from aqueous solution by hypercrosslinked polymeric adsorbent, Acta Polym. Sin. 2 (2006) 213. https://doi.org/10.3724/SP.J.1105.2006.00213

[126] W. Zhang, J. Chen, B. Pan, Q. Chen, M. He, Q. Zhang, F. Wang, B. Zhang, Synergistic effect on phenol/aniline mixture adsorption on nonpolar resin adsorbents from aqueous solution, React. Funct. Polym. 66 (2006) 395–401. https://doi.org/10.1016/j.reactfunctpolym.2005.08.016

[127] H.C. A. Yildiz, A. Gür, Adsorption of aniline, phenol, and chlorophenols on pure and modified bentonite, Russ. J. Phys. Chem. A, 80 (2006) S172. https://doi.org/10.1134/S0036024406130279

[128] Q.X.Z. W. M. Zhang, J. L. Chen, B. C. Pan, J. Environ. Sci. 17 (2005) 529.

[129] C. Jianguo, L. Aimin, S. Hongyan, F. Zhenghao, L. Chao, Z. Quanxing, Equilibrium and kinetic studies on the adsorption of aniline compounds from aqueous phase onto bifunctional polymeric adsorbent with sulfonic groups, Chemosphere. 61 (2005) 502–509. https://doi.org/10.1016/j.chemosphere.2005.03.001

[130] G.R. Reddy, V. V. Mahajani, Insight into wet oxidation of aqueous aniline over a Ru/SiO2 catalyst, Ind. Eng. Chem. Res. 44 (2005) 7320–7328. https://doi.org/10.1021/ie050438d

[131] K. László, Adsorption from aqueous phenol and aniline solutions on activated carbons with different surface chemistry, Colloids Surfaces A Physicochem. Eng. Asp. 265 (2005) 32–39. https://doi.org/10.1016/j.colsurfa.2004.11.051

[132] T.J. Rockey, M. Yang, H.-L. Dai, Aniline on Ag(111): Adsorption configuration, adsorbate–substrate bond, and inter-adsorbate interactions, Surf. Sci. 589 (2005) 42–51. https://doi.org/10.1016/j.susc.2005.05.048

[133] T.F.O. J. O'Brien, T. Curtin, An investigation into the adsorption of aniline from aqueous solution using H-beta zeolites and copper-exchanged beta zeolites, Adsorpt. Sci. Tech. 23 (2005) 255. https://doi.org/10.1260/0263617054353582

[134] L.O. R. Wang, Z. Shi, R. Shi, J. Zhang, The study of adsorption of phenol and aniline on aminated-macroporous hypercrosslinked resins, Acta Polym. Sin. 3 (2005) 339.

[135] G. Liu, J. Yu, QSAR analysis of soil sorption coefficients for polar organic chemicals: Substituted anilines and phenols, Water Res. 39 (2005) 2048–2055. https://doi.org/10.1016/j.watres.2005.03.030

[136] A.A. Gürten, S. Uçan, M.A. Özler, A. Ayar, Removal of aniline from aqueous solution by PVC-CDAE ligand-exchanger, J. Hazard. Mater. 120 (2005) 81–87. https://doi.org/10.1016/j.jhazmat.2004.11.031

[137] O. Duman, E. Ayranci, Structural and ionization effects on the adsorption behaviors of some anilinic compounds from aqueous solution onto high-area carbon-cloth, J. Hazard. Mater. 120 (2005) 173–181. https://doi.org/10.1016/j.jhazmat.2004.12.030

[138] C. Causserand, P. Aimar, J.P. Cravedi, E. Singlande, Dichloroaniline retention by nanofiltration membranes, Water Res. 39 (2005) 1594–1600. https://doi.org/10.1016/j.watres.2004.12.039

[139] S. Angioi, S. Polati, M. Roz, C. Rinaudo, V. Gianotti, M.C. Gennaro, Sorption studies of chloroanilines on kaolinite and montmorillonite, Environ. Pollut. 134 (2005) 35–43. https://doi.org/10.1016/j.envpol.2004.07.018

[140] C.T. N. Kannan, Studies on the removal of aniline blue and acid violet by commercial activated carbon, Indian J. Environ. Prot. 25 (2005) 1.

[141] L. Zhu, B. Lou, K. Yang, B. Chen, Effects of ionizable organic compounds in different species on the sorption of p-nitroaniline to sediment, Water Res. 39 (2005) 281–288. https://doi.org/10.1016/j.watres.2004.11.003

[142] O'Brien, Curtin, O'Dwyer, Adsorption of aniline from aqueous solution using copper-exchanged ZSM-5 and unmodified H-ZSM-5, Adsorpt. Sci. Technol. 22 (2004) 743–754. https://doi.org/10.1260/0263617043026488

[143] K. Inumaru, Y. Inoue, S. Kakii, T. Nakano, S. Yamanaka, Molecular selective adsorption of dilute alkylphenols and alkylanilines from water by alkyl-grafted MCM-41: Tunability of the cooperative organic–inorganic function in the nanostructure, Phys. Chem. Chem. Phys. 6 (2004) 3133–3139. https://doi.org/10.1039/B403124E

[144] B. Ersoy, M.S. Çelik, Uptake of aniline and nitrobenzene from aqueous solution by organo-Zeolite, Environ. Technol. 25 (2004) 341–348. https://doi.org/10.1080/09593330409355467

[145] J. Eriksson, S. Frankki, A. Shchukarev, U. Skyllberg, Binding of 2,4,6-trinitrotoluene, aniline, and nitrobenzene to dissolved and particulate soil organic matter, Environ. Sci. Technol. 38 (2004) 3074–3080. https://doi.org/10.1021/es035015m

[146] J.F.J. C. T. Yan, Determination of aniline in silica gel sorbent by one-step in situ microwave-assisted desorption coupled to headspace solid-phase microextraction and GC–FID, Chromatographia. 59 (2004) 517. https://doi.org/10.1016/j.talanta.2004.03.053

[147] D.M. Nevskaia, E. Castillejos-Lopez, A. Guerrero-Ruiz, V. Mu-oz, Effects of the surface chemistry of carbon materials on the adsorption of phenol–aniline mixtures from water, Carbon N. Y. 42 (2004) 653–665. https://doi.org/10.1016/j.carbon.2004.01.007

[148] T. Tanaka, A. Nakajima, A. Watanabe, T. Ohno, Y. Ozaki, Surface-enhanced Raman scattering spectroscopy and density functional theory calculation studies on adsorption of o-, m-, and p-nitroaniline on silver and gold colloid, J. Mol. Struct. 661 (2003) 437–449. https://doi.org/10.1016/j.molstruc.2003.09.006

[149] G.A. Eimer, M.B. Gómez Costa, L.B. Pierella, O.A. Anunziata, Thermal and FTIR spectroscopic analysis of the interactions of aniline adsorbed on to MCM-41 mesoporous material, J. Colloid Interface Sci. 263 (2003) 400–407. https://doi.org/10.1016/S0021-9797(03)00038-9

[150] F.I. Talens-Alesson, Redox phenomena during adsorption of aniline on Fe- and Zn-bound SDS micelles, Chem. Eng. Technol. 26 (2003) 684–687. https://doi.org/10.1002/ceat.200390104

[151] J. Niu, B.E. Conway, Adsorptive and electrosorptive removal of aniline and bipyridyls from waste-waters, J. Electroanal. Chem. 536 (2002) 83–92. https://doi.org/10.1016/S0022-0728(02)01206-8

[152] T. Alam, H. Tarannum, S.R. Ali, Kamaluddin, Adsorption and oxidation of aniline and anisidine by chromium ferrocyanide, J. Colloid Interface Sci. 245 (2002) 251–256. https://doi.org/10.1006/jcis.2001.7968

[153] M. Uçan, A. Ayar, Sorption equilibria of chlorinated anilines in aqueous solution on resin-bound cobalt ion, Colloids Surfaces A Physicochem. Eng. Asp. 207 (2002) 41–47. https://doi.org/10.1016/S0927-7757(02)00134-6

[154] B.E. Conway, E. Ayranci, H. Al-Maznai, Use of quasi-3-dimensional porous electrodes for adsorption and electrocatalytic removal of impurities from waste-waters, Electrochim. Acta. 47 (2001) 705–718. https://doi.org/10.1016/S0013-4686(01)00751-4

[155] M.W. Kowalska, J.D. Ortego, A. Jezierski, Transformation of 2-(trifluoromethyl)aniline over ion-exchanged montmorillonites: formation of a dimer and cyclic trimer, Appl. Clay Sci. 18 (2001) 233–243. https://doi.org/10.1016/S0169-1317(01)00036-9

[156] E. Alonso, S. Lucas, J. Arevalo, M.J. Cocero, Supercritical extraction of aniline from polluted soil: Effect of operational variables, Chemie Ing. Tech. 73 (2001) 725–725. https://doi.org/10.1002/1522-2640(200106)73:6<725::AID-CITE7253333>3.0.CO;2-C

[157] V.N. Maistrenko, S. V. Sapel'nikova, F.K. Kudasheva, F.A. Amirkhanova, Isomer-selective carbon-paste electrodes for the determination of nitrophenol, nitroaniline, and nitrobenzoic acid by adsorption-stripping voltammetry, J. Anal. Chem. 55 (2000) 586–589. https://doi.org/10.1007/BF02757819

[158] M. Ilic, E. Koglin, A. Pohlmeier, H.D. Narres, M.J. Schwuger, Adsorption and polymerization of aniline on Cu(II)-montmorillonite: vibrational spectroscopy and ab initio calculation, Langmuir. 16 (2000) 8946–8951. https://doi.org/10.1021/la000534d

[159] L. Zhu, B. Chen, X. Shen, Sorption of phenol, p -nitrophenol, and aniline to dual-cation organobentonites from water, Environ. Sci. Technol. 34 (2000) 468–475. https://doi.org/10.1021/es990177x

[160] P. Bharathi, R.C. Deka, S. Sivasanker, R. Vetrivel, Diffusional characteristics of substituted anilines in various zeolites as predicted by molecular modeling methods, Catal. Letters. 55 (1998) 113–120. https://doi.org/10.1023/A:1019074626941

[161] R.-M. Rummel, C. Ziegler, Room temperature adsorption of aniline (C6H5NH2) on Si(100)(2×1) observed with scanning tunneling microscopy, Surf. Sci. 418 (1998) 303–313. https://doi.org/10.1016/S0039-6028(98)00726-2

[162] S. Ardizzone, H. Høiland, C. Lagioni, E. Sivieri, Pyridine and aniline adsorption from an apolar solvent: The role of the solid adsorbent, J. Electroanal. Chem. 447 (1998) 17–23. https://doi.org/10.1016/S0022-0728(98)00007-2

[163] S. Narayanan, R. Pillai Unnikrishnan, Comparison of hydrogen adsorption and aniline hydrogenation over co-precipitated Co/Al2O3 and Ni/Al2O3 catalysts, J. Chem. Soc. Faraday Trans. 93 (1997) 2009–2013. https://doi.org/10.1039/a608074j

[164] T. Bitzer, T. Alkunshalie, N.V. Richardson, An hreels investigation of the adsorption of benzoic acid and aniline on Si(100)-2 × 1, Surf. Sci. 368 (1996) 202–207. https://doi.org/10.1016/S0039-6028(97)80025-8

[165] V.G. Gaikar, T.K. Mandal, R.G. Kulkarni, Adsorptive separations using zeolites: Separation of substituted anilines, Sep. Sci. Technol. 31 (1996) 259–270. https://doi.org/10.1080/01496399608000694

[166] F. Fiçicioğlu, S. Kuliyev, F. Kadirgan, Electrochemical studies of the adsorption of aniline on a smooth polycrystalline platinum electrode, J. Electroanal. Chem. 408 (1996) 231–236. https://doi.org/10.1016/0022-0728(95)04476-0

[167] S.X. Huang, D.A. Fischer, J.L. Gland, Aniline adsorption, hydrogenation, and hydrogenolysis on the Ni(100) surface, J. Phys. Chem. 100 (1996) 10223–10234. https://doi.org/10.1021/jp951868s

[168] R.J.F. N. Gnanapragasam, B. A. G. Lewis, Microstructural changes in sand-bentonite soils when exposed to aniline, J. Geotechnol. Engg. - ASCE. 121 (1995) 119. https://doi.org/10.1061/(ASCE)0733-9410(1995)121:2(119)

[169] P. Fox, Effects of molecular oxygen and pH on the adsorption of aniline to activated carbon, in: Pinisetti, Kamalesh, National Conf. Environ. Eng., 1994: p. 617.

[170] U. Schmiemann, Z. Jusys, H. Baltruschat, The electrochemical stability of model inhibitors: A dems study on adsorbed benzene, aniline and pyridine on mono- and polycrystalline Pt, Rh and Pd electrodes, Electrochim. Acta. 39 (1994) 561–576. https://doi.org/10.1016/0013-4686(94)80102-9

[171] O. P. Homenauth, M. B. McBride , Adsorption of aniline on layer silicate clays and an organic soil, Soil Sci. Soc. Am. J. 58 (1994) 347. https://doi.org/10.2136/sssaj1994.03615995005800020014x

[172] D.L.S. Peng-Chu Zhang, Kinetics of phenol and aniline adsorption and desorption, Soil Sci. Soc. Am. J. 57 (1993) 340. https://doi.org/10.2136/sssaj1993.03615995005700020009x

[173] F.R. Chen, J.J. Fripiat, Formation of radical ion pairs in aniline adsorption on zeolites, J. Phys. Chem. 96 (1992) 819–823. https://doi.org/10.1021/j100181a054

[174] L. Werner, F. Marlow, W. Hill, U. Retter, Optical second harmonic generation at Hg electrodes. the adsorption of aniline, Chem. Phys. Lett. 194 (1992) 39–44. https://doi.org/10.1016/0009-2614(92)85739-W

[175] M.K.N. Yenkie, G.S. Natarajan, Adsorption equilibrium studies of some aqueous aromatic pollutants on granular activated carbon samples, Sep. Sci. Technol. 26 (1991) 661–674. https://doi.org/10.1080/01496399108049907

[176] A.B.E. C. J. Murphy, G. C. Lisensky, L. K. Leung, G. R. Kowach, Photoluminescence-based correlation of semiconductor electric field thickness with adsorbate Hammett substituent constants. Adsorption of aniline derivatives onto cadmium selenide, J. Am. Chem. Soc. 112 (1990) 8344. https://doi.org/10.1021/ja00179a019

[177] C.N. Van Huong, Adsorption of aniline on a polycrystalline gold electrode: Determination of thermodynamic parameters and electron reflectance investigation, J. Electroanal. Chem. Interfacial Electrochem. 264 (1989) 247–258. https://doi.org/10.1016/0022-0728(89)80160-3

[178] R. Holze, Raman spectroscopic investigation of aniline: adsorption and polymerization, J. Electroanal. Chem. Interfacial Electrochem. 224 (1987) 253–260. https://doi.org/10.1016/0022-0728(87)85096-9

[179] H. Shindo, C. Nishihara, Raman spectra of aniline adsorbed on an Ag electrode in acidic solutions, J. Chem. Soc. Faraday Trans. 1 Phys. Chem. Condens. Phases. 84 (1988) 433.

[180] N.E. Zoulis, C.E. Efstathiou, Voltammetric determination of N-alkylated anilines by adsorption/extraction at a carbon-paste electrode, Anal. Chim. Acta. 204 (1988) 201–211. https://doi.org/10.1016/S0003-2670(00)86359-X

[181] R. Holze, Potential- and pH-dependent adsorption of aniline on silver as evidenced with surface enhanced Raman spectroscopy, Electrochim. Acta. 32 (1987) 1527–1532. https://doi.org/10.1016/0013-4686(87)85097-1

[182] J. Gorse, M.F. Burke, G.K. Vemulapalli, Isotherm studies of benzene and aniline on chemically modified silica surfaces, Langmuir. 3 (1987) 179–183. https://doi.org/10.1021/la00074a006

[183] L.I.D. M. I. Urbakh, Renorm-group approach to the calculation of the effective dielectric constant of two-dimensional inhomogeneous systems, Sov. Electrochem. 20 (1984) 962.

[184] V. Amicarelli, G. Baldassarre, V. Balice, L. Liberti, Thermoanalytical study of activated carbon regeneration part IV. adsorption equilibria for phenol, aniline and their nitro-derivatives on granular activated carbon, Thermochim. Acta. 36 (1980) 107–111. https://doi.org/10.1016/0040-6031(80)87001-8

[185] M.A. El-Dib, O.A. Aly, Removal of phenylamide pesticides from drinking waters—II. Adsorption on powdered carbon, Water Res. 11 (1977) 617–620. https://doi.org/10.1016/0043-1354(77)90095-1

[186] M.A. El-Dib, O.A. Aly, Persistence of some phenylamide pesticides in the aquatic environment—II. Adsorption on clay minerals, Water Res. 10 (1976) 1051–1053. https://doi.org/10.1016/0043-1354(76)90034-8

[187] P.C. C. Defosse, Preliminary ESCA study of aniline adsorption on HY zeolites heated at various temperatures, React. Kinet. Catal. Lett. 3 (1975) 161. https://doi.org/10.1007/BF02187509

[188] G.C. Barker, D. McKeown, Kinetics of aniline adsorption-desorption on mercury, J. Electroanal. Chem. Interfacial Electrochem. 59 (1975) 295–302. https://doi.org/10.1016/S0022-0728(75)80184-7

[189] T. Furukawa, Adsorption and oxidation of benzidine and aniline by montmorillonite and hectorite, Clays Clay Miner. 21 (1973) 279–288. https://doi.org/10.1346/CCMN.1973.0210503

[190] M. Tanaka, Infrared study of adsorbed state of aniline on alumina and HCl-treated alumina, J. Catal. 25 (1972) 111–117. https://doi.org/10.1016/0021-9517(72)90207-2

[191] L.R. Snyder, Adsorption from solution. III. derivatives of pyridine aniline and pyrrole on alumina, J. Phys. Chem. 67 (1963) 2344–2353. https://doi.org/10.1021/j100805a021

Chapter 4

Adsorption of proteins onto non-soluble polysaccharides matrixes: a friendly strategy to isolate enzymes with potential application for downstream processes

Nadia Woitovich Valetti, M. Emilia Brassesco and Guillermo Alfredo Picó*

Institute of Biotechnological and Chemical Processes (IPROBYQ - CONICET) and Technology Department. Faculty of Biochemical and Pharmaceutical Sciences. National University of Rosario. Suipacha 570 (S2002RLK) Rosario. Argentina.

* pico@iprobyq-conicet.gob.ar

Abstract

Polysaccharides with electrically charged groups, polyelectrolytes (PE), have the capacity to form hydrogels under different experimental conditions, which act as ionic exchanges with high affinity to adsorb proteins. Presented in this chapter is a description of the current state of the technic on the use of these matrixes. Alginate, carrageenan, chitosan, pectin, etc. are the most used PE to make beds with the capacity of proteins adsorption. The presence of electrically charged residual groups in these PE allows their use as beds for the ion exchange chromatography in stirred tank of packed bed or expended bed. We demonstrated the adsorption of lysozyme, a model protein positively charged, onto Alginate-Guar gum matrixes cross-linked with epichlorohydrin (negatively charged). Their physical characterization, equilibrium isotherms and adsorption kinetics were carried out. Successive cycles of adsorption-washing-elution were performed. The results demonstrate the reversibility of the process and the capacity of this enzyme purification method.

Keywords

Chromatography, Adsorption, Alg-guar gum, Polyelectrolytes, Bioseparation

Abbreviations

GG; guar gum, Alg; Alginate, LZ; lysozyme, Alg-GG; Alg guar gum bed, PE; polyelectrolyte

Contents

1. Introduction

The main problem in the scaling up process of enzymes purification is the use of large volumes of biomass where the target enzyme is present (homogenates or microbial suspensions) and the need to reduce these volumes immediately. These homogenates are formed by suspensions of cells, cell debris, membranes and proteases released from the destruction of tissues. All these give homogenates with low stability and high degradation rate within hours, it is necessary that the isolation methods reach a concentrated extract of the enzyme in a short time. About 90% of the protein purification methods use ammonium sulphate as precipitant in a range of about 40% to 70% of saturation level, which correspond to about 400 g/L of this salt. When this amount is multiplied by the thousands of liters forming a homogenate, the final amount of waste to be treated due to the toxicity of ammonium cation becomes very large.

The goal of a bioseparation process is this it recovers a final product, for example, macromolecules. The bioseparation of a target molecule from a biomass can account for 50–80% of the total cost production, so the development of new materials with adsorption capacity for macromolecules is important since both the raw material and the method become cheaper. Our laboratory has been exploring the interaction between proteins and polysaccharides with net electrical charge, in order to use it as the first step in the concentration and purification of enzymes. Adsorption has been widely used as one of the main steps of downstream operations in several fields, such as biology, medicine, biotechnology and food processing [1]. Within this context, the development of low-cost adsorbents with high adsorption capacity and selectivity has been a great challenge. Moreover, adsorption should be perfectly reversible to optimise recovery, while preserving the activity of the desorbed (recovered) enzyme. The adsorption process of a molecule in a liquid phase may be carried out if using a solid phase (matrix) it has a high affinity for the target molecule. The contacts between the solid and the liquid phase can be done by different ways: -Stirred tank; -Packed bed and - Expanded bed. In any of these ways it is necessary to have a matrix with a high capacity to adsorb the target molecule. Commercial beds are expensive such as is the case of StreamLine ® (General Electric), so they increase, in a significant manner, the cost of the bioseparation process. Thus, as mentioned above, it is necessary to develop new adsorption matrixes, with the same adsorption capacity, minor cost and applicable in scaling up to replacing currently available commercial beds. One alternative is the use of natural PE which under desired experimental conditions forms non-soluble matrixes with high capacity to absorb proteins due to their residual acid or basic electrical charged groups. Such is the case of Alginate (Alg) (COOH), carrageenan (Carr) (SO_3H) and chitosan (Chi) (NH_3) which act as ionic exchanger [2-5]. The advantage of these matrixes is their easy preparation method [6, 7], the possibility of being discarded in the environment without negative impact, and their low cost. PE can form non-soluble beds in different manner:

1) Only one polyelectrolyte: natural PE form non-soluble complexes in the presence of cations such is the case of alginate, carrageen, pectin, etc. [8-10].

2) Two PE of opposite electrical charge: by the formation of columbic bound between as is the case of a mixture of Alginate-Chitosan [11, 12].

3) Two polysaccharides: one polyelectrolyte and a non-electrical charged polysaccharide: in this case, the last one is used to increase the mechanical resistance of the matrix or increase its porosity [13, 14].

A great number of works have reported the adsorption capacity of PE non-soluble matrixes to adsorb small molecules like heavy metal and aromatic contaminant in water [15-18]. However, there are few reports about enzyme adsorption capacity onto these beds, especially the potential application of these systems to scaling up to a level for protein isolation. Rodriguez *et al.* [19] studied the adsorption of cellulase in batch and column using a mixture of two PE of opposite electrical charge, Chi and Alg and reported a 29% of enzymatic recovery. The observed low yield can be due to a decrease in the number of positive charge of Chi by the interaction with the negative Alg charge. Godim et al. [20] recovered human IgG using a matrix of dye- epoxide ligand Chi/Alg. Roy et al. [3] expanded the structure of Alg non-soluble matrix using guar gum (GG), a non-charged polysaccharide, reporting the affinity increase among the matrix and the protein jacalin. However, at the present there are not many studies regarding the molecular mechanism of proteins adsorption in Alg matrixes, which include the analysis of the adsorption kinetic and their respective isotherms, the determination of thermodynamic parameters, and by last a correlation of these result with those obtained using dynamic adsorption (in bed) that should give information very useful to apply these matrixes for the enzymes bioseparation.

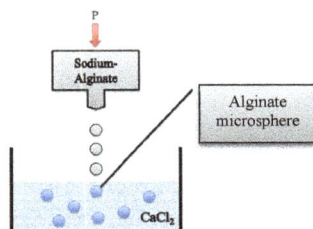

Figure 1. Dropping method to obtain Alg microspheres.

2.   Alginate hydrogels

Sodium Alginate is a particularly attractive biopolymer because it has a wide range of applications, mainly in: biomedicine, industry, food and pharmaceutical fields. It has the capacity to hold water, form gels and form stable emulsions. Alg is a water-soluble linear polysaccharide composed of 1,4-linked β-D-mannuronic and α-L-glucuronic acid residues, which are found in varying composition and sequence [18]. This polymer has the advantage of being of natural origin, having a friendly behaviour when it is discarded in the environment and being low-priced. The gelation of Alg can be carried out under an extremely mild environment using non-toxic reactants. The most important property of Alg is its ability to form gels by reaction with divalent cations, this is called the ionotropic gelation, and it is an effective and facile technique that enables the formation of spherical beads with a regular shape, size and smooth surface. Alg spherical beads can be prepared by extruding a solution of sodium alginate as droplets into a divalent cations solution such as Ca^{2+} or Ba^{2+} (Fig.1) [21].

There are a great number of papers where Alg is transformed into a non-soluble matrix by adding Ca^{2+} to the medium. When a solution of Alg is dropping on to $CaCl_2$, a solid sphere is obtaining due to the viscosity of their solutions. The shape of it is conserved because of Alg-Ca^{2+} complex formation. However, the working pH range of the matrix is limited and when Ca^{2+} is lost, the Alg-Ca^{2+} complex is destroyed, being the Alg soluble [22].

The most common method for the formation of Alg gels is by ionic cross-linking with multivalent cations. This method can take place under gentle conditions, making it ideal for entrapment of sensitive materials. The gelation of Alg occurs by exchange of sodium ions from the guluronic acids blocks with multivalent cations, and the stacking of these blocks to form a characteristic "egg-box" structure (Fig. 2) [21, 22].

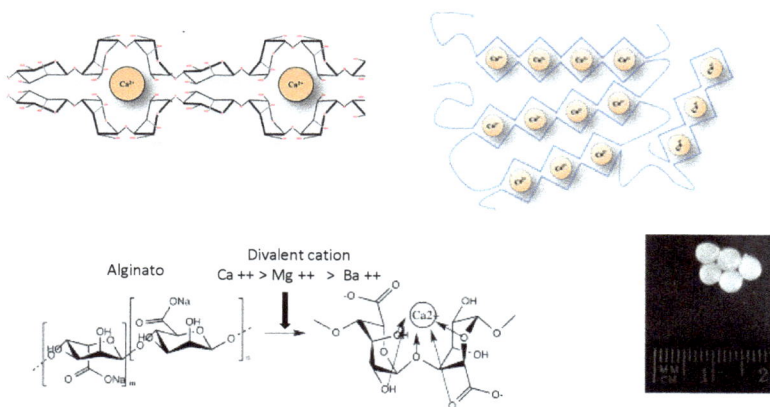

Figure 2. The complex formation between Alg and Ca^{2+} ions. Upper: The egg-box system of the complex formation. Bottom: The interaction of Alg OH- with Ca^{2+}.

Each chain can be linked with many other chains, resulting in the formation of a three-dimensional gel network. These gels can have water contents greater than 95% and can be heat treated without melting [23]. Various cations show a different affinity for Alg, where Ca^{2+} is the most frequently used for Alg gelation.

Gelation of Alg solutions with cations can occur through so-called external gelation, internal gelation and gelation upon cooling [24]. For external gelation, Alg is often dropped into a bath containing cations, such as a $CaCl_2$ solution. The cations diffuse from the continuous phase into the interior of the Alg droplets, and form a gelled Alg matrix, from the outside migrating towards the centre of the Alg droplet [25, 26]. The method is, therefore, also referred to as "diffusion method". Internal gelation also referred to as "internal setting" or "in situ gelation", makes use of a water insoluble Ca^{2+} salt, like $CaCO_3$, which is mixed with the Alg solution. Ca^{2+} ions are subsequently released from the interior of the Alg phase by lowering the pH of the system and/or increasing the solubility of the Ca^{2+} source, resulting in the formation of Alg gel.

The matrix of Alg has used alone or in combination with another polysaccharide to increase the performing and stability of the matrix using the property to form spheres which diameter can be varied between 1 to 2000 μm. The addition of a second polysaccharide non-electrically charge allows obtaining a matrix with other physical and chemistry properties, such as more mechanical resistance and more capacity to bound a solute, because the presence of a second polysaccharide induces the open of the matrix, exposing the negative electrical groups to the interaction with more protein molecules.

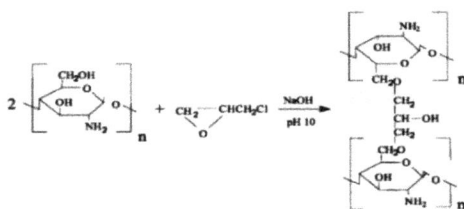

Figure 3. The general chemical reaction between Epi and OH groups of polysaccharides.

A non-soluble matrix can be obtained through a chemical treatment by covalent bonding between a reactant and the chemical groups of the polysaccharide. After this treatment, the matrix gains the properties to work in any pH and independent from the presence of

Ca^{2+} [3, 4]. The crosslinked reagent most used is epichlorohydrin (Epi), which has the capacity to react with the OH of polysaccharides to form epoxy bonds between the polysaccharide chains of a polymer as shown Fig. 3.

Another crosslinker reagent used is the glutaraldehyde, which reacts with COOH groups, inducing a loss of these groups, so the capacity to bind molecules positively charged on the matrix is decreased. This does not happen in Alg beads cross-linked with Epi where the number of COOH groups remains unaltered, having the best capacity to bind positively charged molecules.

3.   Exploring the adsorption protein capacity of Alg-GG beds: its potential use in chromatography

There are a great number of possibilities to used Alg as an adsorption bed: it can be used alone, with a non-charged electrically polysaccharide (as is the case of GG) or, mixing with other negatively charged PE such as Carr, pectin, etc. [27]. The Alg matrix has a negative net electrical charge at pH above 4.0, so it can bind macromolecules having an isoelectric pH (pI) above 5.

Here, we describe the experimental steps to explore the useful application of this system to the recovery of a macromolecule from a biomass. We have selected the lysozyme (LZ) (pI 11.35) as a model protein with the goal to apply this method to a downstream process. The steps are the following:

3.1   Optimisation of enzymes adsorption by cross-linked Alg-GG beads

The beds were obtained according to the Roy et al. method [3]. In able to determine the matrix composition that has the best adsorption capacity, many systems were designed in which the following variables were applied: medium pH, total Epi concentration and Alg-GG ratio being the dependent variable the % of enzyme adsorbed per unit of mass of matrix. Table I shows the results obtained when LZ adsorption was assayed onto the different beds.

The best system for the adsorption of LZ was the system 10. However, the systems 3 and 4 showed a very good percentage of LZ adsorbtion and have the advantage that only a minor initial concentration of Alg is needed. For this reason, the rest of the measures of this work were made with the conditions of the systems 3 and 4.

Table 1. *Systems designed for the LZ adsorption onto Alg-GG beds*

System	Epi (mL/5 g of system)	Alg initial concentration (% p/v)	pH	% of LZ adsorbed
1	1.5	0.6	6	13.0±0.1
2	3.0	0.6	6	56±2
3	1.5	0.6	7	80±2
4	3.0	0.6	7	77±2
5	1.5	0.6	8	28.6±0.4
6	3.0	0.6	8	24.7±0.3
7	1.5	1.0	6	74±2
8	3.0	1.0	6	26±3
9	1.5	1.0	7	62±1
10	3.0	1.0	7	82.5±3.0
11	1.5	1.0	8	26±3
12	3.0	1.0	8	57±14

3.2   Physicochemical characterization of the Alg-GG matrix

For the matrix characterization, different parameters are determined. First, the pH value at which the protein net electrical charge is zero (pHzpc) was determined. This acid-base titration curve is very useful because it allows to know the number of COOH available, which can interact with positively charged groups of protein to be adsorbed.

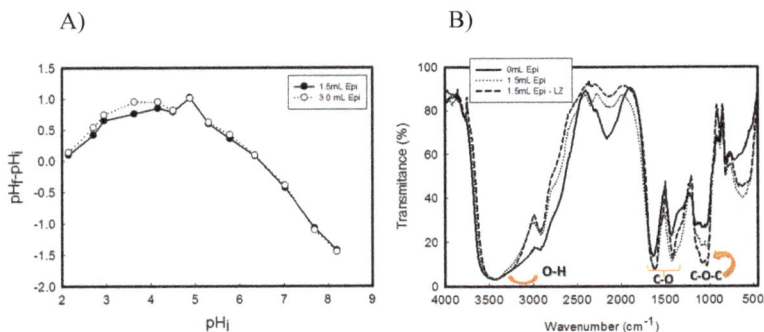

Figure 4. A) pHzpc of the adsorbent (0.6:0.5 ratio Alg-GG beads treated with different Epi volume). Temperature: 25°C. B) IR spectrum of the Alg-GG matrixes (0.6:0.5 ratio) without and with Epi cross-linked.

Fig. 4-A shows a pHzpc value of 6.4 obtained for 0.6:0.5 ratio of Alg:GG matrixes with both Epi volumes. Below the pHzpc value, the solid material has a positive surface charge, promoting the adsorption of anions, and above the pHzpc value, the surface is negatively charged, favouring adsorption of cations [28].

The FTIR spectrum of Alg-GG and cross-linked Alg-GG beads are shown in Fig. 4-B. The peak that mostly characterises epoxide groups is the one that is associated with axial deformation of C–O–C observed at 1120 cm^{-1}, has high intensity for cross-linked beads when compared with the control without treatment. The bands around 1030 cm^{-1} (C-O-C stretching) are attributed to the polysaccharide structure. In addition, the bands at 1617 and 1417 cm^{-1} are assigned to asymmetric and symmetric stretching peaks of carboxyl groups of the alginate. The peak at 3000 cm^{-1} corresponds to the hydroxyl groups of the both polysaccharide [3, 19, 20]. These results confirm the success of the cross-linking reaction.

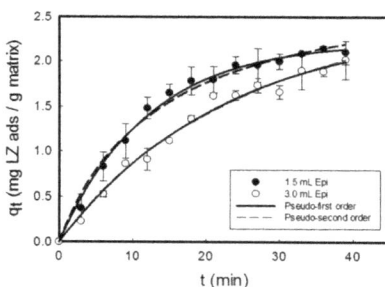

Figure 5. Kinetic of LZ adsorption onto Alg-GG matrixes. Medium: 20 mM phosphate buffer pH 7.0. Initial LZ concentration: 0.16 mg/mL. Temperature: 25°C.

3.3 Adsorption studies of LZ onto Alg-GG bed in batch

3.3.1 Adsorption kinetics

A desired amount of matrix is put in contact with the solute solution at initial concentration, so under continuous shaking the solute concentration is measured vs. time. Fig. 5 shows the experimental data obtained at 25°C, for the adsorption of LZ onto Alg - GG bed (systems 3 and 4 - Table 1). It can be seen that in both cases, the adsorbed LZ (q(t)) increases with the contact time until it reaches a plateau. The point where the plateau begins is the equilibrium time required to achieve the maximum adsorption q(e) in these conditions. Also, it can be seen that the major amount of Epi-induced a decreasing in the LZ adsorption.

To analyse the adsorption kinetic mechanism, the experimental data at different temperatures and initial concentrations of LZ were fitted with two models, namely pseudo-first-order and pseudo-second-order [29, 30] respectively as shown in equations 1 and 2.

$$q_t = q_e(1 - e^{-k_1 t}) \tag{1}$$

$$q_t = \frac{k_2 q_e^2 t}{1 + k_2 q_e t} \tag{2}$$

Where: k_1 and k_2 are the first and second order kinetics constant, respectively, and qt and qe are the amount of LZ adsorbed per g of the matrix in a time "t" and at the equilibrium, respectively. The parameters of the kinetic models and the regression correlation coefficients (R^2) are listed in Table 2. It can be seen that SS values obtained from fitting to pseudo first-order kinetic are low and the correlation coefficient values were found to be greater than 0.99, which shows the applicability of the pseudo first order as a model of LZ adsorption. Besides, qt values obtained for the first order are coincident with the visual value gives in Fig. 5.

Table 2. Kinetics values parameters for the adsorption of LZ onto Alg-GG matrixes cross-linked with two different Epi volumes (1.5 mL and 3.0 mL). Medium: 20 mM phosphate buffer, pH 7.0. Initial LZ concentration: 0.16 mg/mL. Temperature: 25°C.

EpiVolume (mL)		Qe (mg.g-1)	k1 (min-1)	k2 (g.mg-1.min-1)	R²	SS
1.5	Pseudo-first order	2.20 ± 0.04	(8.3 ± 0.5).10-2	NA	0.9934	0.0398
	Pseudo-second order	0.4 ± 22.7	NA	0.16 ± 3.87.10-4	0.9871	0.0779
3.0	Pseudo-first order	2.4 ± 0.15	(4.4 ± 0.5).10-2	NA	0.9877	0.0676
	Pseudo-second order	0.4± 6.4	NA	(7.5 ± 0.08).10-2	0.9861	0.0759

The activation energy of the adsorption process was calculated using the Arrhenius equation as shown in equation 3.

$$ln\frac{k_{279K}}{k_{298K}} = \frac{E_a}{R}\left(\frac{1}{T^{298K}} - \frac{1}{T^{279K}}\right) \tag{3}$$

Where: R is the universal gas constant (1.9817 cal.mol^{-1}.K^{-1}), T is the absolute temperature (K) and Ea is the adsorption activation energy (kcal.mol^{-1}). The magnitude of Ea may give an idea about the type of adsorption. The obtained values of Ea were 4.34 and 7.3 kcal.mol^{-1} for a cross-linked system with 1.5 and 3.0 ml of Epi, respectively. These values are low and hence it can be concluded that the process is governed by interactions of physical nature [31].

The solute transfer is usually characterized by external mass transfer (boundary layer diffusion), intra-particle diffusion or both. The adsorption dynamics can be described by the following three consecutive steps which are as follows [19]: -transport of the solute from bulk solution through liquid film to the adsorbent surface; solute diffusion into the pore of adsorbent except for a small quantity of sorption on the external surface; parallel to this is the intraparticle transport mechanism of the surface diffusion; - adsorption of solute on the interior surfaces of the pores and capillary spaces of the adsorbent. The last step is assumed to be rapid and considered to be negligible.

The overall rate of adsorption will be controlled by the slowest step, which would be either film diffusion or pore diffusion. The most commonly used technique for identifying the mechanism involved in the adsorption processes is to fit the experimental data in an intra-particle diffusion plot. Previous studies by various researchers showed that the plot of qt versus t0.5represents multilinear, which characterizes the two or more steps involved in an adsorption process. The intra-particle diffusion model is given in a simplified form by Dogan et al. [32]:

$$q_t = K_{id} t^{0.5} + C \tag{4}$$

Where: K_{id} is the intra-particle diffusion coefficient (mg.g^{-1}.min$^{-0.5}$) and C is a constant related to the extent of the boundary layer effect. Thus, the K_{id} (mg.g^{-1}.min$^{-0.5}$) value can be obtained from the slope of the plot qt (mg.g^{-1}) versus t$^{0.5}$ (min$^{0.5}$). A linear relationship between qt and t$^{0.5}$ was found, so the slope of the strength line (K_{id}) remained constant value (Fig. 6).

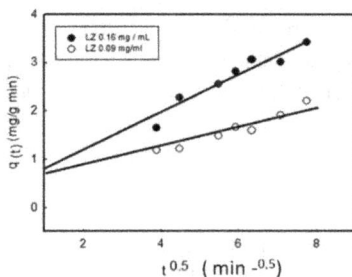

Figure 6. Intra-particle diffusion for LZ adsorption process onto Alg-GG beds. Data were taken from Fig. 5.

This effect was observed for the two Epi relations assayed, suggesting that the diffusion coefficient is not dependent of the matrix cross-linked grade. Because the intra particles diffusion remained constant the adsorption is driven only by the boundary layer, this is represented by the intercept of the plot with the y-axes (zero time) (C) which reflects the solute initial concentration on the boundary layer. As the C values were not dependent on the initial LZ concentration it can be confirmed that the only process to drive the kinetics of the adsorption is the solute diffusion throws the layer [29]. This finding can be due to the small size of the LZ molecule which can diffuse into the pore matrix. The low Ea observed is coincident with the presence of a simple solute diffusion process. The protein kinetics diffusion was also assayed in the presence of different cations of the Hofmeister series as chloride form because of their known effect on the modification of water structure.

Table 3. Relative diffusion coefficient (Kid) calculated from equation (4) in the presence of 100 mM of monovalent cation in chloride form.

Epi Volume (mL)	K_{id} (g.mg^{-1}.min^{-1})				
	Li^+	Control	K^+	Rb^+	Cs^+
1.5	0.38±0.04	0.37±0.02	0.09± 0.04	0.08± 0.04	0.01±0.04
3.0	0.52± 0.02	0.36±0.02	0.13± 0.02	0.10± 0.01	≅ 0

Table 3 shows the effect of the Hofmeister cations series on the diffusion coefficient of LZ value (K_{id}). Also, a linear relationship was found according to equation [5] in presence of the cations assayed (data not shown). As shown Table 3, K_{id} is slightly increased in the presence of Li^+, while the other cations induced a significant decreasing in the following sense: $K^+ > Rb^+ > Cs^+$, this effect can be not attributed to an electrostatic effect because the ionic strength of the medium was constant. Because the matrix is formed by polysaccharides the number of water molecules ordered in the environment of the polysaccharides chains is high, the presence of it is fundamental for the diffusing of the LZ to the pores of the matrix to interacts with the COO^- of Alg. The C value remained constant for all the cations assayed, (data not shown), because this parameter is related to the extent of the boundary layer, it can be concluded that the Hofmeister series do not modify it. So the water layer of hydration on the matrix surface and in the pore wall is not modified. Each Alg saccharine unit has capacity to bound 11-12 water molecules [33], the effect of Alg on the dynamics of water is restricted only to the first hydration shell, the extension of this barriers is great due to the high matrix porosity, so this layer of water is a barrier which it should be overwhelmed by the molecule of LZ as the previous process to interacts with Alg COO^- monomer. Lutter et al. [34] suggested that hydration of cations plays a key role in determining the interactions between the cations and a polymer. Larger monovalent cations such as Rb^+ and Cs^+ have weaker interactions with water molecules through electrostatic attraction with induces salting-in effects on the polymer as compared to smaller monovalent cations (Li^+) [34, 35].

It has been demonstrated that the electrophoretic mobility of an electrical charge particle is influenced by the Hofmeister series of cations [36]. In basis to an interaction between the negative electrical charge of the surface and the opossitive electrical charged of mobile particles such the case of a protein. Our finding showed that cations of the Hofmeister series modified the diffusion coefficient in the following sense: $Cs^+ < Rb^+ < K^+ < Li^+$. This trend in the mobility reflects the affinity of the different cations to the negatively charged electrical surface of the Alg-GG matrix. Therefore, poorly hydrated cation such is the case of Cs^+ interacts more strongly with the charged electrical surface of Alg-GG and, hence, will reduce the magnitude of the charged surface and therefore the LZ diffusion to the surface of the matrix. The strongly hydrated Li^+ ion interacts weakly on the Alg-GG surface and wall pores, so, the LZ diffusion will be not modified.

3.3.2 Adsorption isotherm

Fig. 7 illustrates the experimental equilibrium isotherm data for the Alg-GG matrix (systems 3 and 4). It can be seen that the higher concentration of Epi induces a minor capacity of matrix adsorption, in agree with a more closed structure of the polysaccharide chain. To optimize the design of an adsorption system for the removal of adsorbate, it is important to establish the most appropriate correlation for the equilibrium data. Various isotherm equations have been used to describe the isotherm curve [37]. In order to estimate the validity of isotherm models with experimental data, three equations were used namely Freundlich, Langmuir and Hill as shown equations 5 to 7.

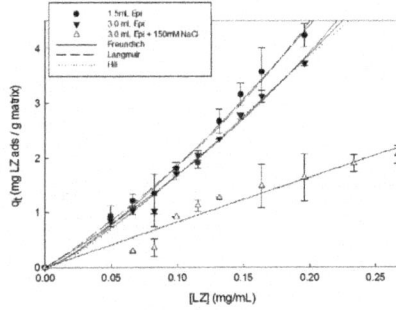

Figure 7. Adsorption isotherm of LZ onto non-soluble Alg-GG beads. Medium 20 mM phosphate buffer pH 7.0.Temperature 25°C.

$$q_e = K_F C_e^{1/n_F} \tag{5}$$

$$q_e = \frac{Q_0 K_L C_e}{1 + K_L C_e} \tag{6}$$

$$q_e = \frac{Q_0 C_e^{n_H}}{K_H + C_e^{n_H}} \tag{7}$$

Where q_e denotes the adsorbate adsorbed per gram of the adsorbent at equilibrium, Ce is the equilibrium concentration of adsorbate in liquid phase, Q_0 is the maximal adsorption capacity and K (K_F, K_L and K_H) and n (n_F and n_H)are empirical constants that indicate the extent of adsorption and the adsorption effectiveness, respectively [38].

Freundlich (F) isotherm is an empirical equation used for the description of multilayer adsorption with the interaction between adsorbed molecules, where K_F is the Freundlich constant related with adsorption capacity (mol.kg^{-1}) and n_F is the Freundlich exponent (dimensionless). The Langmuir (L) model is valid for monolayer adsorption to a surface with a finite number of identical sites. The Hill (H) isotherm, describes a model associated to cooperative adsorption [39].

The statistical analysis of the error function shows that the equation which has the best data fit for both systems is the Freundlich model (see Table 4) at the two temperatures assayed (data not shown). The Freundlich model is in agree with the low activation energy found associated with the adsorption process. The K_F value was decreased with the increase of the polysaccharide chain cross-linked by Epi action, as shown Table 4. The Langmuir and Hill models assayed yielded anomalous parameters demonstrating that the Freundlich is the correct model to fit the experimental data.

Table 4. Isotherms parameters and thermodynamics constants of LZ adsorption onto Alg-GG matrix. Temperature: 25° C

Model	Epi Volume (mL)	K_F	K_L	K_H	$1/n_F$	Q_0 (mg.g⁻¹)	n_H	R^2	SS
Freundlich	1.5	33.5 ± 4.6	N.A	N.A	1.3 ± 0.07	N.A	N.A	0.9878	0.1895
	3.0	27.7 ± 3,5	N.A	N.A	1.2 ± 0.07	N.A	N.A	0.9891	0.1318
Langmuir	1.5	N.A	-1.5 ± 0.3	N.A	N.A	-10.5 ± 3.4	N.A	0.9875	0.1939
	3.0	N.A	-1.3 ± 0.3	N.A	N.A	-11.6 ± 3.0	N.A	0.9875	0.1521
Hill	1.5	N.A	N.A	N.A	N.A	N.A	N.A	N.A	N.A
	3.0	0.5 ± 1.05	N.A	N.A	N.A	22.0 ± 42.2	1.4 ± 0.3	0.9895	0.1277

		ΔH (kcal.mol⁻¹)	ΔS (cal.mol⁻¹.K⁻¹)
Thermodynamics functions value	1.5	7.3	27.0
	3.0	1.0	5.0

Table 4 shows the thermodynamic variables values associated to the LZ adsorption at 25°C. The enthalpy change was positive in agreeing to a physical adsorption mechanism or to a slight coulombic interaction with the disorder of the water ordered in the pores of the matrix, induced by the diffusion of the LZ to them. The positive entropic change observed agree with this mechanism. The salt presence (NaCl 150 Mm) induced a decreasing of the LZ adsorption in agreeing with the presence of a coulombic component participating in the adsorption process. It can be seen that the increasing in the Epi volume for the step of cross-linked induce a decrease in both thermodynamic parameters, due to a more closed structure of the matrix treated with 3.0 mL of Epi.

3.4   Protein desorption conditions and regeneration assay study

The conditions of desorption were assayed by adding NaCl to the system, after the previews step of adsorption and washing of the matrix, to break the coulombic interaction. The presence NaCl 0.3 M induced a total displacement of the LZ from the matrix. Under this condition, the percentage of desorbed protein was 95 ± 2 %. Moreover, any adsorbent is economically viable if it can be regenerated and reused in many cycles of operation. In this case, six cycles during twenty days were carried out. The mean of the LZ adsorption in all the cycling was newly around 50 % of the total protein present, but the desorption with NaCl 0.3 M was approximately in all the cycles closer to 100 %. It can be seen that the adsorption capacity of the matrix was not modified and the enzyme recovery is total in this short period during the recycling prove.

3.5   Dynamic adsorption

Fixed bed column is the most efficient arrangement for conducting adsorption process for industrial applications in wastewater treatment. The design of an adsorption column depends on various important parameters such as flow rate, initial concentration and bed height (mass of adsorbent). The performance of fixed bed is usually described using the breakthrough curve: solute concentration in the eluate-time curves are commonly referred to as the breakthrough curves, and the time at which the effluent concentration reaches the threshold value is called the breakthrough time. The rational design of adsorption systems is based on accurate predictions of breakthrough curves for specific conditions. Despite the usefulness of fixed-bed mode, its analysis is usually complex. The fixed-bed operation is influenced by equilibrium (isotherm and capacity), kinetic (diffusion and convention coefficients), and hydraulic (liquid holdup, geometric analysis, and mal-distribution) factors. A typical breakthrough curve for the LZ adsorption onto Alg-GG is shown in Fig. 8 (insert).

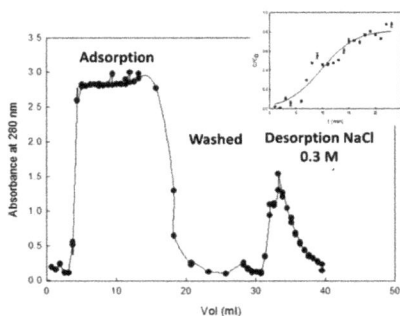

Figure 8. Adsorption and elution curve for the egg white (dilution 1/5) proteins onto Alg-GG matrix, pH 7.0 buffer phosphate 20 mM, diameter 7 mm x 100 mm. Flow rate 0.62 mL/min. (Insert): experimental breakthrough curve for adsorption of LZ.

When an adsorption process is driven only by diffusion through of the layer and the adsorption mechanism follows the Freundlich isotherm model, the adsorption process in a packed bed is described by the Clark model [40, 41]. This model is based on the use of a mass-transfer concept in combination with the Freundlich isotherm [24]. The equation (8) describes the Clark model:

$$\frac{C}{C_o} = \left(\frac{1}{1 + Ae^{-rt}}\right)^{1/n-1}$$

(8)

Where n is the Freundlich constant (from the isotherm) while A and r are the Clark constants. The shown experimental data in the Fig 8 (insert) were fitted to the eq (8) and is demonstrated that our experimental results follow this model. (data not shown).

4.   Fixed bed column using Alg-GG beads for the separation of LZ from white egg

LZ represent 3% of chicken egg white total protein, a dilution 1/5 with water was load into a column of Alg-GG until the absorbance of the eluate was constant, the column was washed with phosphate buffer 20 mM, pH 7.00 until zero absorbance, and the bound LZ was eluted by phosphate buffer 20 mM – 300 mM NaCl. A significant peak was observed of due to the LZ elution (Fig. 8). From the integration of the curve Absorbance vs. time of the eluted peak, the mass balance results in a 75 % recovery of LZ present in the egg white with a purification factor around 5.0. The eluted solution content LZ was analyzed by sodium dodecyl sulphate (SDS) polyacrylamide gel electrophoresis (PAGE). Fig. 9 shows the purification of LZ using the Alg-GG packed bed. It can be seen that only one band is obtained in the zone that corresponds to the LZ marker. This finding demonstrated a good efficient of the Alg- GG bed for the recovery of this protein.

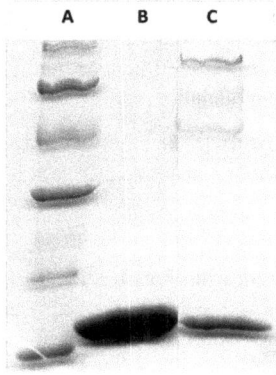

Figure 9. SDS gel electrophoresis. (A) a molecular marker, (B) LZ Sigma and (C) elution from the column.

5. Conclusion

The development of new adsorbents is based on economic issues. Commercial beads for adsorption are formed by polysaccharides chemically cross-linked to make it non-soluble. Agarose generally is used with acid or basic chemical groups covalently attached to it. Besides, several commercial trademarks are available, such as Streamline(™) and Sepharose(™) but their cost is around 2000 USD per L which affects significantly the final cost of the production process in scaling up. Alg-GG beads cross-linked with epichlorohydrin have properties similar to those of commercial adsorbents but are cheaper and easier to prepare. These advantages allow their use at scaling up level. Because the downstream processing costs for enzyme production account for about 50–80% of the total process cost the development of new materials with adsorption capacity for macromolecules is important. Our laboratory has been exploring the interaction between proteins and polysaccharides with net electrical charge, in order to use it as the first step in the concentration and purification of enzymes [20, 21].

The highest binding capacity of commercial beds can be offset by some qualities that have Alg-GG matrix such as:

1) Easy preparation, the matrix may be prepared in the laboratory in only 3 days with readily available materials without the need for specialized equipment or personnel.

2) Shorter purification time due to the faster adsorption and desorption process.

3) Low cost of ~20 USD per L against 2000 USD per L of commercial matrices [39, 42]. Considering also that the matrix can be reused up to 5 times, the cost reduction is even greater. The scaling up a level is strongly influenced by the cost of the unit operations, so it could contribute to increase its application at this level.

4) Minimal waste generation because of non-toxic reagents needed which can be discarded in the environment without a negative effect on it (green Chemistry).

Acknowledgments

This work was supported by a grant from FonCyT PICT 2013-271 .N. Woitovich Valetti thanks to CONICET for their fellowship and M.E. Brassesco to FonCyT. We thank María Robson, Geraldine Raimundo and Mariana DeSanctis for the language correction of the manuscript.

References

[1]   R. K. Scopes, Protein purification: principles and practice / Robert K. Scopes, Springer- Verlag: New York, (1994). https://doi.org/10.1007/978-1-4757-2333-5

[2]   M. -S. Chiou, P.-Y. Ho, H.-Y. Li, Adsorption of anionic dyes in acid solutions using chemically cross-linked chitosan beads, Dyes and Pigments, 60 (2004) 69-84. https://doi.org/10.1016/S0143-7208(03)00140-2

[3]   AL-Othman ZA, Naushad M, Inamuddin (2011) Organic–inorganic type composite cation exchanger poly-o-toluidine Zr(IV) tungstate: Preparation, physicochemical characterization and its analytical application in separation of heavy metals. Chem Eng J 172:369–375. https://doi.org/10.1016/j.cej.2011.06.018

[4]   N. W. Valetti, G. Picó, Adsorption isotherms, kinetics and thermodynamic studies towards understanding the interaction between cross-linked alginate-guar gum matrix and chymotrypsin, Journal of Chromatography B, 1012–1013 (2016) 204-210. https://doi.org/10.1016/j.jchromb.2016.01.027

[5]   Nabi SA, Bushra R, Naushad M, Khan AM (2010) Synthesis, characterization and analytical applications of a new composite cation exchange material poly-o-toluidine stannic molybdate for the separation of toxic metal ions. Chem Eng J 165:529–536. https://doi.org/10.1016/j.cej.2010.09.064

[6]   T. Gotoh, K. Matsushima, K.-I. Kikuchi, Preparation of alginate–chitosan hybrid gel beads and adsorption of divalent metal ions, Chemosphere 55 (2004) 135-140. https://doi.org/10.1016/j.chemosphere.2003.11.016

[7]   A. R. Kulkarni, K.S. Soppimath, T.M. Aminabhavi, A.M. Dave, M.H. Mehta, Glutaraldehyde crosslinked sodium alginate beads containing liquid pesticide for soil application, Journal of Controlled Release 63 (2000) 97-105. https://doi.org/10.1016/S0168-3659(99)00176-5

[8]   M. P. Klein, C. R. Hackenhaar, A. S. Lorenzoni, R. C. Rodrigues, T. M. Costa, J. L. Ninow, P. F. Hertz, Chitosan crosslinked with genipin as support matrix for application in food process: Support characterization and β-d-galactosidase immobilization, Carbohydrate Polymers 137 (2016) 184-190. https://doi.org/10.1016/j.carbpol.2015.10.069

[9]   J. P. Paques, E. van der Linden, C. J. van Rijn, L. M. Sagis, Preparation methods of alginate nanoparticles, Advances in Colloid and Interface Science 209 (2014) 163-171. https://doi.org/10.1016/j.cis.2014.03.009

[10]   D. Spelzini, B. Farruggia, G. Picó, Purification of chymotrypsin from pancreas homogenate by adsorption onto non-soluble alginate beads, Process Biochemistry 46 (2011) 801-805. https://doi.org/10.1016/j.procbio.2010.11.011

[11]   K. C. Aguilar, F. Tello, A. C. Bierhalz, M. G. G. Romo, H. E. M. Flores, C. R. Grosso, Protein adsorption onto alginate-pectin microparticles and films produced by ionic gelation, Journal of Food Engineering 154 (2015) 17-24. https://doi.org/10.1016/j.jfoodeng.2014.12.020

[12]   K. Y. Lee, D. J. Mooney, Alginate: Properties and biomedical applications, Progress in Polymer Science 37 (2012) 106-126. https://doi.org/10.1016/j.progpolymsci.2011.06.003

[13]   K. Dos Santos, J. Coelho, P. Ferreira, I. Pinto, S. G. Lorenzetti, E. Ferreira, O. Z. Higa, M. Gil, Synthesis and characterization of membranes obtained by graft copolymerization of 2-hydroxyethyl methacrylate and acrylic acid onto chitosan, International journal of pharmaceutics 310 (2006) 37-45. https://doi.org/10.1016/j.ijpharm.2005.11.019

[14]   A. G. Sullad, L. S. Manjeshwar, T. M. Aminabhavi, Novel pH-Sensitive Hydrogels Prepared from the Blends of Poly(vinyl alcohol) with Acrylic Acid-graft-Guar Gum Matrixes for Isoniazid Delivery, Industrial & Engineering Chemistry Research 49 (2010) 7323-7329. https://doi.org/10.1021/ie100389v

[15] T. Gotoh, K. Matsushima, K. -I. Kikuchi, Adsorption of Cu and Mn on covalently cross-linked alginate gel beads, Chemosphere 55 (2004) 57-64. https://doi.org/10.1016/j.chemosphere.2003.10.034

[16] J. López-Morales, D. Sánchez-Rivera, T. Luna-Pineda, O. Perales-Pérez, F. Román-Velázquez, Entrapment of Tyre Crumb Rubber in Calcium-Alginate Beads for Triclosan Removal, Adsorption Science & Technology 31 (2013) 931-942. https://doi.org/10.1260/0263-6174.31.10.931

[17] W. W. Ngah, L. Teong, M. Hanafiah, Adsorption of dyes and heavy metal ions by chitosan composites: A review, Carbohydrate Polymers, 83 (2011) 1446-1456. https://doi.org/10.1016/j.carbpol.2010.11.004

[18] F. Tello, R. N. Falfan-Cortés, F. Martinez-Bustos, V. M. da Silva, M. D. Hubinger, C. Grosso, Alginate and pectin-based particles coated with globular proteins: Production, characterization and anti-oxidative properties, Food Hydrocolloids 43 (2015) 670-678. https://doi.org/10.1016/j.foodhyd.2014.07.029

[19] E. Rodrigues, B. Bezerra, B. Farias, W. Adriano, R. Vieira, D. Azevedo, I. Silva, Adsorption of Cellulase Isolated from Aspergillus niger on Chitosan/Alginate Particles Functionalized with Epichlorohydrin, Adsorption Science & Technology 31 (2013) 17-34. https://doi.org/10.1260/0263-6174.31.1.17

[20] D. R. Gondim, N. A. Dias, I. T. Bresolin, A. M. Baptistiolli, D. C. Azevedo, I. J. Silva Jr, Human IgG adsorption using dye-ligand epoxy chitosan/alginate as adsorbent: influence of buffer system, Adsorption 20 (2014) 925-934. https://doi.org/10.1007/s10450-014-9636-6

[21] X. Vecino, R. Devesa-Rey, J. Cruz, A. Moldes, Study of the physical properties of calcium alginate hydrogel beads containing vineyard pruning waste for dye removal, Carbohydrate Polymers 115 (2015) 129. https://doi.org/10.1016/j.carbpol.2014.08.088

[22] Z. Zhang, R. Zhang, L. Zou, D. J. McClements, Protein encapsulation in alginate hydrogel beads: Effect of pH on microgel stability, protein retention and protein release, Food Hydrocolloids 58 (2016) 308-315. https://doi.org/10.1016/j.foodhyd.2016.03.015

[23] J. Venkatesan, I. Bhatnagar, P. Manivasagan, K. -H. Kang, S. -K. Kim, Alginate composites for bone tissue engineering: A review, International Journal of Biological Macromolecules 72 (2015) 269-281. https://doi.org/10.1016/j.ijbiomac.2014.07.008

[24]   A. S. Hoffman, Hydrogels for biomedical applications, Advanced Drug Delivery Reviews 64 (2012) 18-23. https://doi.org/10.1016/j.addr.2012.09.010

[25]   J. Patil, M. Kamalapur, S. Marapur, D. Kadam, Ionotropic gelation and polyelectrolyte complexation: the novel techniques to design hydrogel particulate sustained, modulated drug delivery system: A Review, Digest Journal of Nanomaterials and Biostructures 5 (2010) 241-248.

[26]   J. Yang, J. Chen, D. Pan, Y. Wan, Z. Wang, pH-sensitive interpenetrating network hydrogels based on chitosan derivatives and alginate for oral drug delivery, Carbohydrate Polymers 92 (2013) 719-725. https://doi.org/10.1016/j.carbpol.2012.09.036

[27]   M. George, T. E. Abraham, pH sensitive alginate–guar gum hydrogel for the controlled delivery of protein drugs, International Journal of Pharmaceutics 335 (335) 123-129. https://doi.org/10.1016/j.ijpharm.2006.11.009

[28]   N. Fiol, I. Villaescusa, Determination of sorbent point zero charge: usefulness in sorption studies, Environmental Chemistry Letters 7 (2009) 79-84. https://doi.org/10.1007/s10311-008-0139-0

[29]   E. Daneshvar, M. Kousha, M. Jokar, N. Koutahzadeh, E. Guibal, Acidic dye biosorption onto marine brown macroalgae: Isotherms, kinetic and thermodynamic studies, Chemical Engineering Journal 204 (2012) 225-234. https://doi.org/10.1016/j.cej.2012.07.090

[30]   C. Zhou, Q. Wu, T. Lei, I. I. Negulescu, Adsorption kinetic and equilibrium studies for methylene blue dye by partially hydrolyzed polyacrylamide/cellulose nanocrystal nanocomposite hydrogels, Chemical Engineering Journal 251 (2014) 17-24. https://doi.org/10.1016/j.cej.2014.04.034

[31]   S. -J. Yoon, D. -C. Chu, L. Raj Juneja, Chemical and Physical Properties, Safety and Application of Partially Hydrolized Guar Gum as Dietary Fiber , Journal of Clinical Biochemistry and Nutrition 42 (2008) 1-7. https://doi.org/10.3164/jcbn.2008001

[32]   M. Doğan, Y. Özdemir, M. Alkan, Adsorption kinetics and mechanism of cationic methyl violet and methylene blue dyes onto sepiolite, Dyes and Pigments 75 (2007) 701-713. https://doi.org/10.1016/j.dyepig.2006.07.023

[33]   K. Mazur, R. Buchner, M. Bonn, J. Hunger, Hydration of Sodium Alginate in Aqueous Solution, Macromolecules 47 (2014) 771-776. https://doi.org/10.1021/ma4023873

[34] J. C. Lutter, T. -Y. Wu, Y. Zhang, Hydration of Cations: A Key to Understanding of Specific Cation Effects on Aggregation Behaviors of PEO-PPO-PEO Triblock Copolymers, The Journal of Physical Chemistry B 117 (2013) 10132-10141. https://doi.org/10.1021/jp405709x

[35] Y. Zhang, P. S. Cremer, Interactions between macromolecules and ions: the Hofmeister series, Current Opinion in Chemical Biology 10 (2006) 658-663. https://doi.org/10.1016/j.cbpa.2006.09.020

[36] T. Oncsik, G. Trefalt, M. Borkovec, I. Szilagyi, Specific Ion Effects on Particle Aggregation Induced by Monovalent Salts within the Hofmeister Series, Langmuir 31 (2015) 3799-3807. https://doi.org/10.1021/acs.langmuir.5b00225

[37] S. Rangabhashiyam, N. Anu, M. G. Nandagopal, N. Selvaraju, Relevance of isotherm models in biosorption of pollutants by agricultural byproducts , Journal of Environmental Chemical Engineering 2 (2014) 398-414. https://doi.org/10.1016/j.jece.2014.01.014

[38] V. Vadivelan, K. V. Kumar, Equilibrium, kinetics, mechanism, and process design for the sorption of methylene blue onto rice husk, Journal of Colloid and Interface Science 286 (2005) 90-100. https://doi.org/10.1016/j.jcis.2005.01.007

[39] G. P. Jeppu, T. P. Clement, A modified Langmuir-Freundlich isotherm model for simulating pH-dependent adsorption effects, Journal of contaminant hydrology 129 (2012) 46-53. https://doi.org/10.1016/j.jconhyd.2011.12.001

[40] A. Crhribi, M. Chlendi, Modeling of Fixed Bed Adsorption: Application to the Adsorption of an Organic Dye, Asian J. Textile 1 (2011) 161-171.

[41] H. Nouri, A. Ouederni, Modeling of the Dynamics Adsorption of Phenol from an Aqueous Solution on Activated Carbon Produced from Olive Stones, Journal of Chemical Engineering & Process Technology, 4 (3) (2013). (DOI: 10.4172/2157-7048.1000153) https://doi.org/10.4172/2157-7048.1000153

[42] B. Zhao, W. Xiao, Y. Shang, H. Zhu, R. Han, Adsorption of light green anionic dye using cationic surfactant-modified peanut husk in batch mode, Arabian Journal of Chemistry, (2014). (http://dx.doi.org/10.1016/j.arabjc.2014.03.010) https://doi.org/10.1016/j.arabjc.2014.03.010

Chapter 5

Studies on different desalination processes turning sea water into drinking fresh water

Fakhra Jabeen* and M. Sarfaraz Nawaz

Department of Chemistry, Jazan University, Jazan, Saudi Arabia

* fakhrajabeen@gmail.com

Abstract

Desalination is a practical way of making salty water drinkable and it is used widely around the world, especially in very dry countries and on ships and small islands. The process is quite expensive and it uses a lot of energy. It also produces a very concentrated waste stream of brine which has to be disposed of responsibly. For these reasons, it is generally applied as the last resource, implemented only when all other approches have failed. The most common, modern methods of desalination are thermal processes and reverse osmosis (RO). Although there is an increased trend to RO due to the advances in this technology in the last decade. Humans cannot drink seawater as it contains salt and is saline water. But, luckily saline water can be made into freshwater, so that there is enough water for drinking washing and growing crops and for everyday use. Many parts of the world such as dry desert areas simply do not have enough fresh water from surface water such as rivers, lakes. They have little rainfall and then it may only be seasonal. The scarcity of fresh water and the need for additional fresh water is already critical in many arid regions of the world and will be increasingly important in the future. It is very likely that the need for fresh water will soon be considered, in the same category as oil and energy resources, and to be one of the determining factors of world stability and the prosperity of nations. So a regular dependable supply of water is needed in these areas to enable them to have water to grow crops and build their countries to prosper. As our world populations grow we need to be able to guarantee this basic need and avoid any shortages of fresh water. The solution is to look to the abundant supply of sea water and turn salty water into fresh water for drinking.

Keywords

Desalination, Reverse Osmosis, Distillation, Pretreatment, Saline Water, Membrane Separation

Contents

1.   Introduction

One of the major problems our society faces today is the shortage of fresh water. Increasing world population, rapid industrialization, and periodical droughts have resulted in higher water demand throughout the world. To complicate the issue, freshwater resources are diminishing year after year. Desalination plants have become key components of reliable water supply and guarantee supply even through droughts. Many human activities, such as drinking, agriculture, sanitation, and electricity generation, among others, require significant amounts of water. Fortunately, in many cases, centers of population are located near sources of useable water. However, oceans, which cover more than 70 % of the earth's surface and contain 97 % of the earth's water, have salt water. Since this salty water is unsuitable for many applications, it must be desalinated before it can be used. Desalination means any process that removes the excess salt and other minerals from the sea water in order to obtain fresh water suitable for animal consumption or irrigation [1]. This process is now being used all around the world to provide people with a much needed dependable supply of fresh water. In ancient times, many civilizations used this process on their ships to convert sea water into drinking water. In nature the sun causes water to evaporate from surface sources such as lakes, oceans, and streams. The water vapor eventually comes in contact with cooler air, where it re-condenses to form dew or rain. This process can be copied artificially more rapidly than nature, using man-made processes of heating and cooling. Desalination is not a modern science, it is a natural process. Sun or solar desalination is used by nature to produce rain which is the main source of fresh water on earth. All available man-made distillation systems are duplication on a small scale of this natural process. If almost all of the salt is removed for human consumption, sometimes the process produces table salt as a by-product. In soil desalination, this also happens to be a major issue for agricultural production [2]. The most common desalination methods employ reverse-osmosis in which salt water is forced through a membrane that allows water molecules to pass but blocks the molecules of salt and other minerals. Most of the modern interest in desalination is focused on developing cost-effective ways of providing fresh water for human use. Along with recycled wastewater, this is one of the few rainfall-independent water sources [3]. Due to relatively high energy consumption, the costs of desalinating sea water are generally higher than the alternatives (fresh water from rivers or groundwater, water recycling and water conservation), but alternatives are not always available and rapid overdraw and depletion of reserves is a critical problem worldwide. The most important users of desalinated water are in the Middle East, (mainly Saudi Arabia, Kuwait, the United Arab Emirates, Qatar and Bahrain), which uses about 70% of worldwide capacity; and in North Africa (mainly Libya and Algeria), which uses about 6% of worldwide capacity. Among industrialized countries, the United States is one of the most important users of desalinated water (6.5%), especially in California and parts of Florida. Quoting *Christopher Gasson* of Global Water Intelligence, "At the moment, around 1% of the world's population is dependent on desalinated water to meet their daily needs, but by 2025, the UN expects 14% of the world's population to be encountering water scarcity. Unless people get radically better at water conservation, the desalination industry has a very strong future indeed" [4]. Desalination is particularly relevant in dry countries such as Australia, which traditionally have relied on collecting rainfall behind dams to provide their drinking water supplies. According to the International Desalination Association, in June 2011, 15,988 desalination plants operated worldwide, producing 66.5 million cubic meters per day, providing water for 300 million people [5]. This number has been updated to 78.4 million cubic meters by 2013 or 57% greater than just 5 years prior. The single largest desalination project is Ras Al-Khair in Saudi Arabia, (Figure 1) which produced 1,025,000 cubic meters per day in 2014 although this plant in Saudi Arabia is expected to be surpassed by a desalination plant in California [6]. The largest percent of desalinated water used in any country is in Israel, which produces 40% of its domestic water use from seawater desalination [7]. Desalination remains energy intensive, however, and future costs will depend on the price of both energy and desalination technology [8]. Desalination of brackish water is done in the United States in order to meet treaty obligations for river water entering Mexico. Several Middle Eastern countries have energy reserves so great that they use desalinated water for agriculture. Saudi Arabia's desalination plants account for about 24% of total world capacity. Another desalination plant is the Jebel Ali Desalination Plant (Phase 2) in UAE. It uses multi-stage flash distillation, dual-purpose and it is capable of producing 300 million cubic meters of water per year. Locations relying on desalinated water including Saudi Arabia, Algeria, Maldives, Bahamas, Mauritius, Bahrain, Oman, Balearic Islands, Qatar, California, Canary Islands, Seychelles, Florida, Spain, Gibraltar, Tanzania, Kuwait, UAE, Libya, Western Australia, Malta, Yemen.

Fig.1 Ras Al-Khair, Desalination Plant in Saudi Arabia.

The traditional process used in these operations is vacuum distillation, essentially the boiling of water at less than atmospheric pressure and thus a much lower temperature than normal. This is because the boiling of a liquid occurs when the vapour pressure equals the ambient pressure and vapour pressure increases with temperature. Thus, because of the reduced temperature, low-temperature "waste" heat from electrical power generation or industrial processes can be minimised. The principal competing processes use membranes to desalinate, principally applying reverse osmosis technology [9]. Membrane processes use semipermeable membranes and pressure to separate salts from water. Reverse osmosis plant membrane systems typically use less energy than thermal distillation, which has led to a reduction in overall desalination costs over the past decade. Seawater desalination has the potential to reliably produce enough potable water to support large populations located near the coast.

2. Methods of desalination and their shortcomings

At least three principle methods of desalination exist: thermal, electrical, and pressure. The oldest method, thermal distillation, has been around for thousands of years. In thermal distillation, the water is boiled and then the steam is collected, leaving the salt behind. However, the vaporisation phase change requires significant amounts of energy. More modern methods of distillation make use of various techniques such as low-pressure vessels to reduce the boiling temperature of the water and thus reduce the amount of energy required to desalinate.

A second major type of desalination utilises electric current to separate the water and salt. Typically, electric current will be used to drive ions across a selectively permeable membrane, carrying the dissociated salt ions with it. A key characteristic of this method is that the energy requirement depends on how much salt is initially present in the water. Consequently, it is suitable for water with initial salt concentrations but to energy intensive for sea water.

A third principle method of desalination is reverse osmosis, in which pressure is used to drive water through a selectively permeable membrane, leaving the salt behind. Similarly to electrically-driven separation, the amount of energy required for desalination depends on the initial salt content of the water.

Pretreatment is important when working with reverse osmosis and nanofiltration membranes due to the nature of their spiral-wound design. The material is engineered in such a fashion as to allow only one-way flow through the system. As such, the spiral-wound design does not allow for back pulsing with water or air agitation to scour its surface and remove solids. Since accumulated material cannot be removed from the membrane surface systems, they are highly susceptible to fouling (loss of production capacity). Therefore, pretreatment is a necessity for any reverse osmosis or nanofiltration system. Pretreatment in sea water reverse osmosis systems has four major components:

- Screening of solids: Solids within the water must be removed and the water treated to prevent fouling of the membranes by fine particle or biological growth, and reduce the risk of damage to high-pressure pump components.

- Cartridge filtration: Generally, string wound polypropylene filters are used to remove particles of 1–5 μm diameters.

- Dosing: Oxidising biocides, such as chlorine, are added to kill bacteria, followed by bisulphite dosing to deactivate the chlorine, which can destroy a thin-film composite membrane. There are also biofouling inhibitors, which do not kill bacteria, but simply prevent them from growing slime on the membrane surface and plant walls.

- Prefiltration pH adjustment: If the pH, hardness and the alkalinity in the feedwater result in a scaling tendency when they are concentrated in the reject stream, acid is dosed to maintain carbonates in their soluble carbonic acid form.

$$CO_3^{2-} + H_3O^+ = HCO_3^- + H_2O \quad hCO_3^- + H_3O^+ = H_2CO_3 + H_2O$$

Carbonic acid cannot combine with calcium to form calcium carbonate scale. Adding too much sulfuric acid to control carbonate scales may result in calcium sulphate, barium sulphate, or strontium sulphate scale formation on the reverse osmosis membrane.

Post-treatment and/or polishing steps are required to condition the water after the reverse osmosis membrane process to make it suitable to our application. Brine disposal can be an environmental and economic issue in some areas where the fauna and flora are sensitive to local seawater salinity increase. Brine disposal should be studied and engineered case by case. The art of desalination is to determine and combine available technologies to optimise water production costs and quality.

2.1 Shortcomings

These methods pose a serious environmental threat.

- Discharge concentrated brine back into the ocean
- Affects marine life including the flora & fauna in the marine ecosystem
- Ultimately disrupts the earth's ecosystem
- Degenerate over time, need to be replaced frequently

2.1.1 All types of water can be produced from a desalination plant:

- WHO or EU drinking water
- Irrigation water
- Process water: boiler feed water, cooling water
- Ultrapure water

2.1.2 All types of natural seawater source can be treated:

- Shallow surface seawater
- Deep seawater
- Brackish river water
- Beach well seawater

3. Energy of desalination

Despite the innovative refinements of desalination, the energy requirements are still tremendous. State-of-the-art desalination still requires 7 to 30 kW-h of energy per 1000 gallons of desalinated water. The energy required can vary significantly based on the type of desalination used as well as the initial salt content of the water. Thus, to desalinate 12 billion gallons of water daily, the world uses at least 84 million kW-h of energy; the actual number is likely significantly higher as many plants use older technology that requires more energy per 1000 gallons of purified water. Since a gallon of gasoline contains about 33 kW-h, the world uses the equivalent of at least 2.5 million gallons of gasoline daily to desalinate water.

4.   Product water

The desalinated product water is usually purer than drinking water standards, so when product water is intended for municipal use, it may be mixed with water that contains higher levels of total dissolved solids. Pure desalination water is highly acidic and is thus corrosive to pipes, so it has to be mixed with other sources of water that are piped onsite or else adjusted for pH, hardness, and alkalinity before being piped offsite. In the desalination, process calcium is added to the water to stabilise it, and chlorine is added to kill any bacteria. In the future more and more people will be looking to the sea as a natural resource for water and many areas and islands are now working on desalination projects.

5.   From salt water to tap water humans cannot drink seawater as it contains salt and is saline water. But, luckily saline water can be made into freshwater, so that there is enough water for drinking washing and growing crops and for everyday use. Many parts of the world such as dry desert areas simply do not have enough fresh water from surface water such as rivers, lakes. They have little rainfall and then it may only be seasonal. The scarcity of fresh water and the need for additional fresh water is already critical in many arid regions of the world and will be increasingly important in the future. It is very likely that the need for fresh water will soon be considered, in the same category as oil and energy resources, and to be one of the determining factors of world stability and the prosperity of nations. So a regular dependable supply of water is needed in these areas to enable them to have water to grow crops and build their countries to prosper. As our world populations grow we need to be able to guarantee this basic need and avoid which shortages of fresh water. The solution is to look to the abundant supply of sea water and turn salt water into fresh water for drinking. The cost is very high and so it cannot be afforded by everyone who needs it, but because the desalinization technology is improving fast, so the costs are beginning to fall, making it more affordable to countries and islands that need it.

The two main water sources for desalination are seawater and brackish water. The five key elements of a desalination system are largely the same for both sources. They consist of:

(1) Intake or seawater supply - getting the water from its source to the processing facility.

(2) Pretreatment system - removing <u>suspended solids</u> to prepare the water for further processing.

(3) Desalination or reverse osmosis process-removing <u>dissolved solids,</u> primarily salts and other inorganic matter, from a water source.

(4) Post-treatment - adding chemicals to the desalinated water to prevent corrosion of downstream infrastructure pipes.

(5) <u>Concentrate</u> management and freshwater storage - handling and disposing or reusing the waste from the desalination, and storing freshwater before it's provided to consumers.

Water from the sea is saline contains high levels (referred to as "concentrations") of dissolved salts. In this case, the concentration is the amount (by weight) of salt in water, as expressed in "parts per million" (ppm). If water has a concentration of 10,000 ppm of dissolved salts, then one percent (10,000 divided by 1,000,000) of the weight of the water comes from dissolved salts. One good method used to desalinate seawater is the "reverse osmosis" method. here are some parameters for saline water:

- Freshwater - Less than 1,000 ppm
- Slightly saline water - From 1,000 ppm to 3,000 ppm
- Moderately saline water - From 3,000 ppm to 10,000 ppm highly saline water - From 10,000 ppm to 35,000 ppm

By the way, ocean water contains about 35,000 ppm of salt.

All desalination processes use chemical engineering technology in which a stream of saline water is fed to the process equipment, energy in the form of heat, water pressure or electricity is applied, and two outlet streams are produced; a stream of desalinated (fresh) water and a stream of concentrated brine which must be decomposed of. This is shown in Figure 2.

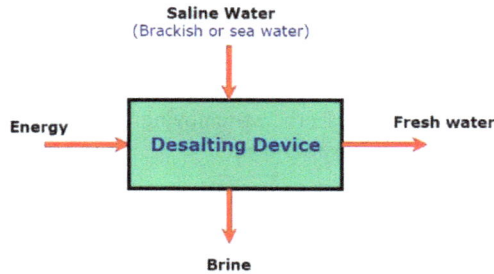

Fig. 2 Basic principle of desalination

There are many processes which can be used to reduce the concentration of dissolved solids in brackish or seawater. The worldwide installed capacity for the different processes is shown in Figure 3.

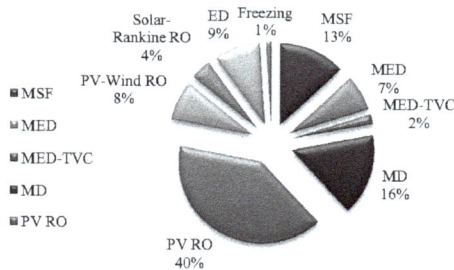

Fig.3 Worldwide installed desalination capacity for the different processes

6. Types of desalination processes

There are different types of desalination processes, which turn seawater into fresh water shown in Figure 4.

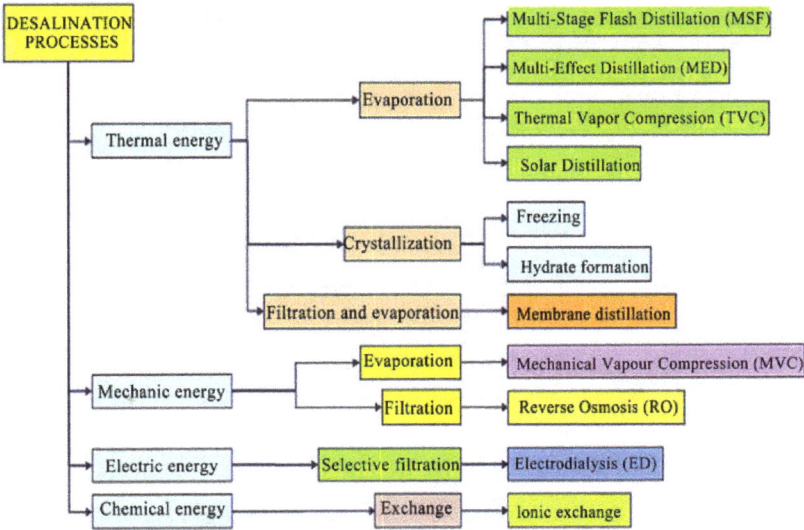

Fig.4 Different types of desalination processes

6.1   Multi-stage flash distillation (MSF)

Multi-stage flash distillation (MSF) is a water desalination process that distills sea water by flashing a portion of the water into steam in multiple stages of what are essentially countercurrent heat exchangers. In this process, seawater is heated and evaporated, after which the steam is condensed to produce desalinated water. An evaporator consists of several consecutive stages (evaporating chambers) maintained at decreasing pressures from the first stage (hot) to the last stage (cold). Seawater flows through the tubes of the heat exchangers where it is warmed by condensation of the vapour produced in each stage. When the water is heated in a vessel both the temperature and pressure increase; the heated water passes to another chamber at a low pressure which causes vapour to be formed; the vapour is led off and condensed to pure water using the cold sea water which feeds the first heating stage. The concentrated brine is then passed to a second chamber at a still lower pressure and more water evaporates and the vapour is condensed as before. The process is repeated through a series of vessels or chambers until atmospheric pressure is reached. Sea water slightly concentrates from stage to stage and builds up the brine flow which is extracted from the last stage. Typically, an MSF plant can contain from 4 to about 40 stages. The water vapour being condensed is used as a thermal energy source to heat incoming seawater. Evaporation and condensation are split into many stages that are repeated several times, thereby increasing overall efficiency. This principle is illustrated in Figure 5 which shows just three stages; in a commercial plant, many more stages are used.

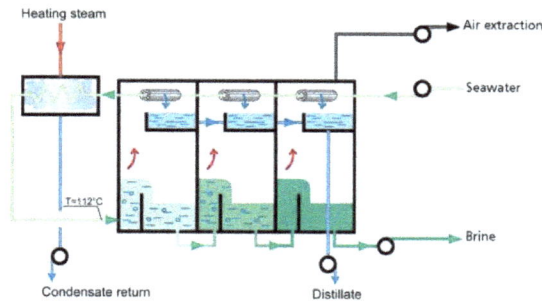

Fig.5 Diagram of the multi-stage flash distillation desalination process

The process employs proven once-through and recirculation technologies, saline water being heated by steam and then led into a series of effects (stages) where reduced pressure results in immediate boiling (flash) without the need for additional heat. This process is mainly employed for large-scale, thermal desalination plants where thermal energy is available in the form of low-pressure steam (>2 bar), e.g. in combination with thermal power plants or industrial complexes. This process operates with maximum efficiency at temperatures of up to 115° C, which results in large plant sizes and low energy consumption. It is considered to be the most reliable and is probably the most widely used, of the three principal distillation processes. One of the good features of this process is its ability to produce large amounts of water at a time. Since it uses the low temperature/pressure evaporator as a thermal energy source, while also making desalination possible with waste heat from the power plant, the MSF type is generally employed when building desalination plants in the Middle East. Multi-stage flash distillation plants produce about 60% of all desalinated water in the world [10].

Advantages

- MSF plants can be constructed to handle large capacities
- The salinity of the feedwater does not have much impact on the process or costs

- It produces very high-quality product water
- There is only a minimal requirement for pretreatment of the feedwater
- The strict operational and maintenance procedures for other processes are not as rigorous for MSF
- There is a long history of commercial use and reliability
- It can be combined with other processes, for example, using the heat energy from an electricity generation plant

Disadvantages

- They are expensive to build and operate and require a high level of technical knowledge highly energy intensive due to the requirement to boil the feedwater, although energy efficiency is substantially enhanced via the heat recovery process
- The recovery ratio is low, therefore more feed water is required to produce the same amount of product water
- The plant cannot be operated below 70-80% of the design capacity
- Blending is often required when there is less than 50 mg/l TDS in the product water

6.2 Multi-effect distillation (MED)

Multi-effect distillation (MED) is a distillation process often used for sea water desalination. It is also known as long-tube vertical distillation (LTV) and is in principle similar to multi-stage flash evaporation, except that steam is used to heat up the seawater in the first stage and the resulting vapor is used in subsequent stages to evaporate the water, and the seawater is used to cool and condense the vapor in each successive stage to that the temperature gradually falls across each stage of the process. Some of the feedwater is flash evaporated, but most of the seawater is dispersed over an evaporator tube bundle and boiled. Steam then condenses to produce fresh water, and this process is continuously repeated. Since the evaporation takes place in a vacuum, the sprayed seawater is able to reach a boiling point even at low temperatures. Since MED uses a progression of stages with ever-dwindling temperatures (at $60{\sim}70^0$C), it is comparatively smaller than MSF in terms of output, but higher heat efficiency and more economical operation. This principle is illustrated in Figure 6 which shows just three stages.

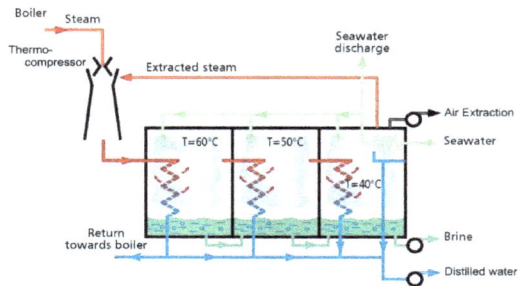

Fig.6 Diagram of the multi-effect distillation desalination process

A MED evaporator consists of several consecutive cells (or effects) maintained at decreasing levels of temperature (and pressure) from the first (hot) cell to the last one (cold). Each cell mainly consists of a horizontal tubes bundle. The top of the bundle is sprayed with the sea water make-up that then flows down from tube to tube by gravity. Brine and distillate are collected from cell to cell till the last one, where from they are extracted by centrifugal pumps.

Advantages

- Low energy consumption compared to other thermal processes
- Operates at low temperature (< 70°C) and at low concentration (< 1.5) to avoid corrosion and scaling
- Does not need pre-treatment of seawater and tolerates variations in sea water conditions highly reliable and simple to operate
- Low maintenance cost
- 24-hour-a-day continuous operation with minimum supervision
- Can be adapted to any heat source, including hot water, waste heat from power generation, industrial processes, or solar heating.
- Small to medium-sized plant sizes
- Reduced scaling risk

- **Disadvantages** high operating costs when waste heat is not available for the distillation process
- The multi-effect flash system operates at high temperatures that increase corrosion and scale formation
- The product water is at an elevated temperature and can require cooling before it can be used as potable water
- The recovery ratio is low, although not as low as for MSF

6.3   Thermal vapour compression (TVC)

Thermal vapour compression (TVC) is used for the production of both potable and process water from the sea and brackish water shown in Figure 7. In this process steam is generated from the seawater using a source of heat and the vapour is then compressed using a compressor. As a result of this compression, the temperature and pressure of the steam are increased. The incoming seawater is used to cool the compressed steam which then condenses into distilled (fresh) water and at the same time the seawater is heated further producing more steam. This process is normally used for small to large scale thermal desalination plants where thermal energy is available in the form of medium-pressure steam (>3 bar). TVC is the most economically advantageous steam heated process for medium-sized plants and it always represents an especially interesting alternative where medium-pressure steam is generated as an industrial by-product or is available in the course of waste heat utilisation. TVC installations are characterised by low investment and operating costs.

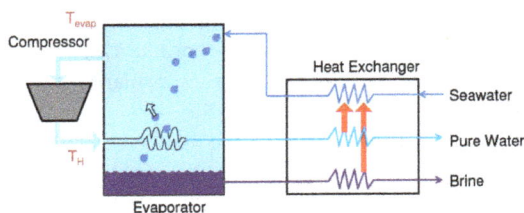

Fig.7 Diagram of the thermal vapour compression desalination process

As in the MED process, the TVC plant consists of several effects (stages), each of which is equipped with heat exchanger tube bundles. The first effect is heated by steam from the thermo-compressor, where an ejector using medium-pressure, motive steam (>3 bar a) recompresses part of the vapour produced in the last effect. The steam produced in each effect heats up the tubes in the subsequent effect. The condensate from the first effect is recycled to the steam generation system. The condensate from the downstream effects and the final condenser is the source of the desalinated product water (distillate).

Advantages

- Small to large plant sizes
- Low investment costs
- Minimised corrosion risk
- Reduced scaling risk
- Lower thermal energy consumption
- Low operating costs
- Efficient use of plant volume

Disadvantages

- Maintenance on compressors and heat exchangers is greater than those of other systems
- Energy consumption is high
- Capital costs are high

6.4   Solar distillation

Solar distillation is the use of solar energy to evaporate water and collect its condensate within the same closed system which can turn salt or brackish water into fresh drinking water shown in Figure 8. The process is known as a solar still and although the size, dimensions, materials, and configuration are varied, all rely on the simple procedure wherein an influent solution enters the system and the more volatile solvents leave in the effluent leaving behind the salty solute [11]. It differs from other forms of desalination that are more energy-intensive, such as reverse osmosis, or simply boiling water due to its use of free energy [12,13]. A very common and, by far, the largest example of solar distillation is the natural water cycle. Here the sun evaporates the water from lakes, rivers, oceans and other surface waters leaving salts and other minerals behind. This evaporated water eventually reaches the upper atmosphere where it recondenses as clouds and precipitates back to the land. This is the basic principle behind the use of solar energy for distillation.

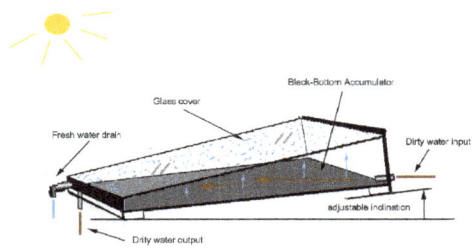

Fig.8 Diagram of the solar distillation desalination process by direct method

Saline water is supplied either continuously or intermittently to a pool ranging in depths of approximately 1 inch to 1 foot. The bottom of the pool has a black surface which absorbs solar energy. The discards salts exit through a drain. A transparent cover composed of glass sheets or plastic film is supported above. These are arranged so that the surfaces slope downward small troughs at their lower edges. These troughs are connected to channels or piping which transport the condensate to storage. A majority of the solar energy is absorbed in the basin bottom with a small amount being absorbed by the salt water itself. Heat is absorbed by the salt water from the basin bottom, raising the temperature and vapour pressure of the water. Partial vaporisation occurs and these vapours are transported upward to the transparent cover by convection covers. The cover is generally 10 to 30 degrees F cooler than the vapours and therefore condensation occurs. The condensation flows down the slope and collects in the troughs. The heat of condensation is transported through the cover and into the atmosphere. Only about half of the original feed is evaporated to prevent salt deposition on the bottom of the tank and the rest goes to waste.

Solar desalination is a technique to desalinate water using solar energy. There are two basic methods; direct and indirect. In the direct method, a solar collector is coupled with a distilling mechanism and the process is carried out in one simple cycle [14]. Water production by direct solar distillation is proportional to the area of the solar surface, incidence angle and has an average estimated value of $3-4L/m^2/day$. Because of this proportionality and the relatively high cost of property and material for construction the direct method distillation tends to favour plants with production capacities less than $200m^3/day$ [15]. Indirect solar desalination employs two separate systems; a solar collection array, consisting of photovoltaic or fluid based thermal collectors, and a separate conventional desalination plant shown in Figure 9. Production by the indirect method is dependent on the efficiency of the plant and the cost per unit produced is generally reduced by an increase in scale. Indirect solar desalination systems using photovoltaic (PV) panels and reverse osmosis (RO) have been commercially available and in use since 2009.

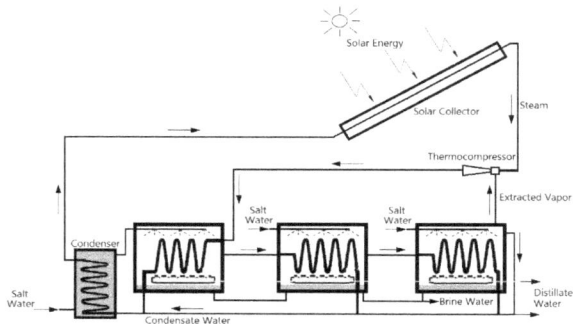

Fig.9 Diagram of the solar distillation desalination process by indirect method

For thousands of years, people have used the sun's heat to desalinate water - largely through the use of relatively simple evaporative processes. So-called 'direct' solar distillation methods have a long history, with evidence that they were used by ancient Greek mariners and Persian alchemists. Basic solar stills are still used in many small desalination and distillation plants. However, while useful at a very small scale, such methods are of limited use in modern agricultural, industrial and urban environments. In recent years, an increasing amount of attention has been paid to 'indirect' solar desalination using modern solar photovoltaic technology alongside methods such as reverse osmosis (RO) and Multiple Effect Distillation, which has the potential to operate at a much larger scale.

Advantages

- Produces pure water [No prime movers required
- No conventional energy required
- No skilled operator required
- Local manufacturing/repairing
- Low investment
- Can purify highly saline water (even sea water)

- No energy costs due to the use of solar panels
- Costs of solar panels are decreasing
- Very low maintenance

Disadvantages

- Requires a lot of space
- Initial investment cost due to the price of land
- Current costs of panels are expensive
- No economy of scale

6.5   Freezing

Freezing, or solidification, is a phase transition in which a liquid turns into a solid when its temperature is lowered below its freezing point. This process is an energy efficient method to produce clean water from aqueous streams and can be applied to a very broad range of water streams and is, therefore, an interesting water treatment option for many industries. Salt waters have a certain critical temperature which is a function of its salinity. When the salt water is reduced to this temperature, ice crystals composed of fresh water are formed. It is then possible to mechanically separate the ice crystals from the solution and re-melt them to get fresh water. This is the basic principle on which freezing desalination methods are based on (Figure 10).

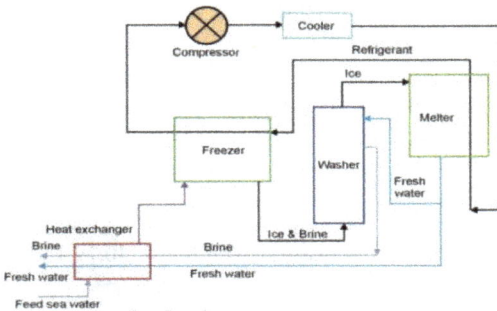

Fig.10 Diagram of the freezing desalination process

It is obvious that this method is not the most practical to use. It can be easily demonstrated on a small-scale basis but when applied to a large scale process, problems arise. The main problem lies in the economic aspects of the initial capital costs, maintenance costs and thermodynamic efficiency relative to upscaling the process. In open waters, some of the salts initially begin to freeze with the water. However, as time passes mass transport takes place throughout the semisolid layer as solidification continues to take place and salinity throughout the layer decreases with time. As a result, glacial and sea ice are relatively salt-free.

There are some advantages to this process. The operating temperature of this type of process is obviously at or below the freezing temperature of water. At these temperatures, scaling and corrosion are greatly reduced. This is one of the drawbacks of other conventional methods. Scaling is due to the buildup of precipitates due to calcium, magnesium, bicarbonate, sulphate, sodium, and chlorine ions in the water. The hard scale formed on the inside of the pipes requires costly maintenance practices to remove. Corrosion of steel pipes in contact with salt water is also increased with temperature. Lower temperatures permit plastics and protective coatings on the steel pipes to prevent the corrosive attack. The thermodynamic efficiency can also be increased due to the lack of heat exchangers required in heat driven desalination processes.

There are several different methods for the separation of the ice crystals from the liquid. The chosen method is dependent on the characteristics of the ice crystals. Two items of concern are crystal size and specific gravity. Filtration seems to be an obvious approach but it is actually impractical. It requires a slow, complicated filtration system and has not yet been applied to any system. A wash-separation method is a more reasonable approach. It takes the specific gravity into account. The solution flows up a screened or perforated column and a floating column of ice crystals is formed. Since the salt gets trapped in this ice column, it must be counter-washed by process water. This is one of the methods used for desalination. Although different methods, like reverse osmosis and distillation, are used commercially, but this can be used for remote areas, cold regions and for small and remote societies.

Advantages

- The main advantage of freezing is its low energy consumption as compared with distillation. The heat of vaporisation of water is 40.79 KJ/mol, whereas that of fusion is only 6.01 KJ/mol

- It can produce very pure potable water, and it has special advantages to produce water for irrigation

Disadvantages

- The major disadvantages of freezing are associated with the slow growth of ice crystals and with washing the salt deposits off the crystals

- it involves handling ice and water mixtures that are mechanically complicated to move and process

6.6   Hydrate formation hydrates are formed when natural gas components, for instance, methane, ethane, propane, isobutene, hydrogen sulphide, carbon dioxide, and nitrogen, occupy empty lattice positions in the water structure shown in Figure 11. Natural gas hydrates are solid crystalline compounds of snow appearance with densities smaller than that of ice. In this case, it seems like water solidifying at temperatures considerably higher than the freezing point of water.

The key circumstances that are essential for hydrate formation can be summarised as:

- The presence of "free" water. No hydrate formation is possible if "free" water is not present.

- Low temperatures, at or below the hydrate formation temperature for a given pressure and gas composition. high operating pressures. high velocities, or agitation, or pressure pulsations, in other words, turbulence can serve as a catalyst.

- The presence of H_2S and CO_2 promotes hydrate formation because both these acid gases are more soluble in water than the hydrocarbons.

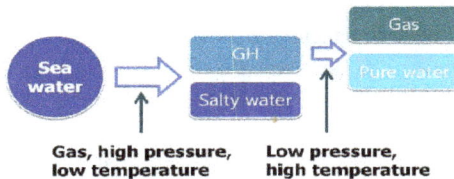

Fig. 11 Diagram of the hydrate formation desalination process.

In this process, the saline water is mixed with a hydrocarbon which forms hydrates or clathrates. In a clathrates a hydrocarbon molecule is enclosed in a molecular "cage" of water molecules forming a solid ice-like phase as shown in Figure 12 which shows methane (CH_4) molecule held in a "cage" of water (H_2O) molecules by Van der Waals forces (Van der Waals forces are weak inter-molecular forces arising from the electrical charges on molecules) [16].

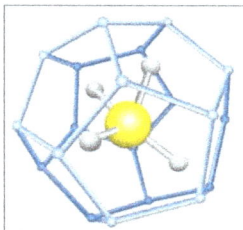

Fig.12 A methane (CH$_4$) molecule in a "cage" of water (H$_2$O) molecules.

The "cage" or hydrate forms ice-like crystals which contain none of the salts present in the seawater in which the hydrate forms. The crystals can be warmed in order to release water molecules; clathrates could be used in cold deep seawater to "entrap" water molecules and then warmed at shallower depths to release the water which would contain no dissolved salts.

Structurally, gas hydrates are inclusion compounds (clathrates) formed by trapping of gas molecules (M) in the voids of crystalline structures consisting of H$_2$O molecules. Gas hydrates have the general formula M nH$_2$O, with n varying from 5.75 to 17 depending on gas composition and conditions under which the hydrates are formed. A range of hydrocarbons such as methane, ethane and butane all have the ability to form clathrates at a temperature in the range of approximately 0^0C to 15^0C. For industrial desalination applications, the preferred chemical would be a refrigerant such as a hydrofluorocarbon which will form clathrates (hydrate) at temperatures above 0^0C at which ice forms. The crystals of hydrates are separated from the brine and melted to release the refrigerant and water free of salts. Although in theory this process has an efficient use of energy and should be able to compete with other desalination processes, the crystals formed have tended to be small and difficult to separate satisfactorily from the brine [17].

Advantages

- The energy is free- no fuel needed, no waste produced
- Not expensive to operate and maintain

Can produce a great deal of energy

Disadvantages

- Depends on the waves
- Needs a suitable site, where waves are consistently strong
- Some designs are noisy, but then again, so are waves, so any noise is unlikely to be a problem
- Must be able to withstand

6.7 Membrane distillation

Membrane distillation (MD) is a thermally-driven separation process, in which only vapour molecules transfer through a microporous hydrophobic membrane shown in Figure 13. The driving force in the MD process is the vapour pressure difference induced by the temperature difference across the hydrophobic membrane. A temperature difference of 5 to 10 K is sufficient to drive the MD process. The selectivity of membranes used for membrane distillation (MD) is based on the retention of liquid water with at the same time- permeability for free water molecules and thus, for water vapour. These membranes are made of a hydrophobic synthetic material (e.g.polytetrafluoroethylene (PTFE), polyvinylidene fluoride (PVDF), polypropylene (PP) and polyethylene (PE)) and offer pores with a standard diameter between 0.1 to 0.5 μm. These are used in MD process and are available in tubular, capillary, or flat sheet forms. As water has strong dipole characteristics, whilst the membrane fabric is non-polar, the membrane material is not wetted by the liquid. Even though the pores are considerably larger than the molecules, the liquid phase does not enter the pores because of the high water surface tension. The driving force which delivers the vapour through the membrane, in order to collect it on the permeate side as product water, is the partial water vapour pressure difference between the two bounding surfaces. This partial pressure difference is the result of a temperature difference between the two bounding surfaces. MD was investigated in the late 1960s but was not commercialised at that time for desalination because of the unavailability of suitable membranes and its relatively high cost. With the availability of new types of the membrane in the 1980s research was again undertaken on membrane distillation and many novel MD modules were designed based on a better understanding of the mass and heat transfer principles of MD [18]. The development of ceramic membranes has made it possible to use higher temperatures which may, in the future, make it possible to develop larger plants than have been manufactured so far. Advantages of the process include lower temperatures than the conventional thermal processes and lower pressures than conventional membrane processes.

Membrane distillation

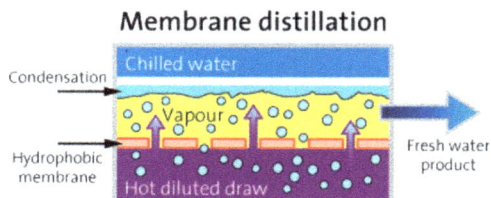

Fig.13 Diagram of the membrane distillation desalination process.

The potential applications of MD are the production of high purity of water, the concentration of ionic, colloid or other nonvolatile aqueous solutions, and removal of trace volatile organic compounds from wastewater. Various applications are involved in MD such as desalination of seawater or brackish water, environmental cleanup, water reuse, food, and medical. All these characteristics of MD process received worldwide attention from both academia and industry in the last decade.

Applications

- Seawater desalination
- Brackish water desalination
- Process water treatment
- Water purification
- Removal/Concentration of ammonium
- Resource concentration

Advantages

- It can be performed at a lower operating pressure and lower temperatures than the boiling point of feed solution.
- It requires lower vapour space.
- It is unlimited to high osmotic pressure.
- It permits very high separation factor of a nonvolatile solute.
- It has potential applications for concentrating aqueous solutions or producing high-purity water.

- It can use any form of low-grade waste heat or be coupled with solar energy systems which make it attractive for production of potable water from brackish water in arid regions.

Additionally, the possibility of using waste heat and renewable energy sources enable MD technique to cooperate in conjunction with other processes on an industrial scale. Hence, MD is a promising, yet still emerging technology for water treatment.

Disadvantages

- The MD process requires that the feed water should be free of organic pollutants; this explains the limited use of this method high initial cost
- Down performance with time
- Sediment clogging
- Reduce power penetration in the effect of sediment
- Energy consumption is approximately the same as that of MSF and MED plants

6.8   Mechanical vapor compression (MVC)

Mechanical vapor compression (MVC) is the evaporation method by which a compressor is used to compress and thus increase the pressure of the steam produced shown in Figure 14. Since the pressure increase of the steam also generates an increase in the steam temperature, the same steam can serve as the heating medium for the liquid. This makes this evaporation method very energy efficient. When this compression is performed by a mechanically driven compressor, the evaporation process is referred to as mechanical vapor compression (MVC) or mechanical vapor recompression (MVR).In case of compression performed by high pressure motive steam ejectors, the process is usually called thermo-compression or steam compression. Evaporation and distillation are the most energy demanding unit operations in the industry. Several methods have been developed to reduce the energy required in these processes, including multiple effect evaporation (MEE) and thermal vapor compression (TVC) but typically the most energy efficient approach will be using mechanical vapor compression (MVC).In this process, the mechanical vapor compressor is the key component in providing the mass and energy transfer required for evaporation. Moreover, the MVC loop does not, contrarily to MEE's and TVC's need a heat sink, and is therefore the optimal solution where cooling water is at a premium. The mechanical work required to drive the vapor compressor is provided either by an electric motor or a back pressure steam turbine. Usually the feed stream is evenly distributed onto a heat transfer surface of the evaporator heat exchanger. Evaporation takes place when the feed stream is heated to boiling by a heat source on the distillate side of the heat exchanger. In the MVC cycle, the water vapor drawn from the feed stream is compressed in the vapor compressor to a pressure level where it can be condensed and reused as heating medium. The vapor condenses on the distillate side of the heat exchanger and can be reused as such as process water, and after a conditional polishing used as boiler feed water.

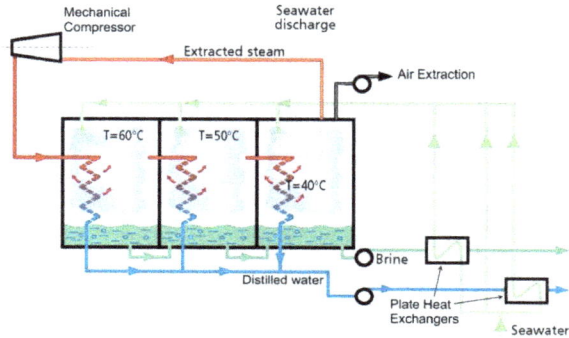

Fig.14 Diagram of the mechanical vapor compression desalination process

Mechanical vapor compression with multi-effect distillation plants (MVC-MED) are used for the production of potable and industrial process water from sea- and brackish water. The MVC desalination process functions like the MED-TVC process with the singular difference that the steam required for evaporation derives from an electric or diesel powered, mechanical steam compressor. Following recompression, the steam generated in the final effect is used to heat the first.MVC represents an economic alternative in the evaporation technology sector for small and medium-sized plants in stand-alone operation. The great advantage of this technology lies in the fact that it does not require cooling water, which means that complicated and cost-intensive feed installations are unnecessary. Therefore, MVC is also extremely suitable for inland applications such as the distillation of brackish water, industrial wastewater and reverse osmosis concentrate. It operates at low temperatures with high thermal efficiency, which leads to the following:

Advantages

- Customized for economic performance over the longer-term, delivering lower operation and maintenance costs

- Eco-friendly solutions – recycle waste steam to generate electricity
- Compact size for reduced footprint and capex
- Trusted partnership for solutions built to last
- Reduced corrosion risk
- Minimized scaling risk
- Process stability
- No need for external thermal energy

Disadvantages

- Limited capacity of the units high maintenance and spare parts requirements highly skilled operators for the operation of the unit
- System required large heat transfer area, which increase capital cost

Mechanical vapor compression desalination refers to a distillation process where the evaporation of sea or saline water is obtained by the application of heat delivered by compressed vapor. Since compression of the vapor increases both the pressure and temperature of the vapor, it is possible to use the latent heat rejected during condensation to generate additional vapor. The effect of compressing water vapor can be done by two methods. The first method utilizes an ejector system motivated by steam at manometric pressure from an external source in order to recycle vapor from the desalination process. The form is designated ejector or thermo-compression. Using the second method, water vapor is compressed by means of a mechanical device, electrically driven in most cases. This form is designated mechanical vapor compression. The MVC process comprises two different versions: vapor compression (VC) and vacuum vapor compression (VVC). VC designates those systems in which the evaporation effect takes place at manometric pressure, and VVC the systems in which evaporation takes place at sub-atmospheric pressures (under vacuum).The compression is mechanically powered by something such as a compression turbine. As vapor is generated, it is passed over to a heat exchanging condenser which returns the vapor to water. The resulting fresh water is moved to storage while the heat removed during condensation is transmitted to the remaining feedstock. The VVC process is the more efficient distillation process available in the market today in terms of energy consumption and water recovery ratio. As the system is electrically driven, it is considered a "clean" process, it is highly reliable and simple to operate and maintain. This plant delivers continuous uninterrupted desalination, independent of waste heat cycle. A mechanical vapor-compression evaporator, like most evaporators, can make reasonably clean water from any water source. In a salt crystallizer, for example, a typical analysis of the resulting condensate shows a typical content of residual salt not higher than 50 ppm or, in terms of electrical conductance, not higher than 10 µS/cm. This results in drinkable water, if the other sanitary requirements are fulfilled. While this cannot compete in the marketplace with reverse osmosis or demineralization, vapor compression chiefly differs from these thanks to its ability to make clean water from saturated or even crystallizing brines with total dissolved solids (TDS) up to 650 g/L. The other two technologies can make clean water from sources no higher in TDS than approximately 35 g/L.

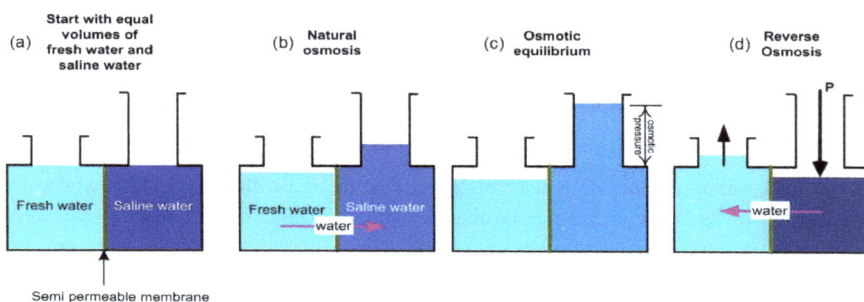

Fig.15 Diagram of the principle of osmosis and reverse osmosis through a semi-permeable membrane.

6.9   Reverse osmosis (RO)

Osmosis is the process in which water passes through a semi-permeable membrane from a low concentration solution into a high concentration solution. It is a process which occurs in plant and animal tissue including the human body. If a pressure is applied to the high concentration side of the membrane the reverse process occurs, namely water diffuses through the semi-permeable membrane from the high concentration solution into the low concentration solution, i.e. reverse osmosis, commonly referred to as RO. This is shown in the Figure 15.

Reverse osmosis is a water purification technology that uses a semipermeable membrane to remove ions, molecules, and larger particles from drinking water. In this process, an applied pressure is used to overcome osmotic pressure, a colligative property, that is driven by chemical potential differences of the solvent, a thermodynamic parameter. This process can remove many types of dissolved and suspended species from water, including bacteria, and is used in both industrial processes and the production of potable water. The result is that the solute is retained on the pressurised side of the membrane and the pure solvent is allowed to pass to the other side. To be "selective", this membrane should not allow large molecules or ions through the pores (holes) but should allow smaller components of the solution (such as solvent molecules) to pass freely. In the normal osmosis process, the solvent naturally moves from an area of low solute concentration (high water potential), through a membrane, to an area of high solute concentration (low water potential). The driving force for the movement of the solvent is the reduction in the free energy of the system when the difference in solvent concentration on either side of a membrane is reduced, generating osmotic pressure due to the solvent moving into the more concentrated solution. Applying an external pressure to reverse the natural flow of pure solvent, thus, is reverse osmosis. The process is similar to other membrane technology applications. However, key differences are found between reverse osmosis and filtration. The predominant removal mechanism in membrane filtration is straining, or size exclusion, so the process can theoretically achieve perfect efficiency regardless of parameters such as the solution's pressure and concentration. RO also involves diffusion, making the process dependent on pressure, flow rate, and other conditions [19]. This is most commonly known for its use in drinking water purification from seawater, removing the salt and other effluent materials from the water molecules.

Reverse osmosis membranes are manufactured from modern plastic materials in either sheets or hollow fibres. In a modern RO plant, the membranes are put together in modules which can be linked together according to the size of plant required. Modern RO plants use four alternative configurations of the membrane, namely, tubular, flat sheets, hollow fibre and the spiral-wound configuration are shown in Figure 16. Membranes manufactured from ceramic materials have been investigated but have not yet been exploited commercially for desalination.

Tubular membrane   Flat sheets membrane   Hollow fibre membrane

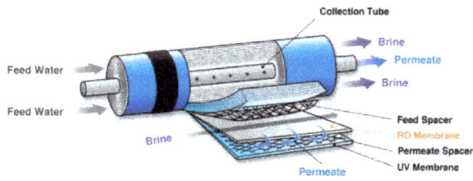

Spiral-wound membrane

Fig.16 The different types of membrane configuration for reverse osmosis (RO) systems.

RO plants vary from small domestic units for use either in a home as shown in Figure 17 or on small ships to large industrial and municipal units for supplying communities with a potable water supply. The world's largest and most advanced seawater reverse osmosis (SWRO) desalination plant has been built in Sorek, **Israel, October 21, 2013, shown in Figure 18**. The Sorek plant is capable of producing 624,000 m³/day of potable water, of which 540,000 m³/day are currently supplied to Israel's water distribution system. The plant sets significant benchmarks in desalination capacity and water cost, resulting in substantial savings for the local water market while alleviating the country's water shortage problem. IDE Technologies, a global water treatment specialist, announced that its seawater reverse osmosis desalination plant in the coastal city of Ashkelon, Israel, Feb. 10, 2015, has reached a world record with its water production process.

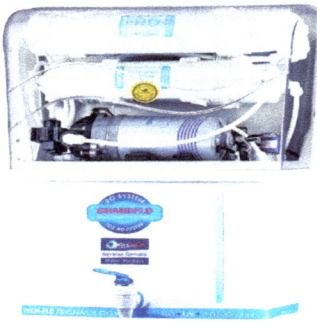

Fig.17 A small domestic RO unit   *Fig.18 Sorek desalination plant*

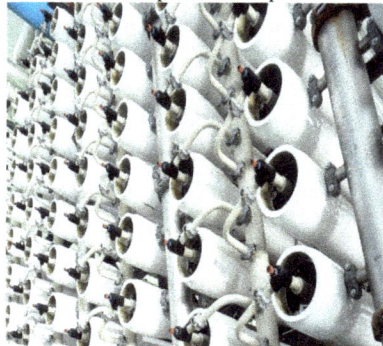

A reverse osmosis system consists of four major components/processes: (1) pretreatment, (2) pressurisation, (3) membrane separation, and (4) post-treatment stabilisation.

Pretreatment: The incoming feedwater is pretreated to be compatible with the membranes by removing suspended solids, adjusting the pH, and adding a threshold inhibitor to control scaling caused by constituents such as calcium sulphate.

Pressurisation: The pump raises the pressure of the pretreated feedwater to an operating pressure appropriate for the membrane and the salinity of the feedwater.

Membrane separation: The permeable membranes inhibit the passage of dissolved salts while permitting the desalinated product water to pass through. Applying feed water to the membrane assembly results in a freshwater product stream and a concentrated brine reject stream. Because no membrane is perfect in its rejection of dissolved salts, a small percentage of salt passes through the membrane and remains in the product water. Reverse osmosis membranes come in a variety of configurations. Two of the most popular are spiral wound and hollow fine fibre membranes. They are generally made of cellulose acetate, aromatic polyamides, or, nowadays, thin film polymer composites. Both types are used for brackish water and seawater desalination, although the specific membrane and the construction of the pressure vessel vary according to the different operating pressures used for the two types of feedwater.

Post-treatment stabilisation: The product water from the membrane assembly usually requires pH adjustment and degasification before being transferred to the distribution system for use as drinking water. The product passes through an aeration column in which the pH is elevated from a value of approximately 5 to a value close to 7. In many cases, this water is discharged to a storage cistern for later use.

Advantages

- The processing system is simple; the only complicating factor is finding or producing a clean supply of feedwater to minimise the need for frequent cleaning of the membrane.

- Systems may be assembled from prepackaged modules to produce a supply of product water ranging from a few litres per day to 750 000 l/day for brackish water, and to 400 000 l/day for seawater; the modular system allows for high mobility, making RO plants ideal for emergency water supply use.

- Installation costs are low.

- RO plants have a very high space/production capacity ratio, ranging from 25 000 to 60 000 $l/day/m^2$.

- Low maintenance, nonmetallic materials are used in construction.
- Energy use to process brackish water ranges from 1-3 kWh per 10001 of product water.
- RO technologies can make use of use an almost unlimited and reliable water source, the sea.
- RO technologies can be used to remove organic and inorganic contaminants.
- Aside from the need to dispose of the brine, RO has a negligible environmental impact.
- The technology makes minimal use of chemicals.

Disadvantages

- The membranes are sensitive to abuse.
- The feedwater usually needs to be pretreated to remove particulates (in order to prolong membrane life).
- There may be interruptions of service during stormy weather (which may increase particulate resuspension and a number of suspended solids in the feedwater) for plants that use seawater.
- The operation of an RO plant requires a high-quality standard for materials and equipment.
- There is often a need for foreign assistance to design, construct, and operate plants.
- An extensive spare parts inventory must be maintained, especially if the plants are of foreign manufacture.
- Brine must be carefully disposed of to avoid deleterious environmental impacts.
- There is a risk of bacterial contamination of the membranes; while bacteria are retained in the brine stream, bacterial growth on the membrane itself can introduce tastes and odours into the product water.
- RO technologies require a reliable energy source.
- Desalination technologies have a high cost when compared to other methods, such as groundwater extraction or rainwater harvesting.

RO membrane technology has developed over the past 40 years to a 44% share in world desalting production capacity, and an 80% share in the total number of desalination plants installed worldwide. The use of membrane desalination has increased as materials have improved and costs have decreased. Today, RO membranes are the leading technology for new desalination installations, and they are applied to a variety of saltwater resources using tailored pretreatment and membrane system design. Two distinct branches of reverse osmosis desalination have emerged: seawater reverse osmosis and brackish water reverse osmosis. Differences between the two water sources, including foulants, salinity, waste brine (concentrate) disposal options, and plant location, have created significant differences in process development, implementation, and key technical problems. Pretreatment options are similar for both types of reverse osmosis and depend on the specific components of the water source. Both seawater and brackish water reverse osmosis (RO) will continue to be used worldwide; new technology in energy recovery and renewable energy, as well as innovative plant design, will allow greater use of desalination for inland and rural communities while providing more affordable water for large coastal cities.

6.10   Electrodialysis (ED)

Electrodialysis (ED) is a membrane process, during which ions are transported through a semi-permeable membrane, under the influence of an electric potential shown in Figure 19. The membranes are cation- or anion-selective, which basically means that either positive ions or negative ions will flow through. It is used to transport salt ions from one solution through ion-exchange membranes to another solution under the influence of an applied electric potential difference. This is done in a configuration called an electrodialysis cell. The cell consists of a feed (dilute) compartment and a concentrate (brine) compartment formed by an anion exchange membrane and a cation exchange membrane placed between two electrodes. In almost all practical electrodialysis processes, multiple electrodialysis cells are arranged in a configuration called an electrodialysis stack, with alternating anion and cation exchange membranes forming the multiple electrodialysis cells. This process is different from distillation techniques and other membrane-based processes (such as reverse osmosis) in that dissolved species are moved away from the feed stream rather than the reverse. Because the quantity of dissolved species in the feed stream is far less than that of the fluid, electrodialysis offers the practical advantage of much higher feed recovery in many applications [20-24]. Electrodialysis reversal (EDR) is an advanced electrodialysis process which utilises a flow and polarity reversal to de-scale membrane surfaces and enables high concentration operation. It is used in higher salinity commercial applications where brine volume and scaling is a concern.

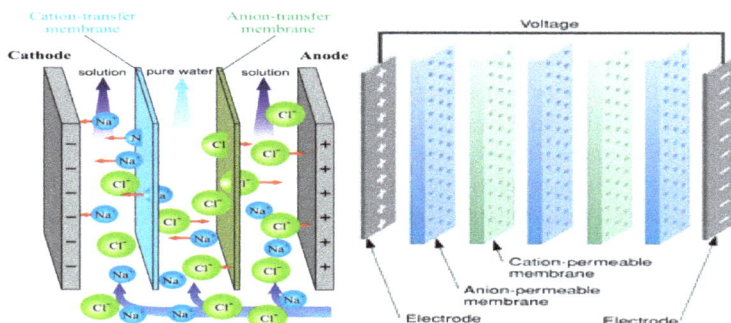

Fig. 19 Diagram of an electrodialysis cell showing how positive and negative ions are removed from saline feed water via ion exchange membranes by means of an electric current.

In an electrodialysis stack, the dilute feed stream, brine or concentrate stream, and electrode stream are allowed to flow through the appropriate cell compartments formed by the ion exchange membranes. Under the influence of an electrical potential difference, the negatively charged ions (e.g., chloride) in the dilute stream migrate toward the positively charged anode. These ions pass through the positively charged anion exchange membrane, but are prevented from further migration toward the anode by the negatively charged cation exchange membrane and therefore stay in the concentrate stream, which becomes concentrated with the anions. The positively charged species (e.g., sodium) in the dilute stream migrate toward the negatively charged cathode and pass through the negatively charged cation exchange membrane. These cations also stay in the concentrate stream, prevented from further migration toward the cathode by the positively charged anion exchange membrane. As a result of the anion and cation migration, electric current flows between the cathode and anode. Only an equal number of anion and cation charge equivalents are transferred from the dilute stream into the concentrate stream and so the charge balance is maintained in each stream. The overall result of the electrodialysis process is an ion concentration increase in the concentrate stream with a depletion of ions in the dilute solution feed stream. The electrode stream is that flows past each electrode in the stack. This stream may consist of the same composition as the feed stream (e.g., sodium chloride) or may be a separate solution containing a different species (e.g., sodium sulphate). Depending on the stack configuration, anions and cations from the electrode stream may be transported into the concentrate stream, or anions and cations from the dilute stream may be transported into the electrode stream. In each case, this transport is necessary to carry current across the stack and maintain electrically neutral stack solutions. By placing multiple membranes in a row, which alternately allow positively or negatively charged ions to flow through, the ions can be removed from wastewater. In some columns concentration of ions will take place and in other columns, ions will be removed. The concentrated salt water flow is circulated until it has reached a value that enables precipitation. At this point, the flow is discharged. This technique can be applied to remove ions from water. Particles that do not carry an electrical charge are not removed. Cation-selective membranes consist of sulfonated polystyrene, while anion-selective membranes consist of polystyrene with quaternary ammonia. Sometimes pre-treatment is necessary before the electrodialysis can take place. Suspended solids with a diameter that exceeds 10 μm need to be removed, or else they will plug the membrane pores. There are also substances that are able to neutralise a membrane, such as large organic anions, colloids, iron oxides and manganese oxide. These disturb the selective effect of the membrane. Pre-treatment methods, which aid the prevention of these effects, are active carbon filtration (for organic matter), flocculation (for colloids) and filtration techniques. In application, electrodialysis systems can be operated as continuous production or batch production processes. In a continuous process, the feed is passed through a sufficient number of stacks placed in series to produce the final desired product quality. In batch processes, the dilute and/or concentrate streams are re-circulated through the electrodialysis systems until the final product or concentrate quality is achieved.

Electrodialysis is usually applied to deionization of aqueous solutions. Some applications of electrodialysis include:

- Large scale brackish and seawater desalination and salt production.
- Small and medium scale drinking water production (e.g., towns & villages, construction & military camps, nitrate reduction, hotels & hospitals)
- Water reuse (e.g., industrial laundry wastewater, produced water from oil/gas production, cooling tower makeup & blowdown, metals industry fluids, wash-rack water)
- Pre-demineralization (e.g., boiler makeup & pretreatment, ultrapure water pretreatment, process water desalination, power generation, semiconductor, chemical manufacturing, food and beverage)
- Food processing
- Agricultural water (e.g., water for greenhouses, hydroponics, irrigation, livestock)

- Glycol desalting (e.g., antifreeze/engine-coolants, capacitor electrolyte fluids, oil and gas dehydration, conditioning and processing solutions, industrial heat transfer fluids, secondary coolants from heating, venting, and air conditioning (HVAC))
- Glycerin purification
- Stabilisation of wine
- Whey demineralization
- Pharmaceutical application

The major application of electrodialysis has historically been the desalination of brackish water or seawater as an alternative to RO for potable water production and seawater concentration for salt production. In normal potable water production without the requirement of high recoveries, reverse osmosis is generally believed to be more cost-effective when total dissolved solids (TDS) are 3,000 ppm or greater, while electrodialysis is more cost-effective for TDS feed concentrations less than 3,000 ppm or when high recoveries of the feed are required. Another important application for electrodialysis is the production of pure water and ultrapure water by electro-deionization (EDI). In EDI, the purifying compartments and sometimes the concentrating compartments of the electrodialysis stack are filled with ion exchange resin. When fed with low TDS feed, the product can reach very high purity levels. The ion exchange resins act to retain the ions, allowing these to be transported across the ion exchange membranes. The main usage of EDI systems is in electronics, pharmaceutical, power generation, and cooling tower applications.

- **Advantages** high water recovery design efficiently processes source water and generates low concentrate volume for disposal
- Rugged systems are less sensitive than reverse osmosis (RO) to particulates and metal oxides
- Targets arsenic, fluoride, radium and nitrate removal for potable water applications
- Long membrane life (typically 20+ years for potable water installations) for a lower cost of ownership
- Chlorine-resistant membranes are tolerant to disinfection, allowing use on challenging waters
- Wastewater reclamation reduces waste discharges and water use charges
- Effectively treats rejects and blowdowns

- Easy to control salt removal and product quality by adjusting amount of electricity applied to membrane stack
- Does not require any of the solutions to change its state of matter
- Simply separates the solute from the solvent, in this case, salt and water respectively

Disadvantages

- Requires massive amounts of electricity to produce the desired effect
- May not be financially feasible for many countries suffering from water shortages
- Chemical usage for pre-treatment is low
- Leaks sometimes occur in the membrane stacks

Electrodialysis (ED) and Electrodialysis Reversal (EDR) processes are driven by direct current (DC) in which ions flow through ion selective membranes to electrodes of opposite charge. In EDR systems, the polarity of the electrodes is reversed periodically. Ion-transfer (perm-selective) anion and cation membranes separate the ions in the feed water. These systems are used primarily in waters with low total dissolved solids (TDS).

Electrodialysis has inherent limitations, working best at removing low molecular weight ionic components from a feed stream. Non-charged, higher molecular weight and less mobile ionic species will not typically be significantly removed. Also, in contrast to RO, electrodialysis becomes less economical when extremely low salt concentrations in the product are required and with sparingly conductive feeds, current density becomes limited and current utilization efficiency typically decreases as the feed salt concentration becomes lower, and with fewer ions in solution to carry current, both ion transport and energy efficiency greatly declines. Consequently, comparatively large membrane areas are required to satisfy capacity requirements for low concentration (and sparingly conductive) feed solutions. Innovative systems overcoming the inherent limitations of electrodialysis (and RO) are available; these integrated systems work synergistically, with each subsystem operating in its optimal range, providing the least overall operating and capital costs for a particular application. As with RO, electrodialysis systems require feed pretreatment to remove species that coat, precipitate onto, or otherwise "foul" the surface of the ion exchange membranes. This fouling decreases the efficiency of the electrodialysis system. Species of concern include calcium and magnesium hardness, suspended solids, silica, and organic compounds. Water softening can be used to remove hardness, and a micrometer or multimedia filtration can be used to remove suspended solids. Hardness, in particular, is a concern since scaling can build up on the membranes.

Various chemicals are also available to help prevent scaling. Also, electrodialysis reversal systems seek to minimise scaling by periodically reversing the flows of dilute and concentrate and polarity of the electrodes.

6.11 Ionic exchange

- Ion exchange (IE) is an exchange of ions between two electrolytes or between an electrolyte solution and a complex. In most cases, the term is used to denote the processes of purification, separation, and decontamination of aqueous and other ion-containing solutions with solid polymeric or mineral 'ion exchangers'. It is a water treatment method where one or more undesirable contaminants are removed from water by exchange with another non-objectionable or less objectionable substance. Both the contaminant and the exchanged substance must be dissolved and have the same type (+,-) of electrical charge. Typical ion exchangers are ion exchange resins, zeolites, clay, and soil humus. Ion exchangers are either cation exchangers that exchange positively charged ions (cations) or anion exchangers that exchange negatively charged ions (anions). There are also amphoteric exchangers that are able to exchange both cations and anions simultaneously. However, the simultaneous exchange of cations and anions can be more efficiently performed in mixed beds that contain a mixture of anion and cation exchange resins or passing the treated solution through several different ion exchange materials. It can be unselective or have binding preferences for certain ions or classes of ions, depending on their chemical structure. This can be dependent on the size of the ions, charge, or structure. Typical examples of ions that can bind to ion exchangers are: h^+ (proton) and OH^- (hydroxide)

- Single-charged monatomic ions like Na^+, K^+, and Cl^-

- Double-charged monatomic ions like Ca^{2+} and Mg^{2+}

- Polyatomic inorganic ions like SO_4^{2-} and PO_4^{3-}

- Organic bases, usually molecules containing the amine functional group $-NR_2H^+$

- Organic acids, often molecules containing $-COO^-$ (carboxylic acid) functional groups

- Biomolecules that can be ionised: amino acids, peptides, proteins, etc.

All natural waters contain, in various concentrations, dissolved salts which dissociate in water to form charged ions. Positively charged ions are called cations; negatively charged ions are called anions. Ionic impurities can seriously affect the reliability and operating efficiency of a boiler or process system. Overheating caused by the buildup of scale or deposits formed by these impurities can lead to catastrophic tube failures, costly production losses, and unscheduled downtime. Hardness ions, such as calcium and magnesium, must be removed from the water supply before it can be used as boiler feedwater. For high-pressure boiler feedwater systems and many process systems, nearly complete removal of all ions, including carbon dioxide and silica, is required. This system is used for efficient removal of dissolved ions from water. Ion exchange resins are used to remove dissolved particles from liquids. Therefore, functional groups are bonded to the polymer beads. These functionalized polymers adsorb particular anions or cations from the liquid and discharge others. Water treatment is the best-known and biggest field of application for ion exchange resins. In the household, such applications are used among others to soften water. They ensure that the calcium ions responsible for water hardness are replaced by sodium ions. Large cation/anion ion exchangers used in water purification of boiler feedwater. It can also be used to remove hardness from water by exchanging calcium and magnesium ions for sodium ions in an ion exchange column. Liquid (aqueous) phase ion exchange desalination has been demonstrated. In this technique anions and cations in salt water are exchanged for carbonate anions and calcium cations respectively using electrophoresis. Calcium and carbonate ions then react to form calcium carbonate, which then precipitates leaving behind fresh water. The desalination occurs at ambient temperature and pressure and requires no membranes or solid ion exchangers. Theoretical energy efficiency of this method is on par with electrodialysis and reverse osmosis. These processes are reversible chemical reactions for removing dissolved ions from solution and replacing them with other similarly charged ions. In water treatment, it is primarily used for softening where calcium and magnesium ions are removed from water; however, it is being used more frequently for the removal of other dissolved ionic species. In a cation exchange process, positively charged ions on the surface of the ion exchanger resin are exchanged with positively charged ions available on the resin surface - typically sodium. Water softening is the most widely used cation exchange process. Similarly, in anion exchange negatively charged ions are exchanged with negatively charged ions on the resin surface - typically chloride. Contaminants such as nitrate, fluoride, sulphate, and arsenic, as well as others, can all be removed by anion exchange. The exchange medium consists of a solid phase of naturally occurring materials (zeolites) or a synthetic resin having a mobile ion attached to an immobile functional acid or base group. Both anion and cation resins are produced from the same basic organic polymers but they differ in the functional group attached to the resin. The mobile ions are exchanged with solute ions having a stronger affinity to the functional group (e.g. calcium ion replaces sodium ion or sulphate ion replaces chloride ion). When the capacity of the resin is exhausted, it is necessary to regenerate the resin using a saturated solution to restore the capacity of the resin and return the resin to its initial condition. Brine, or sodium chloride solution, is most the commonly used regenerate, although others, such as strong acids (hydrochloric acid, sulfuric acid) or strong bases (sodium hydroxide) may also be used. These processes can be operated in a batch or continuous mode. Most water treatment ion exchange processes operate in a continuous mode. Continuous ion exchange processes are usually of the down-flow and packed-bed column type. When the resin capacity is exhausted, it is regenerated.

Applications

Ion exchange is widely used in the food & beverage, hydrometallurgy, metals finishing, chemical & petrochemical, pharmaceutical, sugar & sweeteners, ground & potable water, nuclear, softening & industrial water, semiconductor, power, and a host of other industries. A typical example of application is the preparation of high purity water for power engineering, electronic and nuclear industries; i.e. polymeric or mineral insoluble ion exchangers are widely used for water softening, water purification, water decontamination, etc. This method is widely used in the household to produce soft water. This is accomplished by exchanging calcium Ca^{2+} and magnesium Mg^{2+} cations against Na^+ or H^+ cations. Industrial and analytical ion exchange chromatography is another area of ion exchange chromatography, is a chromatographical method that is widely used for chemical analysis and separation of ions. For example, in biochemistry, it is widely used to separate charged molecules such as proteins. An important area of the application is extraction and purification of biologically produced substances such as proteins (amino acids) and DNA/RNA. These processes are used to separate and purify metals, including separating uranium from plutonium and other actinides, including thorium, and lanthanum, neodymium, ytterbium, samarium, lutetium, from each other and the other lanthanides. This process is also used to separate other sets of very similar chemical elements, such as zirconium and hafnium, which is also very important for the nuclear industry. Ion exchangers are used in nuclear reprocessing and the treatment of radioactive waste.

Advantages

- Purifying water to ensure safe drinking water for homes and communities, urban and rural

- Treating water to meet the quality requirements of any industrial process

- Removing pollutants from waste water

- Recycling and recovering water and valuable products for reuse

- Recovering energy from waste

- Low installation costs
- Low maintenance
- Green technology

Disadvantages

- Parameters should not exceed their limits
- Pretreatment required
- Inefficient for treating high salinity streams

7.   How desalinated water damages our hair (Hair fall)

In desert climates, hair loss is a common complaint affecting both men and women from all nationalities and age groups. Doctors and hairdressers are used to complaints about falling hair. People panic when they see vast amounts of hair on the shower floor and want to know the cause. One of the major reasons for hair loss in both men and women is due to the harmful effects of desalinated water. The main source of tap water in the Arabian Gulf countries is desalinated. However, increasing populations are polluting seawater causing additional bacteria growth. As a result, the treatment plants must add more chlorine to kill bacteria and then add lime (a calcium compound) to help control the chlorine levels. Whilst desalinated water is considered safe to drink, it causes many problems for hair.

7.1   How to prevent hair loss due to desalinated water

- Calcium leaves hair feeling dry and weighted down
- Calcium can build up on the scalp causing flaking of the scalp, giving the appearance of dandruff
- Calcium can damage the hair cuticle causing the hair to break off
- Calcium coats the scalp, blocking new hair growth

7.2   Chlorine in desalinated water

Chlorine is found in tap water and in the swimming pools. It is a harsh oxidizer added to the water to kill bacteria and it adversely affects hair. Any build up at the mouth of the follicle may cause the hair to break off, followed by a coating of the scalp and blocking further new hair growth. Active chlorine in the hair can cause hair to feel gummy when wet and straw-like when dry.

7.3 How the damage is caused

A healthy strand of hair looks like this, with smooth cuticles

 hair damaged by the harsh climate looks like this, with open cuticles

Minerals from desalinated water stick to the hair inside the open cuticles

The mineral crystals grow, causing the protective cuticles to break off

This exposes and damages the inner cortex, causing the hair to break

8. Health and environmental aspects

The number of desalination plants worldwide is growing rapidly, and as the need for fresh water supplies grows acuter, desalination technologies improve and unit costs are reduced. Desalination processes should aim to be environmentally sustainable. Most drinking water applications use WHO drinking water guidelines as water quality specifications. 'WHO Guidelines for Drinking-water Quality' cover a broad spectrum of contaminants, including inorganic and synthetic organic chemicals, disinfection by-products, microbial indicators and are aimed at typical drinking water sources and technologies. Currently, WHO says that existing guidelines may not fully cover the unique factors that can be encountered during the intake, production and distribution of desalinated water. Hence, it imposes stringent guidelines not only for drinking water quality but also for environmental protection issues. This is in order to assist with the optimisation of both proposed and existing desalination facilities and to ensure that nations and consumers will be able to enjoy the benefits of the expanded access to desalinated water with the assurance of quality, safety and environmental protection. Obviously, the intake and pre-treatment of seawater, as well as the discharge of the concentrate reject water produced; have to be adapted to the specific conditions at the site of each desalination plant. Hence, it is necessary to consider and evaluate the criteria that help to select the best available technology and the optimal solution for the intake and outfall system at each plant. These environmental aspects are just as important as the commercial details and must be considered in the design and construction phases and during plant operation. Important elements regarding the environmental impact requirements that should be considered for different options are [25].

- Concentrate discharge standards and locations
- Wastewater discharge standards
- Air pollution control requirements
- Noise control standards
- Land use
- Public services and utilities
- Aesthetics light and glare

9. Advantages of desalination

1. It provides people with potable water - This is perhaps one of the biggest and most important benefits of desalination. By removing dissolved salts and other minerals from seawater, it can be turned to freshwater that is ideal for drinking. It can provide drinking water in areas where no natural supply of potable water exists. Some Caribbean islands get almost all of their drinking water through desalination plants, and Saudi Arabia gets 70 percent of its fresh water via the process. Even in countries where fresh water is plentiful, desalination plants can provide water to drier areas or in times of drought. The United States, for example, uses 6.5 percent of the world's supply of desalinated water.

2. It provides water to the agricultural industry - Desalination does not only produce potable water but also water that can be used for irrigation, which is great for arid regions as well as areas that are going through a drought. Since they will have the chance to produce their own crops, they won't be too dependent on imports and they will get to improve their economy. They can also produce enough food for their residents and keep hunger at bay.

3. It uses tried-and-tested technology - Unlike other water purifying processes, desalination is well past the research stage and has in fact been in use for several decades now. Because of this, it is considered to be a safe, workable, and reliable process that has been tried and tested in many applications by numerous experts over the years.

10. Disadvantages of desalination

1. It consumes a large amount of energy - Opponents of desalination point out that it is not a feasible solution to water supply problems mainly because it requires so much energy. Distillation, for example, requires millions of gallons saltwater to be boiled at high temperatures before it can become potable. Reverse osmosis, meanwhile, uses a lot of energy to overcome the natural osmosis process and remove large particles from seawater using a semipermeable membrane.

2. It can be costly - Desalination plants can be expensive to build because of the equipment and machines they require and can range between $300 million to almost $3 billion. They can also be expensive to keep up since, as mentioned above, they require a lot of energy to complete the desalination process. According to studies, desalinated water is five times more expensive to harvest than freshwater, making it too costly for the average consumer.

3. It can be harmful to the environment - Desalination produces freshwater, but it is also important to remember that it essentially removes salt from seawater - salt that has to be disposed of in one way or another. The problem with this is that chlorine and other chemicals are often added to the water during processing and left behind with the brine which, if dumped back to the ocean, brings with it many harmful substances that can destroy the marine ecology.

Conclusion

As the world's population continues to grow, existing water supplies will become increasingly insufficient. As more and more water is required to meet mankind's needs, desalination of sea water will become an increasingly important source of useable water. Any comprehensive plan addressing mankind's energy usage or ecologic impact must account for the effect of desalination; responsible development requires attention to the most energy-efficient methods of purifying water. Nowadays, desalination has become a very affordable solution to cope with fresh water shortage typically in tropical as well as of off-shore areas. The desalination of brackish water and seawater is proving to be a reliable source of fresh water and is contributing to tackling the world's water shortage problems. This article has reviewed a number of thermal and membrane water desalination processes developed during recent decades. The advantages and disadvantages, including economic, for each desalination technology have also been covered. In recent years, there have been considerable developments in membrane desalination processes, especially in terms of the design of membrane module, energy recovery and pre-treatment methods which have made it cost-competitive with thermal processes. The use of renewable energies for desalination becomes a reasonable and technically mature alternative to the emerging and stressing energy situation and a sustainable solution for water scarcity. Currently, coupling desalination plants with clean environment-friendly energy resources is a pressing issue due to the dramatic increase in fossil fuel prices and the harmful impacts of burning fossil fuels, such as environmental pollution and climate change. The scarcity of drinking water limits the socio-economic development in many areas, where solar resources are abundant. Hence, the use of solar energy for water desalination in countries in Africa and the Middle East region which have plenty of solar energy is a promising issue for meeting water demand and would definitely contribute both towards solving water scarcity problems and reducing carbon dioxide emissions by means of an environmentally friendly process.

Collective research and development programme involving all the stakeholders, governments, industries, universities and research institutions are required to improve and develop seawater desalination technologies to make them affordable worldwide, and especially in countries lacking conventional forms of energy and suffering shortages of water. Accordingly, various water desalination systems require extensive research and analysis for evaluating their potentials of development, applications and performance. The most mature technologies of renewable energy application in desalination are the wind and PV-driven membrane processes and direct and indirect solar distillation. The connection of RO membrane technology would be considered the most cost-competitive solar desalination technologies approaching conventional desalination water costs. Hence the linkage of photovoltaic cells and membrane water desalination processes could be also considered the second most competitive alternative for brackish water desalination because of low water cost, in addition to the current rapid decrease in PV modules prices in the market. However, solar distillation may be advantageous for seawater desalination, as solar-MED is recommended for large-scale solar desalination, other renewable energy resources have to be taken into accounts such as the wind and geothermal energy, which could be suitable for different desalination processes at viable costs. Moreover, the coupling of renewable energy and desalination systems has to be optimised and further technical research development of renewable energy augmented desalination technologies that require little maintenance and waste heat source are recommended which are uniquely suited to provide fresh water in remote areas where water and electricity infrastructure are currently lacking.

The potential benefits of ocean desalination are great for human needs, but the economic, cultural, and environmental costs of worldwide commercialization remain high. In many parts of the world, alternatives can provide the same freshwater benefits of seawater desalination at far lower economic and environmental costs. These alternatives include treating low-quality local water sources, encouraging regional water transfers, improving conservation and efficiency, accelerating wastewater recycling and reuse, and implementing smart land-use planning. Desalination may provide environmental benefits if it can reduce pressures on rivers and streams. Typically, however, the link between desalination and more water for environmental purposes is weak. Unless a water rights order or potential order makes water for the environment mandatory or more water is taken from the environment problematic, there is usually no explicit mechanism to link desalination project approval with environmental water. Without a mechanism, there is no guarantee that water will be used for ecosystem restoration.

References

[1]   "Desalination", The American Heritage Science Dictionary, Houghton Mifflin Company, via dictionary.com. Retrieved August 19, 2007.

[2]   "Australia Aids China in Water Management Project." People's Daily Online, 2001-08-03, via english.people.com.cn. Retrieved August 19, 2007.

[3]   M. Fischetti, Freshwater from the Sea, Scientific American 297 (2007) 118-119. doi:10.1038/scientificamerican0907-118.
        https://doi.org/10.1038/scientificamerican0907-118

[4]   Desalination industry enjoys growth spurt as scarcity starts to bite;
        (https://www.globalwaterintel.com/desalination-industry-enjoys-growth-spurt-scarcity-starts-bite/).

[5]   L. Henthorne, The Current State of Desalination, International Desalination Association. Retrieved 2012.

[6]   http://economictimes.indiatimes.com/slideshows/nation-world/biggest-ocean-desalination-plant-in-california-nears-completion/slideshow/46932071.

[7]   J. Pyper, Israel is creating a water surplus using desalination, EENews, February 7, 2014.

[8]   G. P. Thiel, Salty solutions, Physics Today 68 (6) (2015) 66-67. https://doi.org/10.1063/PT.3.2828

[9]   C. Fritzmann, J. Lowenberg, T. Wintgens, T. Melin, State-of-the-art of reverse osmosis desalination, Desalination 216 (2007) 1-76. doi:10.1016/j.desal.2006.12.009, 2007. https://doi.org/10.1016/j.desal.2006.12.009

[10]   IAEA. The introduction of Nuclear Desalination. IAEA Technical Report Series No. 400, 2000.

[11]   Desalination, a national perspective. National Research Council of the National Academies. 2008.

[12]   M. Abu-Arabi, Status and prospects for solar desalination in the MENA region, Solar Desalination for the 21st Century, (2007) 163-178.

[13]   C. Paton, P. Davies, The seawater greenhouse cooling, fresh water and fresh produce from seawater. In The 2nd International Conference on Water Resources in Arid Environments, Riyadh, 2006.

[14]   L. García-Rodríguez, A. I. Palmero-Marrero, C. Gómez-Camacho, Comparison of solar thermal technologies for applications in seawater desalination, Desalination, 142(2) (2002) 135-142. https://doi.org/10.1016/S0011-9164(01)00432-5

[15]   S. Kalogirou, Solar energy engineering: Processes and systems. Burlington, MA: Elsevier/Academic Press, 2009.

[16]   R. Corfield, Close encounters with crystalline gas, Chemistry in Britain, 38 (5) (2002) 22-25.

[17]   R. A. McCormack, R. K. Andersen, Clathrate desalination plant preliminary research study, U.S. Dept. of the Interior, Bureau of Reclamation, Water Treatment Technology Program Report 5, June 1995.

[18]   L. M. Camacho, L. Dumee, J. Zhang, J-d Li, J. Gomez, M. Duke, S. Gray, Advances in Membrane Distillation for Water Desalination and Purification Applications: A review, Water 5 (2013) 94-196. doi: 10.3390/ w5010094. https://doi.org/10.3390/w5010094

[19]   J. Crittenden, R. Trussell, D. Hand, K. Howe, G. Tchobanoglous, Water Treatment Principles and Design, Edition 2. John Wiley and Sons. New Jersey, 2005. ISBN 0-471-11018-3.

[20] T. A. Davis, Electrodialysis, in Handbook of Industrial Membrane Technology, M.C. Porter, Ed.), Noyes Publications, New Jersey, 1990.

[21] H. Strathmann, Electrodialysis, in Membrane Handbook, W.S.W. Ho, K.K. Sirkar, (Ed.), Van Nostrand Reinhold, New York, 1992.

[22] M. Mulder, Basic Principles of Membrane Technology, Kluwer, Dordrecht, 1996. https://doi.org/10.1007/978-94-009-1766-8

[23] T. Sata, Ion Exchange Membranes: Preparation, Characterization, Modification and Application, Royal Society of Chemistry, London, 2004.

[24] H. Strathmann, Ion-Exchange Membrane Separation Processes, Elsevier, New York, 2004.

[25] T. Peters, D. Pintó. Seawater intake and pre-treatment/brine discharge-environmental issues, Desalination 221 (2008) 576-84. https://doi.org/10.1016/j.desal.2007.04.066

Chapter 6

Adsorption of p-chlorophenol on microporous carbon by microwave activation: isotherms, kinetics and thermodynamics studies

Muthanna J. Ahmed[a,b*], Samar K. Theydan[b]

[a] Department of Chemical and Process Engineering -Faculty of Engineering and Built Environment- Universiti Kebangsaan Malaysia, Bangi 43600, Selangor, Malaysia.

[b] Department of Chemical Engineering -College of Engineering-University of Baghdad, Iraq

*muthanna.ja@gmail.com, samarkarim26@yahoo.com

Abstract

Equilibrium isotherms, kinetics and thermodynamics of p-chlorophenol (PCP) adsorption on microporous activated carbon have been investigated. *Siris* seed pods (SSP), an agricultural solid waste, were utilised as a precursor for preparation of activated carbon (KAC) by microwave induced KOH activation. The yield, surface area, micropores volume, and mesopores volume of KAC were 22.48 %, 1824.88 m^2/g, 0.645 cm^3/g, and 0.137 cm^3/g, respectively. The analysis of pore structure of KAC showed that KOH activation exhibited 82.48% micropores content. The adsorption behaviour was well described by the Langmuir isotherm model, showing a monolayer adsorption capacity of 88.32 and 338.87 mg/g on SSP and KAC, respectively. The investigation of adsorption kinetics indicated that the process closely follows the model of pseudo-second order. Results of thermodynamic studies showed exothermic and spontaneous natures of PCP adsorption under-examined conditions.

Keywords

Activated Carbon, Microwave, Adsorption, Seed Pods Biomass, P-Chlorophenol

Contents

1.   Introduction

Chlorophenols pose a serious ecological problem as environmental pollutants due to their high toxicity, recalcitrance, bioaccumulation, strong odour emission and persistence in the environment and suspected carcinogen and mutagen to the living organisms [1]. Therefore, such pollutants need to be constantly monitored in the aquatic environment, where a value of 0.5 mg L^{-1} is the upper permissible limit of these compounds in publicly supplied water [2]. Among these toxic organic contaminants, p-chlorophenol PCP has been recognised by the environmental protection agency EPA as priority pollutant [3]. PCP is widely used in chemical, pharmaceutical and petroleum industries and thus it can easily enter the natural environment through effluents of factories and can be found in surface water and soil [4]. Thus, different methods have been applied for the removal of PCP from wastewaters, such as photocatalytic degradation, electrochemical oxidation, biodegradation, dechlorination, ozonation, wet oxidation, and adsorption [5-11]. Among these techniques, adsorption technology is currently being used extensively for the removal of low concentrations of organic and inorganic micropollutants from aqueous solutions.

Several adsorbents were used such as metal oxide, resin, polymer, sewage sludge-based adsorbent, chitosan, zeolite, clays and rocks [12-18]. Activated carbon is one of the most effective adsorbents for organic compounds because of its extended surface area, high adsorption capacity, high thermostability, microporous structure and special surface reactivity. Indeed, activated carbon adsorption has been cited by the US Environmental Protection Agency as one of the best existing control technologies [19]. Despite the extensive use of activated carbon for wastewater treatment, its production remains an expensive process due to the application of conventional heating techniques and the use of relatively expensive and nonrenewable precursors like coal and petroleum.

Recently, microwave heating has been widely used in preparation and regeneration of activated carbons. The main difference between microwave devices and conventional heating systems is the heating pattern. The transformation of microwave energy is not by conduction or convection as in conventional heating, but by dipole rotation and ionic conduction inside the particles [20]. Thus, the treatment time along with the energy consumption can be significantly reduced through microwave heating. Also, abundance and availability of lignocelluloses materials make them good sources of precursors for activated carbons. Harvesting and processing of various agricultural crops result in considerable quantities of by-products. Such wastes are usually inexpensive, for which the effective utilisation has been desired.

Some of the agricultural solid wastes such as corn cobs, cane pith, plum kernels, sewage sludge, rattan sawdust, and coconut shell [21-26] have been successfully converted into activated carbons on a laboratory scale. *Siris* or *Albizia lebbek* is a medium to large tree belongs to the family Fabaceae. It is specially planted on both sides of the highways for beautification and shade. The tree is native to tropical Africa, Asia, and northern Australia, and annually produces seed pods which are 20 cm long and 3 cm wide [27]. These pods have little or no economic value and their disposal not only is costly but may also cause environmental problems. Although isotherms and kinetics behaviours of PCP adsorption were extendedly studied, investigations about adsorption thermodynamics of such pollutant on activated carbon are limited. In our previous work, we focused on determination of best preparation conditions for activated carbon from *Siris* seed pods using K_2CO_3 activation [28]. Therefore, the present study focused on adsorption behaviour of PCP onto activated carbon by microwave-assisted KOH activation. It was proved that chemical activation with KOH is a very powerful method for the production of activated carbon with high surface area, narrow pore size distribution, and well developed microporosity, which is preferred for adsorption of phenol and its derivatives.

2. Materials and methods

2.1 Activated carbon preparation

Siris seed pods collected from trees located around Baghdad University premises, Iraq were washed with water, dried at 110 °C for 24 h, crushed using a coffee grinder (Hammer, china), and sieved into a uniform size of 0.8 mm. The proximate analysis of the pods represented in weight percent was as follows: 36.4% cellulose, 18.9% hemicellulose, 13.6% lignin, 5.14% ash and 1.07% moisture. The dried pods with the mass of 3 g were firstly impregnated with 20 ml of potassium hydroxide (supplied by BDH chemicals Ltd Company) solution at an impregnation ratio of 1 g/g and remain for 24 h at ambient temperature. The impregnated precursor was activated in a modified microwave oven (MM717CPJ, china) operated at a radiation power of 620 W for a radiation time of 8 min. Then the cooled product was mixed with 0.1 M HCl solution and left overnight at ambient temperature, then filtered and subsequently was repeatedly washed with distilled water until the pH of washing effluent reached 6.5–7.0. The final product was dried in an oven (Model IH-100, England) at 110 °C to a constant weight and sieved into uniform granules for further use.

2.2 Activated carbon characterization

The characteristics and yield of activated carbon were determined. The yield was calculated as grams of activated carbon per gram of *Siris* seed pods utilised for activation. Ash and moisture contents were determined by standard and oven drying methods, respectively [29, 30]. The surface morphology of the *Siris* seed pods before and after activation was examined using scanning electron microscopy SEM (300 K Pixel CMOS, China). The physical properties of the activated carbon were characterised with an automatic chemisorption & physisorption analyser (Quantachrome Instrument Corp., USA). The surface area was determined from the application of BET equation to the adsorption-desorption isotherm of N_2 at 77 K [31]. Micropores volume was determined by applying the Dubinin-Radushkevich equation. The mesopores volume was determined using BJH desorption branch [32] and the pore size distribution was determined from the density functional theory [33].

2.3 Adsorption isotherm

Batch equilibrium studies were conducted at different temperatures of 303, 313, and 323K in 100 ml Erlenmeyer conical flasks containing 40 ml of PCP solutions with different concentrations (50-250 mg/l). Carbon with 0.02 g and particle size of 0.25 mm was added to each flask and kept on a thermostat shaker (Type TR-1, Germany) at an agitation speed of 200 rpm. The samples were agitated for 24 h to reach equilibrium and the pH of the solution was adjusted to 7 by adding drops of 0.1M HCl or NaOH solutions. After filtration, the concentrations of PCP in the filtrates were determined using UV-Visible Spectrophotometer (Shimadzu UV-160A) at its maximum wavelength of 280 nm. All adsorption trials and sample tests were carried out in triplicates. The PCP uptake at equilibrium, q_e (mg/g), was calculated as follows:

$$q_e = \frac{(C_o - C_e) V}{W} \tag{1}$$

Where C_o and C_e (mg/l) are PCP concentrations at initial and equilibrium, respectively, V (l) is the volume of PCP solution, and W (g) is the weight of activated carbon. To determine the maximum PCP adsorption capacity of prepared carbon, Langmuir, Freundlich, and Temkin isotherm models [34-36] were applied to fit the experimental equilibrium isotherm data and a nonlinear regression analysis was used to fit these equations to experimental data.

2.3.1 Langmuir isotherm

The Langmuir isotherm assumes that monolayer adsorption occurs at binding sites with homogenous energy levels, no interactions between adsorbed molecules and no transmigration of adsorbed molecules on the adsorption surface. The linear form of this isotherm equation is given as:

$$\frac{C_e}{q_e} = \frac{1}{q_m K_L} + \frac{1}{q_m} C_e \tag{2}$$

Where C_e (mg/l) is the equilibrium concentration of the PCP solution, q_e (mg/g) is the adsorption capacity at equilibrium, K_L (l/mg) is the constant related to free energy of adsorption, and q_m (mg/g) is the maximum adsorption capacity at monolayer coverage.

2.3.2 Freundlich isotherm

The Freundlich isotherm is an empirical equation that assumes heterogeneous adsorbent surface with its adsorption sites at varying energy levels. The linear expression of this isotherm is written as:

$$\ln q_e = \ln K_F + \frac{1}{n} \ln C_e \tag{3}$$

Where K_F $(mg/g(1/mg)^{1/n})$ is the relative adsorption capacity and n signifies the affinity of the adsorbate to the adsorbent.

2.3.3 Temkin isotherm

The Temkin isotherm takes into account that the presence of indirect adsorbate/adsorbate interactions affects the heat of adsorption of all adsorbed molecules in the layer would decrease linearly with coverage. The Temkin isotherm is represented linearly as:

$$q_e = B \ln A + B \ln C_e \tag{4}$$

Where R is the gas constant (8.314 J/mol K), T is the absolute temperature (K), A and b are Temkin parameters where b is related to the heat of adsorption and A (l/mg) is the binding energy parameter.

2.4 Adsorption kinetics

The batch kinetic studies were done by following the same procedure of adsorption studies at various PCP initial concentrations of 50, 150, and 250 mg/l. The aqueous samples were taken at preset time intervals and the uptake of PCP at time t, q_t (mg/g), was calculated by the following equation:

$$q_t = \frac{(C_o - C_t) V}{W} \tag{5}$$

Where C_t (mg/l) is PCP concentration at time t (min). Three kinetic models: pseudo-first order model, pseudo-second order model, and intraparticle diffusion model [37-40] were used to analyse the kinetic data of PCP adsorption on prepared carbon.

2.4.1 Pseudo-first order model

Lagergren proposed a method for adsorption analysis which is the pseudo first- order kinetic equation. The linear form of this equation is

$$\ln(q_t - q_e) = \ln(q_e) - K_1 t \tag{6}$$

where qe (mg/g) and qt (mg/g) are the amounts of adsorbed adsorbate at equilibrium and at time t, respectively, and k_1 (min^{-1}) is the rate constant of pseudo-first-order equation.

2.4.2 Pseudo-second order model

The sorption kinetics may be described by a pseudo-second-order model. The linear form of this equation is:

$$\frac{t}{q_t} = \frac{1}{K_2 q_e} + \frac{t}{q_e} \qquad (7)$$

Where K_2 (g/mg.min) is the equilibrium rate constant of pseudo-second-order model.

2.4.3 Intraparticle diffusion model

According to the intraparticle diffusion model proposed by Weber and Morris, the root time dependence may be expressed by the following equation:

$$q_t = K_3 t^{1/2} + C \qquad (8)$$

Where qt (mg/g) is the amount of solute on the surface of the sorbent at time t, K_3 (mg/g. min$^{1/2}$) is the intraparticle diffusion rate constant and C (mg/g) is the intercept and it gives an idea about the thickness of the boundary layer, where a value of C close to zero indicates that diffusion is the only controlling step of the adsorption process. Least-squares regression analysis was used to fit above models to experimental data. Besides the value of correlation coefficient R^2, the applicability of the kinetic models to describe the adsorption process was further validated by the normalised standard deviation, Δq (%), which is defined as:

$$\Delta q \, (\%) = 100 \sqrt{\frac{\sum [(q_{exp} - q_{cal})/q_{exp}]^2}{N - 1}} \qquad (9)$$

Where, N is the number of data points, q_{exp} and q_{cal} (mg/g) are the experimental and calculated adsorption capacities, respectively.

2.5   Adsorption thermodynamics

Adsorption thermodynamics study can provide information on whether the process is spontaneous or not, exothermic or endothermic, chemisorption or physisorption; with the aid of thermodynamic parameters such as a change in Gibbs free energy (ΔG), enthalpy (ΔH), and entropy (ΔS). These parameters are determined by using the following equations:

$$\ln (K_d) = \frac{\Delta S}{R} - \frac{\Delta H}{RT} \qquad (10)$$

$$\Delta G = -RT \ln (K_d) \qquad (11)$$

$$K_d = \frac{q_e \cdot (W/V)}{C_e} \qquad (12)$$

Where R is the universal gas constant (8.314 J/mole.K), T is temperature (K), and K_d is the distribution coefficient for the adsorption.

3. Results and discussion

3.1 Yield and characterization

In this study, 22.48 % maximum yield has been achieved from *Siris* seed pods precursor which is higher than obtained from solidago Canadensis [41] using KOH activator (Table1). This may be due to the high volatile and lignocellulosic contents of *Siris* seed pods compared to solidago Canadensis precursor. Ash and moisture contents were 5.23 and 9.45%, respectively. The surface area, micropore volume and mesopore volume of prepared carbon are 1824 m^2/g, 0.645 cm^3/g, and 0.137 cm^3/g, respectively. These values have been compared with the characteristics of other carbons prepared from various biomass wastes by microwave assisted KOH activation [42-47], as summarised in Table 2. It can be concluded that KOH activation gave carbon with high surface area and high micropores content. The higher volatile content in *Siris* seed pods can allow for forming interconnected pores which allow for better diffusion of activator to the inside of the precursor particle resulting in a well-developed micropores structure. Fig. 1 shows the SEM images for the *Siris* seed pods and activated carbon. It can be seen from Fig. 1a, that the surface of raw precursor has smooth nature and there are very little pores on it. The activation with KOH improves the porous structure and develops many pores (Figs. 1b). The development of porosity is associated with gasification according to the following reduction reactions:

$$6KOH + 2C \rightarrow 2K + 3H_2 + 2K_2CO_3$$
$$K_2CO_3 + 2C \rightarrow 2K + 3CO$$
$$K_2CO_3 \rightarrow K_2O + CO_2$$
$$K_2O + 2C \rightarrow 2K + CO$$

The metallic potassium K produced from the above reactions will diffuse into the internal structure of carbon matrix, widening the existing pores, and created new porosities. Fig. 2 displays the isotherms of N_2 adsorption-desorption and pore size distribution for the prepared carbon. From Fig 2a, it can be observed that the isotherm presents a high adsorption at low relative pressure, revealing microporous nature, where the adsorption branch resembles that of a type I isotherm in the international union of pure and applied chemistry IUPAC classification. From this figure it can be also concluded that the isotherm displayed a small hysteresis loop, indicating the presence of very small mesopores volume. Fig. 2b presents the pore size distribution for the activated carbons prepared in this study. This figure shows that the pore structure concerns mainly of micropores, which are defined by IUPAC as pores size < 20 Å. The percentage of micropores area for KAC structure is 82.48%. This revealed that Siris seed pods precursor produce microporous activated carbon which is useful for many industrial applications.

3.2   Adsorption isotherms

The adsorption isotherm indicates how adsorbate molecules are distributed between the liquid and solid phases when the adsorption process is in an equilibrium state. Analysis of the isotherm data by fitting different isotherm models is an important step in determining a suitable model for design purposes. The equilibrium isotherms at various temperatures were studied by varying the initial concentration of PCP under the conditions of pH 7.0, contact time 120 min, and carbon dose 0.5 g/l. Three models: Langmuir, Freundlich and Temkin isotherms, Eqs. (2)-(4), were applied for fitting the experimental adsorption isotherm data as shown in Fig. 3. The results of this fitting, as summarised in Table 3, show that the R^2 values of the three models descend in the order: Langmuir > Temkin > Freundlich. Thus, the best fitting for experimental data at various temperatures is achieved by the linear plot of C_e/q_e versus C_e for Langmuir isotherm (Fig. 3a). This result revealed the homogeneous nature of PCP adsorption on the carbon surface, as explained by Wu et al. [47] who showed the well fitting of Langmuir isotherm to adsorption isotherm data of PCP on activated carbon from corncob hull by KOH activation. The favourable adsorption nature can be concluded from the value of parameter n> 1 which evaluated from a linear plot of In C_e versus In q_e for Freundlich isotherm fitting (Fig. 3b) and summarised in Table 2. The Temkin isotherm considered a linear decrease of sorption energy as the degree of completion of the sorption sites of an adsorbent is increased. The heat of adsorption of all the molecules in the layer would decrease linearly with coverage due to adsorbent-adsorbate interactions. This isotherm gives an indication of the effect of temperature on adsorption process in terms of K_T, which determined from a linear plot of In C_e versus qe as shown in Fig. 3b. The decrease in values of K_T with increasing of temperature reveals the exothermic nature of PCP adsorption (Table 2). In this study, the maximum adsorption capacity of 338.87 mg/g has been reported according to the Langmuir isotherm (Table 2). This value has been compared with PCP capacity on raw precursor SSP (88.32 mg/g) and those on different adsorbents (Table 3). At adsorption conditions mentioned in Table 3, it can be seen that the modification of SSP to activated carbon improves the capacity for PCP and can be considered as effective adsorbents for removal of PCP from aqueous solutions.

3.3   Adsorption kinetics

Adsorption kinetics provides valuable information about the reaction pathways and mechanism of the reactions. The kinetics of PCP adsorption on prepared carbon were analysed using pseudo-first-order, pseudo second-order, and intraparticle diffusion models. The conformity between experimental data and the model predicted values was expressed by the correlation coefficients (R^2). A relatively high R^2 value indicated that the model successfully describes the adsorption kinetics.

The experimental kinetics data for PCP adsorption at different initial concentrations have been plotted as adsorbed amounts versus contact time, as shown in Fig. 4a. This figure shows that the equilibrium is achieved in about 120 min. These results are in agreements with those observed by many researchers [13, 48] who reported 120 min equilibrium time for adsorption of PCP on various adsorbents. This figure also shows that rapid increase in capacity for PCP is achieved during the first 25 min. The fast adsorption at the initial stage may be due to the higher driving force making the fast transfer of PCPh ions to the surface of carbon particles and the availability of the uncovered surface area and the remaining active sites on the adsorbent. Generally, kinetics models are used to determine the rate of the adsorption process. Three kinetic models: pseudo-first order, pseudo-second order, and intraparticle diffusion, Eqs. (6)-(8), were used to correlate the experimental kinetic data of Fig. 4a. The pseudo-first order is of low R^2 values, as shown in Table 4. Moreover, the high deviation between the experimental and calculated adsorption capacity can be seen from this table. This reflects poor fitting of this model to the experimental data. On the other hand, the linear plot of t/q_t versus t (Fig. 4b) for pseudo-second order equation is of high R^2 values (Table 4). The results of this table show that the adsorption kinetics data are better represented by pseudo-second-order model and the calculated q_e values agree well with the experimental q_e values. This indicates the second-order kinetics for PCP adsorption on prepared carbon. These results are in agreements with those reported by many researchers [21, 48] for adsorption of p-chlorophenol on activated carbons from agricultural wastes. It can be noticed that the increase of initial PCP concentrations causes a decrease the value of rate constant K_2 (Table 4). This may be attributed to the high competition for the sorption surface sites at high concentration which leads to higher sorption rates.

The intraparticle diffusion model is widely used to predict the rate controlling step, which mainly depends on either surface or pore diffusion. R^2 values for this model were lower compared to those of pseudo-first order and pseudo-second order models (Table 4). Also, there is a high deviation between the theoretical and experimental values Δq (%). The plot of q_t versus $t^{1/2}$ for intra-particle diffusion model is shown in Fig. 4c. The first sharper portion is the instantaneous adsorption or external surface adsorption. The second portion is the gradual adsorption stage where the intraparticle diffusion is the rate-limiting. In some cases, the third portion exists, which is the final equilibrium stage where intraparticle diffusion starts to slow down due to extremely low adsorbate concentrations left in the solutions. In order to say that the intraparticle diffusion is the rate controlling step, the plot of q_t versus $t^{1/2}$ should be linear and pass through the origin. As can be noticed from Fig. 4c, the plot did not pass through the origin and this deviation from the origin or near saturation might be due to the difference in mass transfer rate in the initial and final stages of adsorption. From these results, it can be concluded that intraparticle diffusion is not the dominating mechanism for the adsorption of PCP on prepared carbon.

3.4 Adsorption thermodynamics

The values of ΔH and ΔS have been calculated from the slope and intercepts of the linear plot of Eq. (10) represented by In (k_d) versus $1/T$ (Fig. 5). The values of ΔG are determined from Eq. (11) at each temperature. The results are listed in Table 5. The negative ΔG values indicate spontaneous nature for PCP adsorption. The decrease in ΔG values with increasing temperature suggests that higher temperature makes the adsorption easier. Negative ΔH show the exothermic nature of PCP adsorption. The value of ΔH obtained in this study is 3.03 kJ/mole (Table 5) which indicates physisorption nature of PCP adsorption, where physisorption is corresponding to ΔH ranging from 2.1 to 20.9 kJ/mol [49]. The negative ΔS value suggests a decrease in the randomness at sorbate-solution interface during the adsorption process. The spontaneous and exothermic natures of PCP adsorption have also been reported by Monsalvo et al. [50] on sewage sludge-based adsorbents.

4. Conclusions

In this study, p-chlorophenol adsorption from aqueous solutions by activated carbon has been investigated. *Siris* seed pods were used as agricultural waste precursors for the production of activated carbon using microwave-induced KOH activation. The best correlation for equilibrium isotherm data of p-chlorophenol was achieved by Langmuir isotherm, showing a maximum p-chlorophenol capacity of 338.87 mg/g on prepared carbon compared to 88.32 mg/g on *Siris* seed pods. The adsorption kinetic data were well described by the pseudo-second order model. KOH activation produced carbon structure with 82.48% micropores content. The investigation of adsorption thermodynamics revealed exothermic and spontaneous natures of p-chlorophenol adsorption under-examined conditions.

Acknowledgement

We gratefully acknowledge Department of Chemical Engineering and University of Baghdad for assist and support of this work.

References

[1] P.S. Majumder, S.K. Gupta, Degradation of 4-chlorophenol in UASB reactor under methanogenic conditions, Bioresour. Technol., 99 (2008) 4169–4177. https://doi.org/10.1016/j.biortech.2007.08.062

[2] O. Hamdaoui, E. Naffrechoux, Sonochemical and photosonochemical degradation of 4-chlorophenol in aqueous media. Ultrasonics Sonochem., 15 (2008) 981–987 https://doi.org/10.1016/j.ultsonch.2008.03.011

[3] P. Wongwisate, S. Chavadej, E. Gulari, T. Sreethawong, P. Rangsunvigit, Effects of monometallic and bimetallic Au-Ag supported on sol-gel TiO2 on photocatalytic degradation of 4-chlorophenol and its intermediates, Desalination 272 (2011) 154-163. https://doi.org/10.1016/j.desal.2011.01.016

[4] H. Wang, Z.Y. Bian, D.Z. Sun, Degradation mechanism of 4-chlorophenol with electrogenerated hydrogen peroxide on a Pd/C gas-diffusion electrode, Water Science and Technol., 63 (2011) 484-490. https://doi.org/10.2166/wst.2011.247

[5] K. Naeem, F. Ouyang, Influence of supports on photocatalytic degradation of phenol and 4-chlorophenol in aqueous suspensions of titanium dioxide, J. Environm. Sci., 25 (2013) 399–404. https://doi.org/10.1016/S1001-0742(12)60055-2

[6] X. Duan, L. Tian, W. Liu, L. Chang, Study on electrochemical oxidation of 4-Chlorophenol on a vitreous carbon electrode using cyclic voltammetry, Electrochimica Acta 94 (2013) 192– 197. https://doi.org/10.1016/j.electacta.2013.01.151

[7] AL-Othman ZA, Inamuddin, Naushad M (2011) Adsorption thermodynamics of trichloroacetic acid herbicide on polypyrrole Th(IV) phosphate composite cation-exchanger. Chem Eng J 169:38–42. https://doi.org/10.1016/j.cej.2011.02.046

[8] J. Su, S. Lin, Z. Chen, M. Megharaj, R. Naidu, Dechlorination of p-chlorophenol from aqueous solution using bentonite supported Fe/Pd nanoparticles: Synthesis, characterization and kinetics, Desalination, 280 (2011) 167–173. https://doi.org/10.1016/j.desal.2011.06.067

[9] Y. Pi, L. Zhang, J. Wang, The formation and influence of hydrogen peroxide during ozonation of para-chlorophenol, J. Hazard. Mater., 141 (2007) 707–712. https://doi.org/10.1016/j.jhazmat.2006.07.032

[10] C. Catrinescu, D. Arsene, C. Teodosiu, Catalytic wet hydrogen peroxide oxidation of para-chlorophenol over Al/Fe pillared clays (AlFePILCs) prepared from

different host clays, Applied Catalysis B: Environm., 101 (2011) 451–460. htttps://doi.org/10.1016/j.apcatb.2010.10.015

[11]   Naushad M, Ahamad T, Sharma G, et al (2016) Synthesis and characterization of a new starch/SnO2 nanocomposite for efficient adsorption of toxic Hg2+ metal ion. Chem Eng J 300:306–316. https://doi.org/10.1016/j.cej.2016.04.084

[12]   H.S. Wahab, T. Bredow, S.M. Aliwi, A computational study on the adsorption and ring cleavage of para-chlorophenol on anatase TiO2 surface, Surface Sci., 603 (2009) 664–669. https://doi.org/10.1016/j.susc.2009.01.001

[13]   M.S. Bilgili, Adsorption of 4-chlorophenol from aqueous solutions by xad-4 resin: Isotherm, kinetic, and thermodynamic analysis, J. Hazard. Mater., B137 (2006) 157–164. https://doi.org/10.1016/j.jhazmat.2006.01.005

[14]   C. Pa˘curariu, G. Mihoc, A. Popa, S.G. Muntean, R. Ianos, Adsorption of phenol and p-chlorophenol from aqueous solutions on poly (styrene-co-divinylbenzene) functionalized, Chem. Eng. J., 222 (2013) 218–227. https://doi.org/10.1016/j.cej.2013.02.060

[15]   V.M. Monsalvo, A.F. Mohedano, J.J. Rodriguez, Adsorption of 4-chlorophenol by inexpensive sewage sludge-based adsorbents, Chem. Eng. Research and design, 90 (2012) 1807–1814. https://doi.org/10.1016/j.cherd.2012.03.018

[16]   J.M. Li, X.G. Meng, C.W. Hu, J. Du, Adsorption of phenol, p-chlorophenol and p-nitrophenol onto functional chitosan, Bioresour. Technol., 100 (2009) 1168–1173. https://doi.org/10.1016/j.biortech.2008.09.015

[17]   A. Kuleyin, Removal of phenol and 4-chlorophenol by surfactant-modified natural zeolite, J. Hazard. Mater., 144 (2007) 307–315. https://doi.org/10.1016/j.jhazmat.2006.10.036

[18]   B. Koumanova, P. Peeva-Antova, Adsorption of p-chlorophenol from aqueous solutions on bentonite and perlite, J. Hazard. Mater., A90 (2002) 229–234. https://doi.org/10.1016/S0304-3894(01)00365-X

[19]   F. Derbyshire, M. Jagtoyen, R. Andrews, A. Rao, I. Martin-Gullon, E. Grulke, Carbon materials in environmental application. In: L.R. Radovic, ed., Chemistry and Physics of Carbon, 27, Marcel Dekker, New York. 2001, pp. 1–66.

[20]   C.O. Ania, J.B. Parra, J.A. Menéndez, J.J. Pis, Effect of microwave and conventional regeneration on the microporous and mesoporous network and on the adsorptive capacity of activated carbons, Microporous and Mesoporous Mater., 85 (2005) 7–15. https://doi.org/10.1016/j.micromeso.2005.06.013

[21] R.L. Tseng, S.K. Tseng, Pore structure and adsorption performance of the KOH-activated carbons prepared from corncob, J. Colloid and Interface Sci., 287 (2005) 428-437. https://doi.org/10.1016/j.jcis.2005.02.033

[22] R.L. Tseng, F.C. Wu, Analyzing a liquid-solid phase counter current two- and three- stage adsorption process with the Freundlich equation, J. Hazard. Mater., 162 (2009) 237-248. https://doi.org/10.1016/j.jhazmat.2008.05.031

[23] V.M. Monsalvo, A.F. Mohedano, J.J. Rodriguez, Activated carbons from sewage sludge application to aqueous-phase adsorption of 4-chlorophenol, Desalination, 277 (2011) 377-382. https://doi.org/10.1016/j.desal.2011.04.059

[24] B.H. Hameed, L.H. Chin, S. Rengaraj, Adsorption of 4-chloropenol onto activated carbon prepared from rattan sawdust, Desalination, 225 (2008) 185-198. https://doi.org/10.1016/j.desal.2007.04.095

[25] V.K. Gupta, S.K. Srivastava, R. Tyagi, Design Parameters for the treatment of phenolic wastes by carbon column obtained from fertilizer waste material, Water Res., 34 (2000) 1543-1550. https://doi.org/10.1016/S0043-1354(99)00322-X

[26] N. Fernandez, E. Chacin, C. Garcia, N. Alastre, F. Leal, C.F. Forster, The use of seed pods from Albizia lebbeck for the removal of alkyl benzene sulphonates from aqueous solutions, Process Biochem., 31 (1996) 383-387. https://doi.org/10.1016/0032-9592(95)00074-7

[27] H.O. Adubiaro, O. Olaofe, E.T. Akintayo, Chemical composition, calcium, zinc and phytate interrelationships in Albizia lebbeck and Daniella Oliveri seeds, Elect. J. Environ. Agric. Food Chem., 10 (2011) 2523-2530.

[28] M.J. Ahmed, S.K. Theydan, Adsorption of p-chlorophenol onto microporous activated carbon from Albizia lebbeck seed pods by one-step microwave assisted activation, J. Anal. Appl. Pyrolysis, 100 (2013) 253-260. https://doi.org/10.1016/j.jaap.2013.01.008

[29] ASTM standard, standard test method for total ash content of activated carbon, Designation D2866-94; 2000.

[30] F.A. Adekola, H.I. Adegoke, Adsorption of blue-dye on activated carbons produced from Rice Husk, Coconut Shell and Coconut Coir pitch, Ife J. Sci., 7(1) (2005) 151-157. https://doi.org/10.4314/ijs.v7i1.32169

[31] S. Brunauer, P.H. Emmett, E. Teller, Adsorption of gases in multimolecular layers, J. Am. Chem. Soc., 60 (1938) 309-319. https://doi.org/10.1021/ja01269a023

[32]   S.J. Gregg and K.S.W. Sing, Adsorption, surface area and porosity, London, Academic Press., 1966.

[33]   C. Lastoskie, K.E. Gubbins, N. Quirke, pore size distribution analysis of microporous carbons: a density functional theory approach, J. Phys. Chem., 97 (1993) 4786-4796. https://doi.org/10.1021/j100120a035

[34]   I. Langmuir, The constitution and fundamental properties of solids and liquids, J. Am. Chem. Soc., 38 (1916) 2221-2295. https://doi.org/10.1021/ja02268a002

[35]   H.M.F. Freundlich, Über die adsorption in lösungen, Z. Phys. Chem., 57 (1906) 385-470.

[36]   M.J. Temkin and V. Phyzev, Recent modifications to Langmuir isotherms, Acta Physiochim, USSR 12 (1940) 217-222.

[37]   S. Langergen and B.K. Svenska, Zur theorie der sogenannten adsoption geloester stoffe, Veteruskapsakad Handlingar, 24 (1898) 1-39.

[38]   Y.S. Ho and G. Mckay, Pseudo-second order model for sorption processes, Process Biochem., 34 (1999) 451-465. https://doi.org/10.1016/S0032-9592(98)00112-5

[39]   W.J. Weber and J.C. Morris, Kinetics of adsorption on carbon from solution, J. Saint. Eng. Div. Am. Soc. Civil Eng., 89 (1963) 31-60.

[40]   L.S. Luo, Preparation of activated carbon from solidago Canadensis and its adsorption performance for Cd(II), M.SC. Thesis, China, 2012.

[41]   B. Xing, C. Zhang, L. Chen, G. Huang, Preparation of activated carbon from lignite for electrochemical capacitors by microwave and electrical furnace heating, Advanced Mater. Res., 194-196 (2011) 2472-2479. https://doi.org/10.4028/www.scientific.net/AMR.194-196.2472

[42]   K.Y. Foo, B.H. Hameed, Microwave-assisted preparation and adsorption performance of activated carbon from biodiesel industry solid residue: influence of operational parameters, Bioresour. Technol., 103 (2012) 398-404. https://doi.org/10.1016/j.biortech.2011.09.116

[43]   K.Y. Foo, B.H. Hameed, Coconut husk derived activated carbon via microwave induced activation: effects of activation agents, preparation parameters, and adsorption performance, Chem. Eng. J., 184 (2012) 57-65. https://doi.org/10.1016/j.cej.2011.12.084

[44]   K.Y. Foo, B.H. Hameed, Adsorption characteristics of industrial solid waste derived activated carbon prepared by microwave heating for methylene blue, Fuel Processing Tehnol., 99 (2012) 103-109. https://doi.org/10.1016/j.fuproc.2012.01.031

[45]   K.Y. Foo, B.H. Hameed, Porous structure and adsorptive properties of pineapple based activated carbons prepared via microwave assisted KOH and K2CO3 activation, Microporous and Mesoporous Mater., 148 (20120 191-195.

[46]   K.Y. Foo, B.H. Hameed, Utilization of rice husks as a feedstock for preparation of activated carbon by microwave induced KOH and K2CO3 activation, Bioresour. Technol., 102 (2011) 9814-9817. https://doi.org/10.1016/j.biortech.2011.07.102

[47]   F.C. Wu, P.H. Wu, R.L. Tseng, R.S. Juang, Preparation of novel activated carbons from H2SO4-Pretreated corncob hulls with KOH activation for quick adsorption of dye and 4-chlorophenol, J. Environm. Manage., 92 (2011) 708-713 https://doi.org/10.1016/j.jenvman.2010.10.003

[48]   R.L. Tseng, K.T. Wu, F.C. Wu, R.S. Juang, Kinetic studies on the adsorption of phenol, 4-chlorophenol, and 2,4-dichlorophenol from water using activated carbons, J. Environ. Manage., 91 (2010) 2208-2214. https://doi.org/10.1016/j.jenvman.2010.05.018

[49]   Z. Belala, M. Jeguirim, M. Belhachemi, F. Addoun, G. Trouve, Biosorption of basic dye from aqueous solution by date stones and palm-trees waste: kinetics, equilibrium and thermodynamic studies, Desalination, 271 (2011) 80-87. https://doi.org/10.1016/j.desal.2010.12.009

[50]   V.M. Monsalvo, A.F. Mohedano, J.J. Rodriguez, Adsorption of 4-chlorophenol by inexpensive sewage sludge-based adsorbents, Chem. Eng. Res. and design 90 (2012) 1807-1814. https://doi.org/10.1016/j.cherd.2012.03.018

Chapter 7

Removal of phosphates and sulphates in a multi-ion system with nitrates

Patricia A. Terry*, Megan Olson Hunt, Renee Henning

Department of Natural and Applied Sciences, University of Wisconsin-Green Bay, 2420 Nicolet Drive, Green Bay, WI 54311, USA

* terryp@uwgb.edu

Abstract

Eutrophication remains a water quality issue globally, and evidence demonstrates that sulphates in water may interact to release phosphates bound in underlying soil sediments, such that removal of aqueous phosphate may not be adequate to eliminate eutrophication. Further, sulphates promote the formation of cyanobacteria, which creates potentially toxic conditions in affected waterways. This work characterises the effect of phosphates, sulphates, and nitrates on the co-removal of phosphates and sulphates from contaminated water via ion exchange with calcined hydrotalcite, a clay mineral double layer hydroxide. To assess the statistical significance of main effects and interactions between anions on mean residual target anion levels, fixed-effects two- and three-way analyses of variance were used. Langmuir isotherms for single-ion removal are estimated and compared to those for ternary solutions at phosphate, sulphate, and nitrate levels typical for contaminated ground and surface waters. For phosphate removal in the ternary system, the two-way interactions between sulphate and initial phosphate, and between sulphate and nitrate were statistically significant. However, phosphate removal remained high - between 94 and 99% - in all cases, demonstrating that this is a viable removal method. Sulphate removal was also dictated by significant interactions (two- and three-way), but, as with phosphate, the reduction was still successful in general. The findings indicate that while high sulphate levels may not be removed sufficiently so as to prevent eutrophication if phosphates are held in soil sediments, for intermediate and low levels of sulphate, hydrotalcite is a useful material for the co-removal of phosphate and sulphate in eutrophic waters.

Keywords

Ion Exchange, Hydrotalcite, Phosphate Removal, Nitrate, Langmuir Isotherm

Contents

1.   Introduction

Since the middle of the 20th century, lakes globally have experienced eutrophication with increased growth in aquatic plants and subsequent fish kills. In the 1960s, researchers observed the connection between aquatic plant growth and the quantity of nutrients reaching a lake through anthropogenic activities such as agricultural fertiliser application, sewage discharge, and urban run-off. A report prepared in 1968 for the Organization for Economic Co-operation and Development concluded that phosphorus was the limiting nutrient in most lakes (Schindler, 2006). To prevent eutrophication, the United States Environmental Protection Agency (1986) (U.S. EPA) recommends total phosphate concentrations should not exceed 25 μg/L in lakes.

One consequence of algal blooms from eutrophication is fish kills attributed to the low-oxygen conditions created by anaerobic decomposition of dead algae. The blooms also replace the natural vegetation used by organisms and fish as habitats and food, contributing to greater fish mortality (Dodds et al., 2009). Based on recovery plan costs for endangered and threatened species, $44 million/year is spent to prevent biodiversity loss attributed to eutrophication, and the total cost of eutrophic algal blooms could cost an estimated $2.2 billion/year.

In addition to the aforementioned negative effects of algal blooms, some algal species also release cyanobacteria toxins, which produce poison potentially fatal to aquatic animals, and are also commonly known to cause illness, skin irritation, and even death in humans and other non-aquatic animals that enter the water body. Novotny (2011) examined a reservoir in the Czech Republic and a lake in China with excessive nutrient loads that caused harmful cyanobacteria algal blooms, degrading the water quality used for drinking. He attributed the conditions to increases in nutrient load from fertiliser application, concentrated animal farms, agricultural tile drainage, increased urban drainage, and sewage disposal. Additionally, many beach closings on Lake Erie, United

States, have been attributed to the toxin (U.S. Senate Subcommittee on Water and Wildlife, 2011). In 2011, the toxic algal bloom of mycrosystis in Lake Erie was the worst in recorded history, causing beach closings. Schindler (1977) found that having low nitrogen levels compared to phosphorus caused cyanobacteria to proliferate because of their ability to fix nitrogen from the air. The bacteria are able to convert atmospheric nitrogen (N_2) into bio-available ammonia (NH_3), nitrites (NO_2^-), or nitrates (NO_3^-) with the enzyme nitrogenase.

Sulphate is also a concern for surface water contamination. Increases in sulphate occur from atmospheric deposition as a result of anthropogenic emissions of sulphur into the air, leaching as a result of excessively fertilised land and mining run-off. The United States Environmental Protection Agency (1999) conducted a study with the Centers for Disease Control and Prevention (CDC) on sensitive populations, transients, and infants, due to concerns of diarrhoea resulting from exposure to abrupt increases in sulphate concentrations in drinking water. Sulphate was included on the Drinking Water Contaminant Candidate List published on March 2, 1998.

A more recently observed concern about sulphate is its effect on eutrophication. In 2006, Smolders et al. explained how the presence of sulphates caused eutrophication in the absence of excessive external phosphate nutrient loading in Netherland's surface waters. The decay of organic matter requires microbial processes that obtain energy from electron transfer. When oxygen is present, it is the primary electron acceptor in the decay process. If oxygen is not present, bacteria may use nitrate and potentially iron hydroxides (FeOOH) or sulphate (SO_4^{2-}) as the electron acceptor. When iron and sulphate are reduced, they form FeS_x which has less potential for phosphorus sorption than iron hydroxide, increasing phosphate mobility from soils underlying a water body. This release of sediment-bound phosphate into the aqueous system creates eutrophic conditions. The rate of organic decomposition also increases in alkaline conditions created by the reduction of nitrate, iron hydroxides, and sulphate, increasing the availability of phosphate nutrients.

Experiments performed in the Netherlands by Lamers et al. (2006) concluded that an increase in sulphate availability caused an increase in the reduction of sulphate to sulphide, which forms a compound with sediment-held iron that is reduced from iron hydroxide. This decreases the iron available to bond with sediment-held phosphate, increasing its aqueous availability. The authors believe sulphate bonds at the anion site in soils released the phosphate that was adsorbed by the soils and sediment. Hence, they deduced that a reduction in nitrogen and phosphorus loading may not be enough to reduce eutrophic conditions of wetlands, and that sulphate levels should also be considered.

Much work has focused on the removal of anions from aqueous solutions as single ions, the co-removal of nutrients, or the co-removal of heavy metals. Many materials have displayed adsorption and ion exchange capabilities for phosphate removal, including the minerals apatite (Bellier et al., 2006), goethite (Norwack et al., 2006), and Filtrate-P (Adam et al., 2006). Methods found to remove phosphates include adsorption onto a zirconium sulphate micelle mesostructure (Pitakteerathem et al., 2013), recovery by a continuous struvite reaction crystallization process (Hutnik et al., 2013), immobilization using natural calcium-rich sepiolite (Yin et al., 2013), adsorption onto biochar/MgAl-LDH ultra-fine composites (Zhang et al., 2013), and enhanced biological removal (Mielczarek et al., 2013). Recent studies successfully recovered phosphate from wastewater using layered double hydroxide ion exchangers on superparamagnetic microparticles (Mandel et al., 2013; Drenkova-Tuhtan et al., 2013).

Co-removal of phosphate and nitrate has been achieved through adsorption onto ceramic material made from coal ash and metallic iron (Ji et al., 2010), the use of plant-microorganism combined systems (Li et al., 2011), and membrane biological reactors (Galil et al., 2009). In attempts to prevent calcium sulphate formation in desalination, sulphate was selectively removed from seawater using the weak-base anion-exchange resin Relite MG1/P (Zhu et al., 2011).

Layered double hydroxides (LDHs) form a class of ion exchange media that consist of positively charged fixed layers with exchangeable anions, and are either naturally occurring or inexpensive to manufacture. The LDH hydrotalcite (HTC; $[Mg_2Al(OH)_6]_2CO_3 \cdot 3H_2O$) is a strong base ion exchanger, with CO_3^{2-} as the exchangeable ion. Due to the double hydroxide structure of HTC, the macropore size allows the exchange of large anions. Kuzawa et al. (2006) found high removal rates for phosphates using a synthetic hydrotalcite anion exchange with the ability to regenerate. Terry (2009) studied the removal of phosphate and nitrate and showed phosphate removal was successful in single-ion solutions and in the presence of the nitrate ion. Nitrate removal was achieved for the low concentration of 25 mg/L initial nitrate, but the removal rate decreased significantly for higher initial concentrations and in the presence of phosphate.

The goal of this work is to characterise the co-removal of phosphates and sulphates in a ternary system with nitrate. Removal is performed via ion exchange with HTC, a double layer hydroxide clay mineral material. Fixed-effects two- and three-way analyses of variance (ANOVAs) are used to determine statistical significance regarding the effect of sulphate on phosphate removal (and vice-versa) in a binary system, and the effect of nitrate, phosphate, and sulphate on both phosphate and sulphate removal in a ternary system. Note that nitrate removal is not considered, as previous work demonstrated that the presence of phosphate significantly inhibits nitrate removal (Terry, 2009). However, since treatments used to remove one ion are usually influenced by the presence of others, interactions between ions are considered when assessing phosphate and sulphate removal. Lastly, Langmuir isotherm model parameters are estimated based on the experimental results.

2.   Methods and materials

The HTC ($[Mg_2Al(OH)_6]_2CO_3 \cdot 3H_2O$) ion exchange media used in these experiments was purchased from Sigma-Aldrich chemicals. It was calcined by heating at 450 °C for at least four hours. Nitrates were added as sodium nitrate ($NaNO_3$), phosphates as sodium dihydrogen phosphate (NaH_2PO_4), and sulphates as sodium sulphate ($MgSO_4$), all of which were research-grade from Fisher Chemicals. Sulphate and phosphate levels were measured using a calibrated Shimadzu UV-1601 UV-visible spectrophotometer. An ascorbic acid soluble reactive phosphorus procedure, HACH standard procedure 8048 using HACH Phosver 3 reagent powder, was used for phosphate analysis. The test measurement range was 0 to 0.82 mg/L P, reported as 0 to 2.52 mg/L PO_4^{3-}. The sulphate analysis followed the U.S. EPA SulfaVer 4 Method 8051, using HACH SulfaVer 4 reagent powder. The test measurement range was 2 to 70 mg/L.

All experiments were performed in triplicate using 100 mL samples in acid-washed, high-density polyethylene bottles. Samples were prepared by adding the appropriate quantity of nitrates, phosphates, and sulphates, followed by 0.5 g/L HTC. Each sample was then shaken for 20 minutes with an NBS gyratory shaker on high speed. Following ion exchange, the HTC was separated from the suspension by vacuum filtration through Whatman 0.7 mm glass fibre filters, and the residual ion concentrations in the aqueous solution were analysed.

Both the sulfate-phosphate binary system and the ternary system of phosphate, sulphate, and nitrate were characterised. All experiments were performed using the following ranges: 0 to 6.13 mg/L phosphate (0 to 2 mg/L P), 0 to 30 mg/L nitrate, and 0 to 300 mg/L sulfate. These levels are often found in impaired surface waters and wells in agricultural areas and are also associated with contaminated bodies of water that have experienced cyanobacteria blooms (The United States Environmental Protection Agency, 1986).

To assess the statistical significance of main effects and interactions between anions on residual target anion levels, fixed-effects two- and three-way ANOVAs were used with $\alpha = 0.05$. Experimental trials were independent and run in random order, and thus the data were evaluated as a completely randomised design. These analyses were conducted with PROC MIXED in SAS/STAT® version 9.4.

Langmuir isotherm models were used to describe the ion exchange removal of a target anion in both the binary and ternary systems. The parameterization of the Langmuir isotherm model used here was

$$y = bx_{eq}/(1 + Kx_{eq}),$$ (1)

where y is the mg of target anion exchanged per unit mass (g) hydrotalcite, x_{eq} is the residual target anion in solution following ion exchange (mg/L), and b and K are the Langmuir constants. To estimate the parameters of the Langmuir curve, PROC NLIN in SAS was used with the Gauss-Newton algorithm. Given R^2 (the coefficient of determination, a measure of goodness-of-fit) is not defined for non-linear models, a pseudo-R^2 is reported, as suggested by Schabenberger and Pierce (2002). By definition, pseudo-$R^2 = 1 - SSE/SST_{Corrected}$, where SSE is the error sums of squares, and $SST_{Corrected}$ is the total sums of squares, corrected for the mean.

3. Results and discussion

Results are presented in the form of residual concentrations and isotherms for the target species. For each isotherm, estimates of the Langmuir constants (and their approximate 95% confidence intervals) are presented, along with pseudo-R^2 as a goodness-of-fit measure. As described above, fixed-effects ANOVAs were used to determine statistical significance regarding the effect of the initial target species' and non-target species' concentrations on the removal of the target species. In all instances, model assumptions (normality, equal variance) were met.

Phosphate Removal

The effect of sulphate and initial phosphate on ion exchange removal of phosphate in a binary system was measured for low (0.613 mg/L) and high (6.13 mg/L) phosphate initial concentrations, while sulphate ranged from 0 to 300 mg/L. Figure 1 shows residual phosphate as a function of sulphate for both high and low phosphate, and Figure 2 gives the comparative isotherms over a phosphate range of 0 to 6.13 mg/L initial phosphate. Langmuir model estimates are reported below in Table 1.

When assessing the effect of sulphate and initial phosphate on residual phosphate in a two-way ANOVA, all model terms, including the two-way interaction, were significant with $p < 0.0001$. Specifically, as Figure 1 shows, the slope for the high phosphate group was significantly steeper than that for the low. Simple effects analysis indicated that for 50 mg/L of sulphate and higher, mean residual phosphate was significantly different across the low and high initial phosphate groups ($p < 0.0001$ for all). When sulphate was

0 or 25 mg/L, there was no significant difference in the two initial phosphate levels (p = 0.696 in both instances). This finding is also clearly depicted in Figure 1.

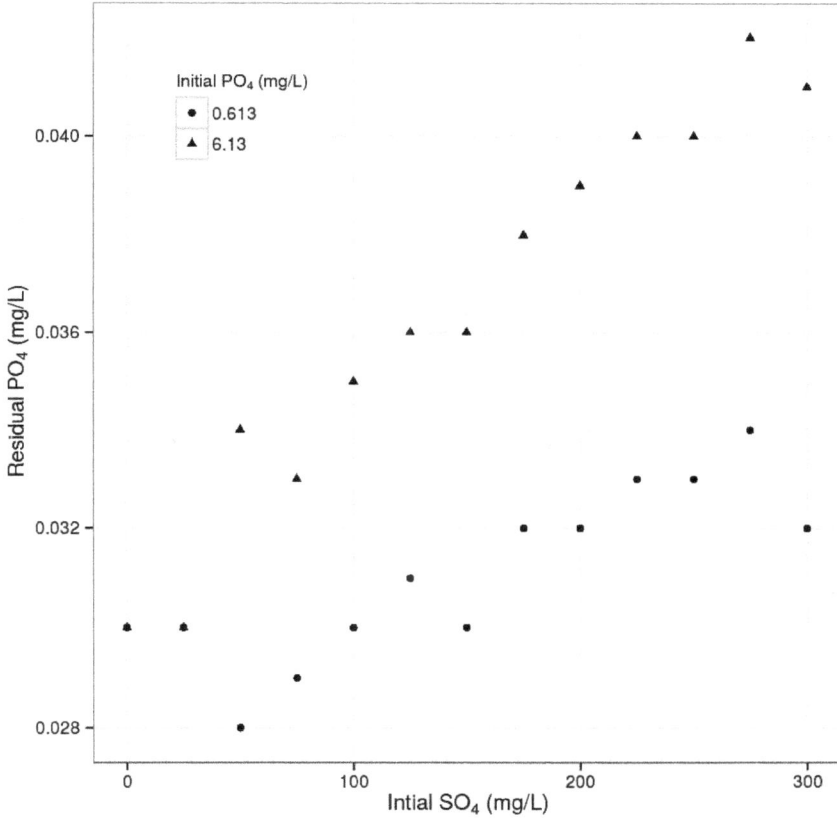

Figure 1. Effect of initial phosphate and sulphate on residual phosphate.

In the absence of sulphate, mean residual phosphate was 0.03 mg/L for both 0.613 and 6.13 mg/L initial phosphate, which corresponds to 95% and 99% removal, respectively. Further, for sulphate concentrations of 25 mg/L or less, mean residual phosphate differed only negligibly across initial phosphate levels. Even though there was a statistically significant interaction indicating lower mean residual phosphate for lower initial levels, percentage removal was still high in all cases. Specifically, even for high sulphate levels (300 mg/L), residual phosphate was only 0.033 mg/L (94.6% removal) for low initial phosphate, and only 0.041 mg/L (99% removal) for high. In conclusion, these phosphate residuals are still low enough to render this a suitable method of phosphate removal despite sulphate and initial phosphate levels.

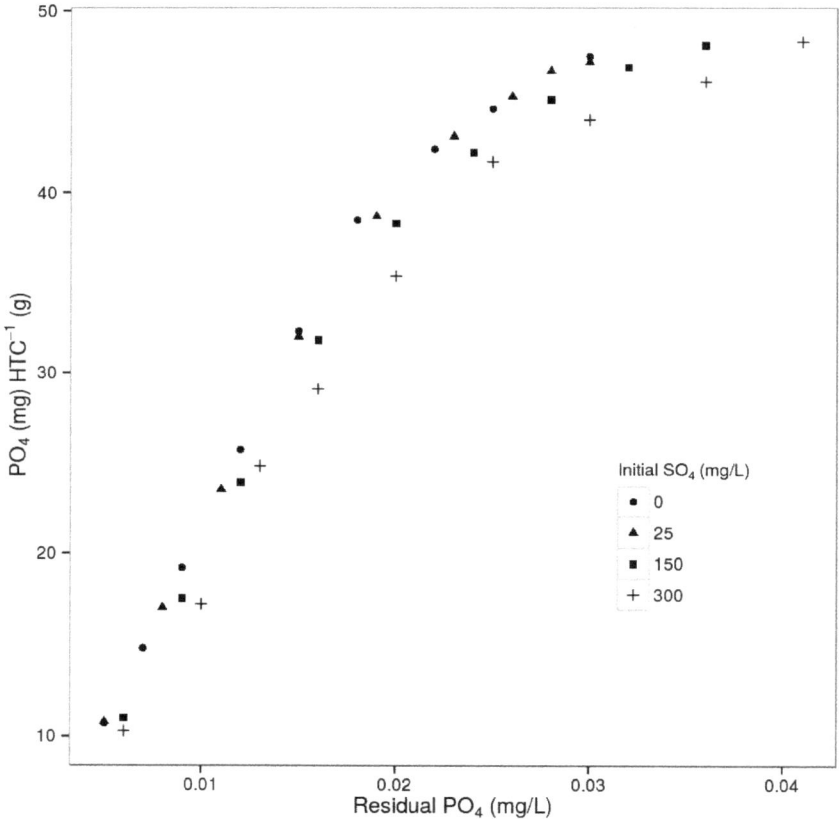

Figure 2. Phosphate isotherms as a function of initial sulfate.

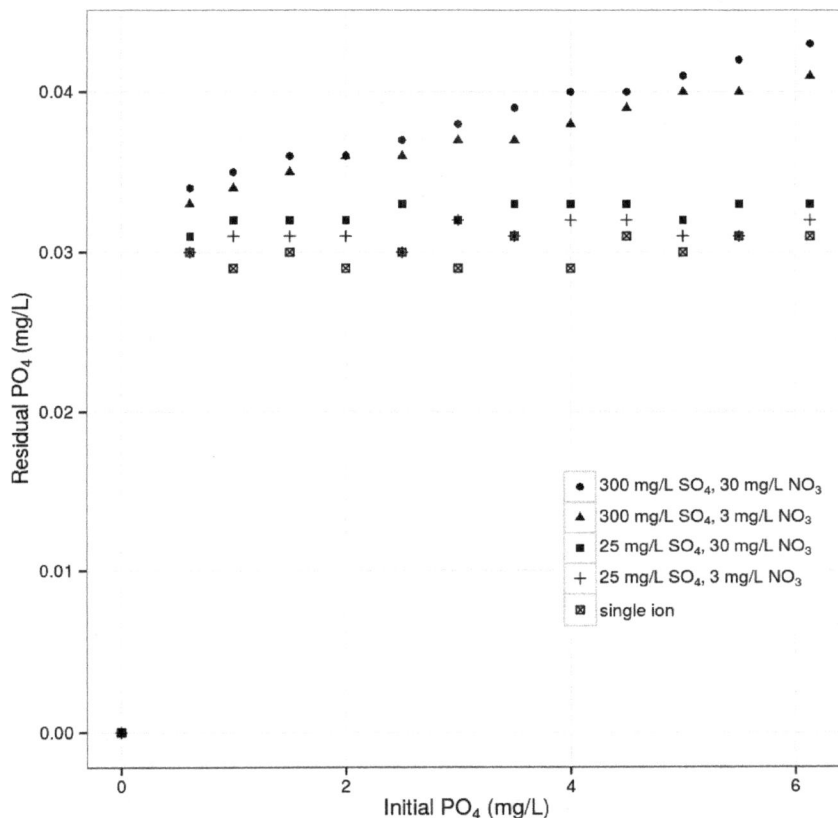

Figure 3. Effect of initial phosphate, sulfate, and nitrate on residual phosphate.

The phosphate isotherms in Figure 2 further confirm the above conclusions in that all curves are quite similar, with the single-ion and 25 mg/L initial sulphate isotherms being essentially the same. The approximate 95% confidence intervals (CIs) for the isotherm parameter estimates (Table 1) support the idea of similar curves, as they largely overlap for all models. While the curves for 150 and 300 mg/L sulphate show somewhat reduced phosphate removal, the process is still highly effective at eliminating phosphate from water overall.

Table 1. Langmuir model parameter estimates for the effect of sulphate on phosphate ion exchange. Approximate 95% confidence intervals are given in parentheses.

Sulphate (mg/L)	b	K	Pseudo-R^2
0	2730.21 (2202.32, 3258.10)	21.33 (8.28, 34.39)	98.18%
25	2755.13 (2280.42, 3229.82)	22.71 (11.49, 33.92)	98.81%
150	2669.70 (2064.72, 3274.69)	24.90 (10.61, 39.19)	97.64%
300	2534.18 (1944.12, 3124.24)	25.87 (11.71, 40.02)	97.35%

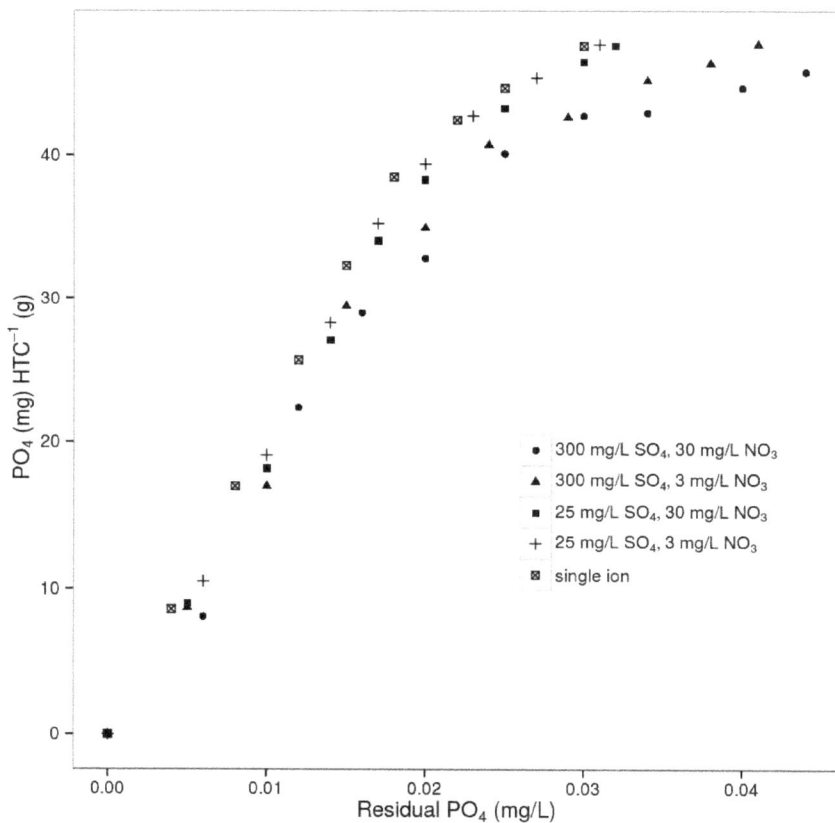

Figure 4. Phosphate isotherms as a function of initial sulphate and nitrate.

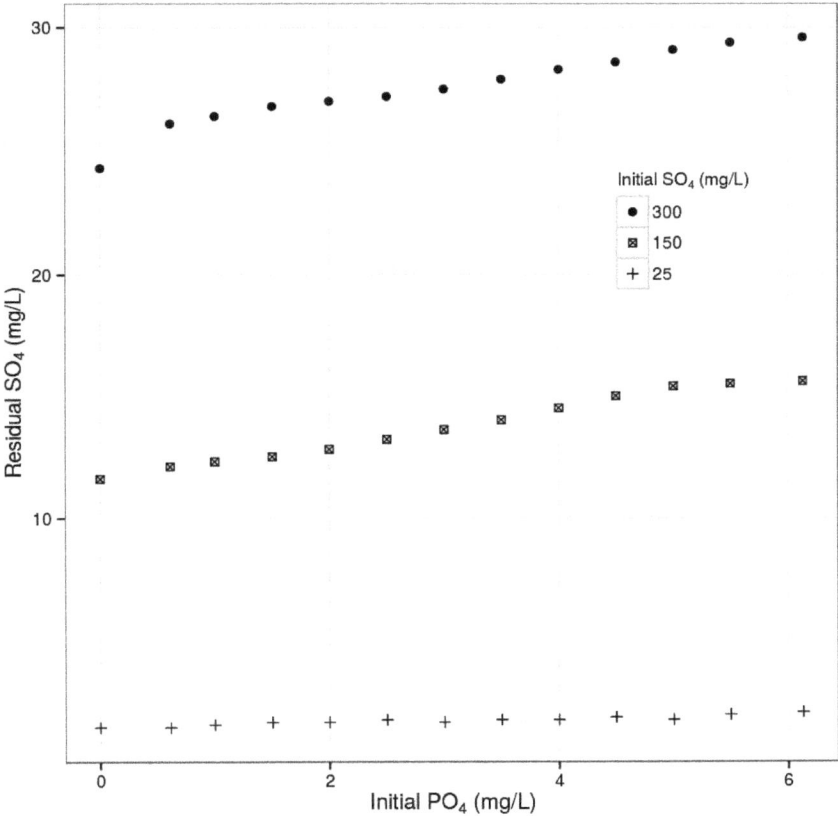

Figure 5. Effect of initial phosphate and sulphate on residual sulfate.

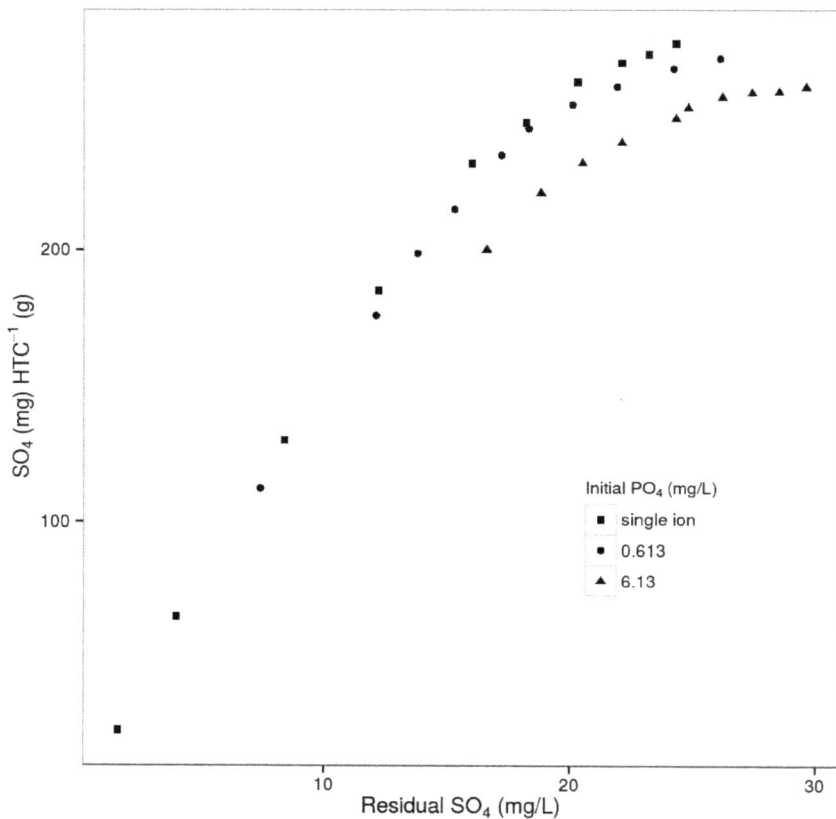

Figure 6. Sulphate isotherms as a function of initial phosphate.

In the ternary system, the effect of initial phosphate, sulphate, and nitrate on phosphate removal was assessed. Figure 3 compares mean residual phosphate as a function of initial phosphate for the following cases: a single-ion phosphate solution, both low sulphate(25 mg/L) and low nitrate (3 mg/L), low sulphate and high nitrate (30 mg/L), high sulphate(300 mg/L) and low nitrate, and both high sulphate and high nitrate. Figure 4 compares isotherms for each of these and Table 2 reports the Langmuir constant estimates for each isotherm in Figure 4.

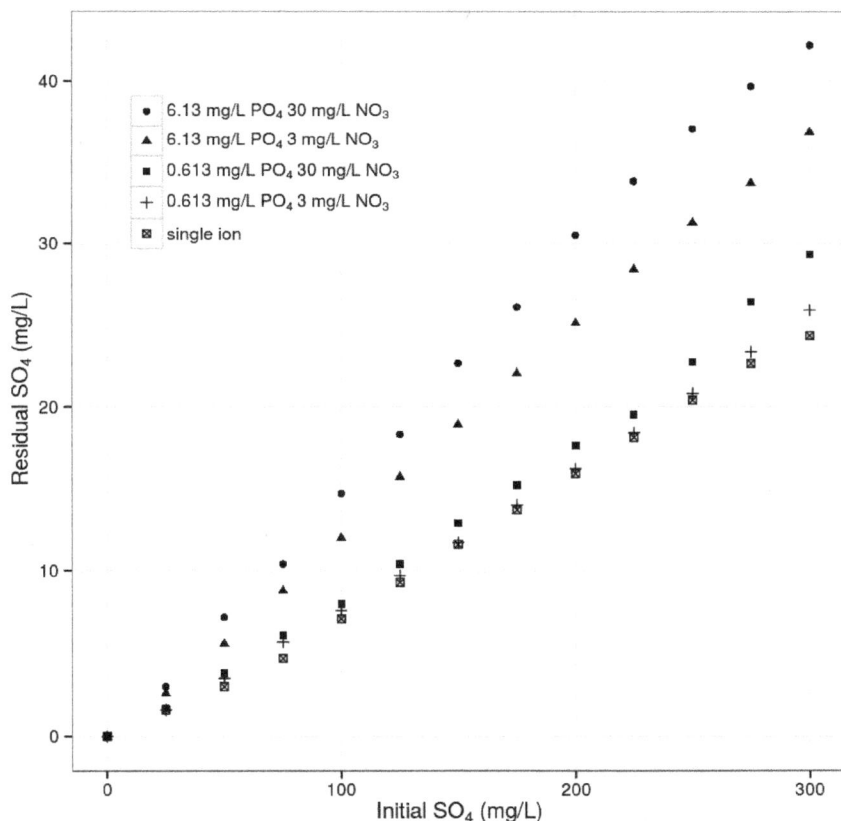

Figure 7. Effect of initial phosphate, sulfate, and nitrate on residual sulfate.

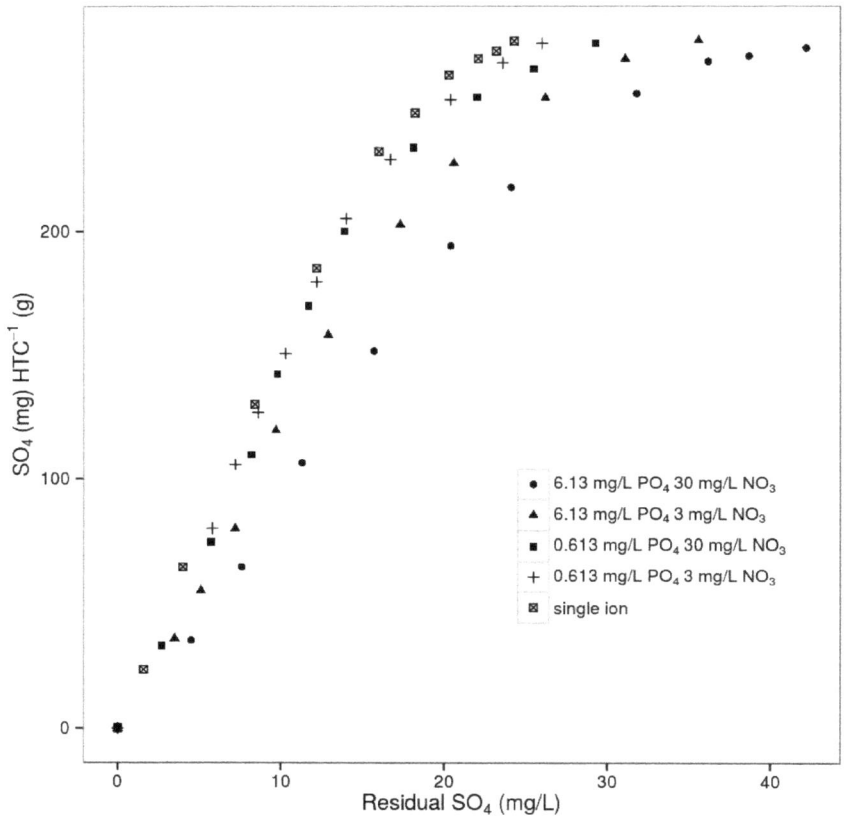

Figure 8. Sulphateisotherms as a function of initial phosphate and nitrate.

The three-way ANOVA for these data indicated the two-way interactions between sulphate and initial phosphate ($p < 0.0001$), and between sulphate and nitrate ($p = 0.0023$) were statistically significant, as is shown in Figure 3. The three-way interaction and two-way interaction between nitrate and initial phosphate were not significant ($p = 0.901$ and 0.978, respectively).

At low sulphate levels, Figure 3 demonstrates that residual phosphate was relatively constant and low for both low and high nitrate across all values of initial phosphate. Even for high initial phosphate, mean residual phosphate only increased from 0.031 mg/L in the single-ion solution to 0.033 mg/L in the low sulphate/high nitrate solution. However, at high sulphate levels, residual phosphate increased with increasing initial phosphate for both nitrate levels.

As with the binary system, even though mean residual phosphate was statistically significantly higher for some ion combinations compared to others, removal was still notable in all cases, indicating HTC is acceptable for removing phosphate in this ternary system involving sulphates, nitrates, and phosphates. Specifically, in the case of 6.13 mg/L initial phosphate, mean residual phosphate increased from 0.031 mg/L in the single-ion phosphate solution to only 0.043 mg/L for both high sulphate and nitrate, which was the highest average observed in the data. Subsequently, at lower initial phosphate levels, the impact of the ternary system is even less pronounced. The isotherms plots in Figure 4 and estimates in Table 2 support this assessment. Namely, there is almost no difference between the isotherm for the single-ion phosphate system and those for low sulphate across both nitrate levels. The isotherms for high sulphate do show a reduction in phosphate removal, but not to a degree that would render the process ineffective. As in the binary system, 95% CIs for the Langmuir isotherm estimates (Table 2) show notable overlap, again indicating similarity amongst the curves.

Table 2. Langmuir model parameter estimates for the effect of sulphate and nitrate on phosphate ion exchange. Approximate 95% confidence intervals are given in parentheses.

Sulphate (mg/L)	Nitrate (mg/L)	b	K	Pseudo-R^2
0	0	2836.12 (2259.54, 3412.70)	23.73 (9.67, 37.78)	98.98%
25	3	2540.21 (1906.54, 3173.89)	18.51 (3.19, 33.84)	98.39%
25	30	2479.72 (1930.31, 3029.13)	19.16 (5.96, 32.35)	98.78%
300	3	2681.17 (2000.91, 3361.42)	29.98 (13.96, 46.00)	98.34%
300	30	2683.63 (1898.14, 3469.11)	33.19 (13.97, 52.40)	97.64%

SulphateRemoval

The effect of phosphate and initial sulphate on the ion exchange removal of sulphate in a binary system was measured for low (25 mg/L), intermediate (150 mg/L), and high (300 mg/L) sulphate initial concentrations. Phosphate ranged from 0 to 6.13 mg/L. Figure 5 shows residual sulphate as a function of phosphate for low, intermediate, and high initial sulphate, and Figure 6 gives the comparative isotherms over a sulphate range of 0 to 300 mg/L initial sulphate. Langmuir model estimates are given in Table 3.

Similar to the findings for the impact of sulphate on phosphate removal, assessment of phosphate and initial sulphate on mean residual sulphate indicated both main effects and the interaction were significant, all with $p < 0.0001$. Here, the pattern of mean residual sulphate across phosphate levels was similar for the intermediate and high values of initial sulphate, while the slope was essentially flat (and thus different from the other two curves) for the low initial sulphate group (Figure 5).

In the absence of phosphate, mean residual sulphate was 1.6 mg/L for 25 mg/L initial sulphate(93% removal), 11.6 mg/L for 150 mg/L initial sulphate(92% removal), and 24.3 mg/L for 300 mg/L initial sulphate(92% removal). As Figure 5 shows, at low initial sulphate, the amount of phosphate did not greatly impact the removal of sulphate. Even at 6.13 mg/L phosphate, sulphate was removed to an average of 2.2 mg/L, a 91.2% reduction.

The impact of phosphate became more pronounced at 150 mg/L initial sulphate, with mean residual sulphate reduced to 15.6 mg/L (89.6% removal) at 6.13 mg/L phosphate. At the same phosphate level, 300 mg/L initial sulphate was reduced to 29.6 mg/L (90.1% removal). However, as was the case with phosphate removal in the binary system, even though statistical results comparing mean residual sulphate indicated significant differences across various ion combinations, percent sulphate reduction for 300 mg/L initial sulphate decreased only from 91.9 to 90.1% across the range of phosphate, which is still high enough to consider this a suitable method of sulphate removal.

Lastly, the sulphate isotherms in Figure 6 (estimates in Table 3) again support previous conclusions. The isotherm for 0.613 mg/L initial phosphate is not notably different from that for single-ion sulphate removal for low sulphate values, although it does show slightly increased residuals after 25mg/L initial sulphate. Although at high phosphate levels the sulphate isotherm is reduced across all sulphate concentrations, the process still removed at least 90% of sulphate from the water. As with the other cases, there is much overlap amongst the CIs for the Langmuir estimates in this system (Table 3).

Table 3. Langmuir model parameter estimates for the effect of phosphate on sulphate ion exchange. Approximate 95% confidence intervals are given in parentheses.

Phosphate (mg/L)	b	K	Pseudo-R^2
0	20.59 (17.22, 23.95)	0.031 (0.018, 0.044)	99.30%
0.613	22.28 (17.92, 26.64)	0.040 (0.022, 0.058)	96.94%
6.13	25.16 (20.63, 29.69)	0.061 (0.043, 0.079)	95.99%

Finally, the effect of initial sulphate, phosphate, and nitrate on sulphate removal was measured. Figure 7 compares residual sulphate as a function of initial sulphate: the single-ion sulphate solution, both low phosphate (0.613 mg/L) and low nitrate (3 mg/L), low phosphate and high nitrate (30 mg/L), high phosphate (6.13 mg/L) and low nitrate, and both high phosphate and high nitrate. Figure 8 compares isotherms for each of these and Table 4 reports estimates of the Langmuir constants for each isotherm in Figure 8.

The three-way ANOVA for this system indicated a significant three-way interaction was present ($p < 0.0001$). In general, high phosphate levels resulted in higher mean residual sulphate compared to low phosphate, despite nitrate level. However, the effect of nitrate interacted with phosphate in that, for high phosphate, there was a larger difference in slopes across nitrate levels compared to the low phosphate group, where the slopes were more similar across the two levels of nitrate (Figure 7).

As Figure 7 shows, across all groups, mean residual sulphate increased as initial sulphate increased. For low phosphate, over 93% of sulphate was removed for both nitrate levels when initial sulphate was 25 mg/L or less. In both of the low phosphate cases, mean residual sulphate increased with increasing initial sulphate concentration, but with higher means for high nitrate than for either low nitrate or the single-ion solution. However, at 300 mg/L initial sulphate, low phosphate, and high nitrate, sulphate removal was still 90.2%.

Compared to low phosphate, high phosphate concentrations resulted in higher mean residual sulphate for both nitrate levels, with the highest means observed for high phosphate/high nitrate across the range of initial sulphate(Figure 7). At 300 mg/L initial sulphate, mean residual sulphate increased from 24.3 mg/L in the single-ion solution to 42.2 mg/L for both high phosphate and nitrate, a reduction in removal from 91.9% to 86%. As with the binary system, even though various mean sulphite residuals were statistically significantly different, removal of 86% in the worst case demonstrates that this process would still be applicable to the removal of sulphate.

Lastly, the isotherms for this ternary system are depicted in Figure 8, with the Langmuir parameter estimates given in Table 4. At low sulphate levels, the isotherms for the single-ion and low phosphate solutions are similar. Only for higher sulphate levels is a reduction noted for the two low phosphate groups. On the other hand, the isotherms for high phosphate are notably reduced compared to the others, with that for high nitrate having the most pronounced effect.

In this instance, the point estimate for the b parameter of the Langmuir model for the high phosphate/high nitrate case falls outside the approximate 95% CIs for all other models. Additionally, the estimate for high phosphate/low nitrate falls outside of the CIs for all models except low phosphate/high nitrate, although it is close to the lower confidence bound in this case as well. Both of these findings align with what was observed in Figure 8.

Table 4. Langmuir model parameter estimates for the effect of phosphate and nitrate on sulphate ion exchange. Approximate 95% confidence intervals are given in parentheses.

Phosphate (mg/L)	Nitrate (mg/L)	b	K	Pseudo-R^2
0	0	20.59 (17.47, 23.72)	0.031 (0.019, 0.043)	99.52%
0.613	3	19.14 (16.31, 21.97)	0.028 (0.016, 0.040)	98.87%
0.613	30	19.03 (15.23, 22.83)	0.032 (0.016, 0.048)	98.39%
6.13	3	15.75 (12.60, 18.89)	0.026 (0.013, 0.039)	98.51%
6.13	30	12.39 (9.90, 14.89)	0.019 (0.0091, 0.029)	98.68%

4.   Summary and conclusions

Experiments were performed to determine the effect of phosphates, sulphates, and nitrates on the removal of either phosphates or sulphates in multi-ion solutions via ion exchange with calcined hydrotalcite, a synthetically made double layer hydroxide. Levels of each anion associated with contaminated ground or surface waters were considered. Langmuir isotherm models were estimated to characterise and compare each system. To assess the statistical significance of main effects and interactions between anions on residual target anion levels, fixed-effects two- and three-way ANOVAs were used.

For phosphate removal, it was determined for the ternary system that the two-way interactions between sulphate and initial phosphate, and between sulphate and nitrate were statistically significant. However, since phosphate removal only decreased from 95% in the single-ion solution to 94% for low initial phosphate, high sulphate, and high nitrate, and remained at 99% for high initial phosphate, ion exchange with hydrotalcite is a viable method of phosphate removal in this system.

Sulphate removal in this ternary system was dictated by both two- and the three-way interactions. However, as with phosphate, sulphate removal remained high overall: For 300 mg/L (high) initial sulphate, removal was reduced from 91.9% in the single-ion solution to 86% in a ternary system with high phosphate and nitrate. In conclusion, the data indicate that only for very high sulphate levels may sulphate not be removed at a level sufficient to prevent eutrophication if phosphates are held in soil sediments. However, for intermediate and low sulphate levels, co-removal of phosphate and sulphate high, showcasing the usefulness of hydrotalcite in the removal of these ions in eutrophic waters.

References

[1]   Adam, K., A. K. Sovik, and T. Krogstad (2006) Sorption of phosphorus to Filtrate-PTM – the effect of different scales. Water Research 40(6): 1143-1154. https://doi.org/10.1016/j.watres.2006.01.009

[2]   Bellier, N., F. Chazarenc, and Y. Comeau (2006) Phosphorus removal from wastewater by mineral apatite. Water Research 40(15): 2965-2971. https://doi.org/10.1016/j.watres.2006.05.016

[3]   Dodds, Walter K., Wes W. Bouska, Jefferey L. Eitzmann, Tyler J. Pilger, Kristen L. Pitts, Alyssa J. Riley, Joshua T. Schloesser, and Darren J. Thornbrugh (2009) Eutrophication of U. S. Freshwaters: analysis of potential economic damages. Environmental Science & Technology 43(1): 12-19. https://doi.org/10.1021/es801217q

[4]   Drenkova-Tuhtan, Asya, Karl Mandel, Anja Paulus, Carsten Meyer, Frank Hutler, Carsten Gellermann, Gerhard Sextl, Matthias Franzreb, and Heidrun Steinmetz (2013) Phosphate recovery from wastewater using engineered superparamagnetic particles modified with layered double hydroxide ion exchangers. Water Research 47(15): 5670-5677. https://doi.org/10.1016/j.watres.2013.06.039

[5]   Galil, N., K. Malachi, C. Sheindorf (2009) Biological nutrient removal in membrane biological reactors. Environmental Engineering Science 26(4):817-824. https://doi.org/10.1089/ees.2008.0234

[6]   Hutnik, N., A. Kozik, A. Mazienczuk, K. Piotrowski, B. Wierzbowska, and A. Matynia (2013) Phosphates (V) recovery from phosphorus mineral fertilizers

industry wastewater by continuous struvite reaction crystallization process. Water Research 47(11): 3635-3643. https://doi.org/10.1016/j.watres.2013.04.026

[7]   Ji, G., Y. Zhou, J. Tong (2010) Nitrogen and phosphorous adsorption behavior of ceramsite material made from coal ash and metallic iron. Environmental Engineering Science 27(10):871-878. https://doi.org/10.1089/ees.2010.0086

[8]   Kuzawa, K., Y. J. Jung, Y. Kiso, T. Yamada, M. Nagai, and T. G. Lee (2006) Phosphate removal and recovery with a synthetic hydrotalcite as an adsorbent. Chemosphere 62(1): 45-52. https://doi.org/10.1016/j.chemosphere.2005.04.015

[9]   Lamers, Leon P. M., Sarah-J. Falla, Edyta M. Samborska, Ivo A. R. van Dulken, Gijs van Hengstum and Jan G. M. Roelofs (2006) Factors controlling the extent of eutrophication and toxicity in sulfate-polluted freshwater wetlands. Limnology and Oceanography 47(2): 585-593. https://doi.org/10.4319/lo.2002.47.2.0585

[10]   Li, H., H. Zhao, H. Hao, J. Liang, F. Zhao, L. Xiang, X. Yang, Z. He, and P. Stoffela (2011) Enhancement of nutrient removal from eutrophic water by a plant-microorganisms combined system. Environmental Engineering Science 28(8):543-554. https://doi.org/10.1089/ees.2011.0026

[11]   Mandel, Karl, Asya Drenkova-Tuhtan, Frank Hutter, Carsten Gellermann, Heidrun Steinmetz, and Gerhard Sextl (2013) Layered double hydroxide ion exchangers on superparamagnetic microparticles for recovery of phosphate from waste water. Journal of Materials Chemistry A 1(5): 1840-1848. https://doi.org/10.1039/C2TA00571A

[12]   Mielczarek, A., H. Nguyen, J. Niellsen, and P. Nielsen (2013) Population dynamics of bacteria involved in enhanced biological phosphorus removal in Danish wastewater treatment plants. Water Research 47(4): 1529-1544. https://doi.org/10.1016/j.watres.2012.12.003

[13]   Norwack, B. and A. T. Stone (2006) Competitive adsorption of phosphate and phosphonates onto goethite. Water Research 40(11): 2201– 2209. https://doi.org/10.1016/j.watres.2006.03.018

[14]   Novotny, Vladimir (2011) The danger of hypertrophic status of water supply impoundments resulting from excessive nutrient loads from agricultural and other sources. Journal of Water Sustainability 1(1): 1-22.

[15]   Pitakteerathem, N., A. Hafuka, H. Satoh, and Y. Watanabe (2013) High-efficiency removal of phosphate from water by zirconium sulfate-surfactant micelle

mesostructure immobilized on polymer matrix. Water Research 47(11): 3583-3590. https://doi.org/10.1016/j.watres.2013.04.006

[16] Schabenberger, O., Pierce, F.J. (2002) Contemporary Statistical Models for the Plant and Soil Sciences. CRC Press, 343.

[17] Schindler, D. W. (1977) Evolution of phosphorus limitation in lakes: Natural mechanisms compensate for deficiencies of nitrogen and carbon in eutrophied lakes. Science 195: 260-262. https://doi.org/10.1126/science.195.4275.260

[18] Schindler, D. W. (2006) Recent advances in the understanding and management of eutrophication. Limnology and Oceanography 51(1, part 2): 356-363. https://doi.org/10.4319/lo.2006.51.1_part_2.0356

[19] Smolders, A. J. P., L. P. M. Lamers, E. C. H. E. T. Lucassen, G. Ven Der Velde, and J. G. M. Roelofs (2006) Internal Eutrophication: How it works and what to do about it – a review. Chemistry and Ecology 22(2): 93-111. https://doi.org/10.1080/02757540600579730

[20] Terry, Patricia A. (2009) Removal of nitrates and phosphates by ion exchange with hydrotalcite. Environmental Engineering Science 26(3): 691-696. https://doi.org/10.1089/ees.2007.0222

[21] United States Environmental Protection Agency (1986) Quality criteria for water 1986. EPA 440/5-86-001. Online available at http://water.epa.gov/scitech/swguidance/standards/criteria/aqlife/upload/2009_01_13_criteria_goldbook.pdf

[22] United States Environmental Protection Agency (1999) Health Effects from Exposure to High Levels of Sulphatein Drinking Water Study. EPA 815-R-99-001. Online available at: http://www.epa.gov/ogwdw/contaminants/unregulated/pdfs/study_sulfate_epa-cdc.pdf.

[23] U. S. Senate Subcommittee on Water and Wildlife. Testimony of Andy Buchsbaum Regional Executive Director, Great Lakes Natural Resources Center National Wildlife Federation (October 4, 2011) Nutrient pollution: An overview of nutrient reduction approaches. Retrieved from: http://epw.senate.gov/public/index.cfm?FuseAction=Files.View & FileStore_id=8d780b54-8647-4f33-b91b-6d15dfb2412e.

[24] Yin, H., M. Kong, and C. Fan (2013) Batch investigations on P immobilization from wastewaters and sediment using natural calcium rich sepiolote as a reactive

material. Water Research 47(13): 4247-4258.
https://doi.org/10.1016/j.watres.2013.04.044

[25] Zhang, M., B. Gao, Y. Yao, and M. Inyang (2013) Phosphate removal ability of biochar/MgAl-LDH ultra-fine composites prepared by liquid-phase deposition. Chemophere 92(8): 1042-1047.
https://doi.org/10.1016/j.chemosphere.2013.02.050

[26] Zhu, Li, Cesar B. Granda, and Mark T. Holtzapple (2011) Prevention of calcium sulphateformation in seawater desalination by ion exchange. Desalination and Water Treatment 36:57-64. https://doi.org/10.5004/dwt.2011.1862

Chapter 8

Application of adsorption techniques for sour and greenhouse gas treatment

Safdar Hossain SK[1]*, Mohammed Mozahar Hossain[2]

[1] Department of Chemical Engineering, King Faisal University, Al-Hasa- 31982, Saudi Arabia

[2] Department of Chemical Engineering, King Fahd University of Petroleum & Minerals, Dhahran- 31682, Saudi Arabia

[a]snooruddin@kfu.edu.sa, [b]mhossain@kfupm.edu.sa

Abstract

Sour (SO_2, NO_X & H_2S) and greenhouse gases (CO_2) are present in various proportions in most of the gaseous effluents from industrial facilities and automobiles. Recently, the rise in their concentration in the atmosphere has begun to show a detrimental effect on humans and other components of the ecosystems. While new, more efficient processes have now been designed with more stringent environmental regulations, but with the prevalent use of current polluting sources, warrants the need for devising efficient abatement and separation techniques for the harmful gases from flue gases, and their subsequent storage or destruction. Capture of harmful gases using solid adsorbent is a commercially promising method for the treatment of the flue gas from conventional power plants. In this chapter, we present an up to date account of the various types of conventional and emerging solid adsorbents for the capture of sour and greenhouse gases from flue gas. Major types of adsorption equipment used in the industry for the gas treatment are briefly discussed.

Keywords

Greenhouse Gases, Sour Gases, Treatment Technologies, Adsorption, Pressure Swing Adsorption

Contents

1. Introduction

Due to a rapid increase in the population in this last few decades, and demand for a high quality life, the demand for energy has increased manifold and, most certainly, it will continue to increase in the near future. Currently, around 80% of the energy production comes from burning fossil fuels (coal, natural gas, and oil), and although this share is in decline, but in most of the scenarios it is expected to remain around 75% by 2030 [1-3]. Combustion of fossil fuels in an electrical energy generating thermal power, and in automobiles for transportation generates a large amount of flue or waste gases. The flue gas composition and flow rates vary depending on the nature of the fuel and the processes. However, carbon dioxide (CO_2), methane (CH_4), and nitrous oxide (N_2O)[4] are present appreciable concentration in the flue gas emitted from power plants and other sources, such as petroleum refineries, steel plants, cement factories and on-road transportation vehicles. These gases are collectively classified as the "Greenhouse" gases.

Other GHGs of minor importance include water vapour and many halocarbon compounds, but their emissions are not very important. By definition, greenhouse gases are gases that absorb infrared or ultraviolet radiation in the atmosphere, trapping heat and thereby increasing the average temperature of the earth. This increased temperature has led to "Global warming" and due to this extreme weather conditions are becoming common. Apart from the GHGs, sulphur dioxide (SO_2), hydrogen sulphide (H_2S), and oxides of nitrogen (NO_X) are also present in minor concentration in the emission from the above-mentioned sources. The natural gas stream which substantial percentage of hydrogen sulphide and or carbon dioxide is called sour gas. Among various sulphur and nitrogen oxide species, SO_2, NO, and NO_2 are considered the most toxic and harmful gases emitted into the atmosphere. These acidic gases are primary sources of atmospheric pollution and are believed to be causing increasingly serious environmental problems, mainly through the formation of acid rain and photochemical smog as well as ozone layer destruction. Furthermore, high concentrations of these undesirable contaminants in the air pose serious health threats to human beings by contributing to a broad range of health issues, including respiratory diseases such as asthma, bronchitis, emphysema, and throat inflammation, among others [4-6]. In particular, these acid gases highly contribute to the formation of secondary organic aerosols (SOAs), which have an impact on the climate as well [4].

Around 37 % of the CO_2 is generated by power plants, 31 % by the combustion of oil in automobiles while 11 % comes from industrial activities. The total CO_2 emission stands at 37.5 gigatonnes in 2014 which is expected to slow down substantially but will rise to 40.6 gigatonnes in 2030[7]. By this emission rate, we will not be able to achieve the 2 °C rise in temperature (as agreed upon in the Kyoto protocol) to limit the impact of global warming which has already started to show its effect. It is speculated that the rate of growth in coal and oil demand slows down but the volumetric demand is not going to decrease in the next few decades, while the natural gas demand is set to increase but the prices for oil and coal remain low and the price for natural gas is set to steadily increase. Therefore, coal powered power generation is expected to increase by an average of 0.2 % per year from 2011 through 2014 and coal will remain the largest source of electricity despite a significant portion of the energy demand is set to be provided by renewable energy[8].

There has been a concentrated effort for the treatment of gas stream containing CO_2, NO_X, and SO_X[7]. The treatment technologies for the removal of CO_2 can be classified into four broad categories namely (1) cryogenic distillation (2) membrane gas separation (3) gas absorption using liquid (4) adsorption using solid [9-11]. Among these treatment processes, gas absorption using a liquid medium is widely established and considred a mature process. Two types of absorption processes are used commercially (a) physical absorption (b) chemical absorption [12]. In the physical absorption process, the gas is physically dissolved in the appropriate solvent at higher pressure and low temperature. This process is economical from an energy point of view only for a concentrated stream of CO_2 with high partial pressure. Hence, CO_2 can be separated from such solvents mainly by reducing the pressure in the desorber, significantly reducing the energy requirements in the desorption process. The main physical solvents that could be employed are cold methanol (Rectisol process), dimethylether of polyethylene glycol (Selexol process), propylene carbonate (Fluor process) and sulpholane [1]. The main advantage of the physical absorption process is that (unlike chemical absorption) physical solvents have no absorption limitation. This technology is well established on a large scale in ammonia production plants but needs to be demonstrated in full-scale power plants. [13]. Chemical absorption combines physical absorption along with a chemical reaction with the solvent. This Chemical absorption has been used successfully for low-pressure gas streams containing between 3% and 25% of CO_2. Most often, an aqueous solution of alkanol amine (mono, di, tri) is used as the liquid media for chemical absorption of the CO_2 gas. Other alternatives like sodium bicarbonate; ammonia was also used for the chemisorption. It is a mature technology and has been used commercially for decades [13]. The advantage of this technology is that it is suitable for retrofitting of existing power plants. However, this technology has several drawbacks including (1) low CO_2 loading capacity; (2) high equipment corrosion rate; (3) amine degradation by SO_2, NO_2, and O_2 in the flue gases which induces a high absorbent makeup rate; (4) high energy consumption during high temperature absorbent regeneration; (5) large equipment size [11, 14].

Adsorption process for gas purification through the selective adsorption of the contaminating gas on solid adsorbent has been well investigated [15-17]. Adsorption offers several advantages over absorption process such as being cheap. An array of adsorbents have been in use such as zeolite, metal oxide, activated carbon, carbon nanotubes [18]. More recently, graphenes, metal-organic frames works have also been explored as novel adsorbents. Adsorbents play a crucial role in the overall effectiveness and feasibility of the overall process. An ideal adsorbent should possess the following key features: (1) it should show high adsorption capacity (2) It should have high selectivity towards the target adsorbate in a flue gas mixture (3) moderate heat of adsorption for low energy requirement (4) fast adsorption and desorption kinetics for easy regeneration (5) excellent chemical and mechanical stability [17]. In this chapter, traditional and emerging adsorbent materials for the treatment of CO_2, NO_X and SO_X from flue gas streams have been briefly presented. Their important properties related to adsorption such as adsorption capacity, selectivity, kinetics and regeneration procedure are also discussed. Finally, major adsorption equipment, their configuration and recent developments in this field have been briefly discussed.

2. Non-carbon materials for gas purification

2.1 Zeolites

Zeolites are porous, highly crystalline aluminosilicates and have the general formula $M_{x/n}[(AlO_2)_x(SiO_2)_y]wH_2O$ where M represents a metal cation such as Li, Na, K or Ca of valence n.; x and y are integers; and w is the number of water molecules per unit cell [19, 20]. There are two types of zeolites; natural and synthetic. Zeolite consists of a three dimensional network of SiO_4 and AlO_4 tetrahedra. The exchangeable metal cations are present to ensure electrical neutrality of the zeolite molecules because each AlO_4 tetrahedron in the crystal lattice carries a negative charge. Zeolites have a uniform pore diameter of 0.3 to 1 nm which depends on the type of the zeolite, the cation present, and the used synthesis procedure. These unique features of zeolites make them ideal candidate for use as adsorbent for an adsorption of various gas molecules, including acidic gases such as CO_2, SO_X and NO_X [21]. They are also called molecular sieve because of their sieving properties. These materials have a well defined pore size distribution, high surface area, and large volume that are suitable for catalysis, gas separation, drying organic liquids, membranes, separation of ethanol and water to break azeotropes, and separation of oxygen from the nitrogen in air [22, 23]. The different types of zeolites and their modifications have been extensively studied as regenerable adsorbent material for the treatment of flue gases containing CO_2, SO_X and NO_X. Several excellent review papers on the use of zeolites as adsorbent materials are available in the open literature [15, 16].

It is well known that the dominant process of CO_2 adsorption on zeolites is physisorption and a very small amount of CO_2 is adsorbed by chemisorption in the form of carbonate or carboxylate. In general, zeolites offer higher adsorption capacities compared to other adsorbents like hydrotalcite and oxides. Physisorption process is influenced by the electric field created by the cations present in the zeolites and by the hydrogen bonding with the silanols groups present on the surface of the zeolites. It is confirmed by the investigation of Barthoeuff of the adsorption of pyrrole on zeolites X, Y, L, mordenite, and ZMS-5 that basic strength of the zeolites increases with increase of aluminium content because aluminium is more electronegative than silicon [24]. In another effect, the highest base strength of the zeolite is exhibited by the cations with low electronegativity. Pawlesa et al. investigated the CO_2 adsorption on MCM-22 and MCM-49 ion exchanged with Li+, K+, Na+, and CS+ ions. Highest adsorption capacity was found to be for the Na-MCM-49 with Al/Si ratio of 15. Lowest adsorption capacity for CS+ was due to the steric hindrance arising from the large size of CS+ [25].

2.1.1 Removal of CO_2

Harlick et al. experimentally investigated the performance of thirteen commercial zeolites (5A, 13X, NaY, NaY-1-, H-Y-30, H-Y-80, HISiv 1000, H-ZSM-5-30, H-ZSM-5-50, H-ZSM-5-80, H-ZSM-5-280, and HiSiv 3000) for CO_2 adsorption in N_2. The highest adsorption capacity of 4.5 mmol/g was found for 13X at 295 K and 1 bar. It is recommended that for a low-pressure CO_2 feed and very low regeneration pressure NaY a 13 X adsorbent should be used [26].

Because a variety of zeolites are available as natural or synthetic ones, it is evident that their adsorptions properties will vary greatly. The adsorption properties depend on the Al/Si ratio, aluminium content, type and number of cations, their charge, other physical properties like pore volume, pore size and surface area [27].

Sriwardane et al. investigated the effect of pore size on zeolites using 4A, 5A, 13X, APG-II, and WEG 592 on the adsorption of CO_2. The highest adsorption was achieved using 13X, which has the highest pore diameter and pore volume among all the zeolites tested. It is hypothesized that the different pore sizes result in a different electric field in the pores leading to the different adsorption capacities [28].

It is well known that CO_2 adsorption is favourable at high pressure and low temperature. Rodrigues et al. carried out CO_2 adsorption on the 13X zeolite at 293, 308, and 323 K from low pressures to around 50 bar. It was found that the CO_2 uptake at any temperature decreases significantly even with small changes in the operating temperature. The amount of CO_2 uptake increases swiftly up to 1 bar and then slightly linear increases up to the pressure studied [27]. Water is present in the flue gas only in small quantity but any attempt to adsorb CO_2 with zeolite, H_2O competes. The detrimental or favourable, the presence of H_2O affects the adsorption depending on the concentration of CO_2. Rege et al. studied the effect of water vapour on the adsorption of CO_2 in 13X zeolite. It was concluded that for very dilute CO_2 flue gas streams (<300 PPM), a small amount of water vapour assists adsorption, while at higher CO_2 concentrations (<1000 PPM), water vapour inhibits CO_2 adsorption on zeolites [29]. Ruthven et al. investigated the effect of water vapour on CO_2 on NaLSX, LILSZ, and CaX. The adsorption of CO_2 on zeolite CaX measured at 0.06 bar CO_2 and 323K decreased from 2.5 to 0.1 mmol/g as the water vapour concentration increased from 0.8 to 16.1 wt% [30].

Adsorption of CO_2 is faster on zeolite compared to other adsorbents. The equilibrium capacity is reached within few minutes in most of the zeolites. Similar to other adsorbents, the adsorption rate is fast initially but flattens out eventually [27, 31, 32]. For any adsorption process to be cost effective and practical for an adsorption design, the adsorbent materials should be regenerable for a very large number of cycles. Zeolites have good regeneration characteristics; that means, it can be reused many times by regenerating it without compromising its activity [32]. Zeolites are regenerated mainly by (pressure swing adsorption (2) temperature swing adsorption (2) Electric swing adsorption. It is concluded by Tenzel et al. that the CO_2 adsorption capacity could be recovered with negligible changes by regenerating the adsorbent at 473 K for 12 h [19]. Siriwardena et al. evaluated the PSA and TSA on the CO_2 adsorption on sodium form of type X zeolites, NaX. After the first regeneration cycle, the adsorption capacity was reduced at a lower temperature but was fully regenerated for up to 4 cycles with regeneration at a higher temperature (623 K)[28]. These results show that the regeneration of the CO_2 adsorbed zeolites can be more easily carried out using TSA compared to PSA.

2.1.2 Removal of SO_X/NO_X

Several non-regenerable and regenerable processes exist for the removal of SO_X from flue gas [33]. However, these processes suffer from several drawbacks, which is the main incentive for exploring other processes. Adsorption of SO_X using zeolite offers such a good process where the adsorbent can be reused after regenerating it. Zeolites have a high surface area, pore volume, porosity and permeability, which favour their use as an adsorbent for SO_2 [12]. Sakizci et al. investigated the adsorption of SO_2 on KL, Na-A, Ca-A and Na-X at room temperature (25 °C). It was found that these zeolites showed good adsorption capacity ranging from 130-170 mg/g with a stream containing 2000 ppm SO_2 stream [34, 35]. After such encouraging results, several other zeolites such as silicate, dealuminated Y zeolites (DAY), Na-Y were investigated for the adsorption of SO_2. They all showed good adsorption capacities [36].

Marcu et al. studied the adsorption of SO_2 on Y-Zeolite suing fixed bed reactor in presence of N_2. They carried out the experiment in the range of 25 to 200 °C. The Y-zeolite showed adsorption capacity of 175 mg/g at 25 °C. As expected the amount of SO_2 adsorbed decreased with increase in temperature. Physical adsorption was the major process. The zeolite was regenerated at 400 °C and it did not altered even after 20 cycles of regeneration. The SO_2 molecules are probably adsorbed by hydrogen bonding to one or two conveniently positioned surface hydroxyl groups [37].

Srinivasan et al. synthesized a variety of zeolites such as zeolite x, y, and Na-P1 using class F fly ash. The adsorption capacity depended on the moisture content and the type of the zeolite. High adsorption capacity (6-7 mg/g) was obtained for very dry zeolite [38].

Kopac et al. studied the adsorption on 13X zeolites in the temperature range 25-400 °C. The material showed good adsorption capacity below 250 °C for dilute streams [36].

Selective catalytic reduction and high-temperature ammonia gas injection non-catalytic methods are used to remove NO_X from flue gas. Because low-concentration hazardous gas (several hundred ppm of NO_X in diesel engines) is the target, reaction efficiency is not so high, which leads to a high-fuel penalty in the after treatment [39]. This process suffers from many drawback (1) disposal of unreacted species NH_3 and hydrocarbons (2) complex multistep process (3) high cost (4) high operating temperature. A large number of early efforts made in the 1960–1970 periods produced two current commercial catalytic systems: noble metal-based three-way catalysts for the purification of automobile emissions and vanadium-titanium-based catalysts for the control of stationary nitrogen monoxide (NO) emissions by selective catalytic reduction with ammonia [40]. Nitrous oxide is formed, which is a greenhouse gas, and ammonia slip at lower reaction temperature are the main drawbacks of this process.

Nitric oxide (NO) can be adsorbed on zeolite either by reversible physical adsorption of the NO to the cation or it can irreversibly combine with the surface to form a nitrosyl complex. The adsorption characteristics are largely defined by the interaction between the NO and the cation present in the zeolite. Adsorption is carried out at a lower temperature. Xing et al. evaluated zeolite A, Y, beta, ZSM-5, 13X and SAPO-34 for adsorption of NO and NO_2 at 40 °C. It was found that a very high adsorption capacity was achieved for 13X in the presence of oxygen whereas H-beta shows maximum capacity in absence of O_2. Copper exchanged ZSM shows good adsorption properties for NO_2 at low temperature but NO is needed to be oxidized to NO_2 for adsorption [41].

2.2   Silica and other oxides

Microporous materials (zeolite, carbon) have pore diameters in the range of 0.5-2 nm. For this reason, they offer mass transfer limitation to the adsorbate molecule because they have to diffuse into the pores for adsorption to take place [42]. They showed low adsorption capacity, poor kinetics, saturates quickly and poor regeneration. Therefore, to improve mass transfer limitation, it is proposed to use mesoporous materials, which have diameter of 1.5-10 nm. Works on ordered mesoporous materials are focused mainly on silica and silica-based solids because silica chemistry offers a flexible state that allows better control of the microstructure [43]. Ordered mesoporous silica (OMS) materials have attracted a great deal of research attention for their wide potential applications as adsorbents. Their outstanding features (high surface areas, large pore volumes, uniform and tunable pore sizes) endow such an adsorbent to have a high adsorption capacity with fast adsorption kinetics for a wide size-range of guest molecules. Several mesoporous silica, such as M41S, MCM-41, SBA-15, SBA-16 have been used. They did not show remarkable properties because still, the phenomena is physical adsorption [44]. Its adsorption capacity decreases with temperature and it shows low selectivity for multi-component systems. Ordered mesoporous silica-based materials have their walls amorphous and present lots of silanols, which can lead to poor hydrothermal stability but can be used advantageously to graft organic functional groups. Modification with amine groups is required for tailoring the sorption properties. Amine functionalized mesoporous silica materials can be classified as physical impregnation of mesoporous material with amine rich polymers (II) chemical [32, 45] amine and host material exist in class I, the amines are covalently bonded to the silanol groups in support. Chemically modified ones have high stability and porosity compared with the physical one [43, 44, 46-49].

3. Activated carbon

Several varieties of carbons are popular as an adsorbent material for a variety of target species. The most famous and widely used form is activated carbon [15]. It has been used in catalysis as support materials, for the treatment of water containing various inorganic and organic pollutants. It is also used in the adsorption of CO_2, SO_X and NO_X. Their popularity as adsorbent is largely due to its large area, high pore volume, the presence of meso and micropores which are primary requirements of a good adsorbent [50]. The cost of raw materials makes it one of the cheapest adsorption materials available today. They can be produced commercially from many low-cost starting materials such as coals, petroleum by-products, woods, coconut shells, tree bark, sawdust and rice husks to name few [51]. A typical synthesis of activated carbon involves two phases, namely carbonization followed by the activation process. The carbonization process involves the pyrolysis of the starting material in an inert atmosphere to make char. The activation process involves chemical or physical treatment of the resultant char to achieve desired pore size distribution and a high surface area. This activation process gives its name to the carbon as activated carbon. Adsorption capacities for CO_2 separation on activated carbon have been extensively studied [52].

4. Nano carbon materials

4.1 Carbon Nanotubes

Carbon nanotubes (CNT) are allotropes of carbon with quasi-one-dimensional tubular structure made by the rolling-up of sp2- hybridized graphene sheets. One single graphene sheet rolled up is called a single-wall carbon nanotube (SWCNT) [53]. A multi-walled carbon nanotube (MWCNT) is a stack of graphene sheets rolled up into concentric cylinders. These nanotubes show a very high aspect ratio (Length over diameter), which can be as high as 1000. This unique structure and morphology have imparted extraordinary properties to the CNTs [54]. It is very natural that they are finding applications in wide areas of science and engineering such as catalysis, energy storage, energy conversion, sensors, & adsorption. CNTs offer a high porous and hollow structure, large surface area, high aspect ratio, and low mass density [55-57]. They also offer strong interaction between the surface and the adsorbent molecules. All these features make them suitable for application adsorbent especially for gas treatment [58, 59].

The first report on the adsorption of CO_2 on a single wall nanotube (SWNT) has been done by Cinke et al. in 2003. They studied the adsorption phenomena on SWCNT in the temperature range of 0-200 °C. They compared the performance of SWCNT and carbon with surface area, 1617 and 1284 m^2/g respectively. The adsorption capacities for SWCNT and carbon were found to be 87 and 44 mg/g at 35 °C. The adsorption process was found to be by physical adsorption and the adsorption capacity decreases with increase in temperature. SWCNT show large pore volume, surface area and are arranged in bundles. This leads to the additional adsorption sites, such as grooves site and interstitial channels [60].

Su et al. studied the adsorption of CO_2 on the pure and modified MWCNT. They used different amino group introducing precursors such as 3-aminopropyl-triethoxysilane (APTS), ethylene diamine (EDA), and polyethyleneimine (PEI). It is understood that the adsorption capacity of pristine CNT can be increased by the incorporation of various additional types of function groups on the surface of the CNTs by chemical treatment or thermal treatment. Therefore, chemical modification is expected to have a positive impact on the adsorption of CO_2 on CNT. The found equilibrium adsorption capacity (qe) at 20 °C for the APTS modified MWCNT and pristine MWCNT were 75.9 and 114 mg/g, respectively. The adsorption capacities of CNT modified by EDA and PEI showed negligible improvement [61].

Omidfar et al. (2015) studied the effect of diameter of MWCNTs and the effect of amine modification using urea on the adsorption capacity of MWCNT. It was observed that the adsorption capacities increased with the increase in diameter of the MWCNT. It was concluded that the adsorption took place not only on the surface of the nanotubes but also in the interplanar spacing between the individual CNTs. It was also confirmed that the amine modified CNT showed increased adsorption capacity. The adsorption capacities were found to be 64.1 and 61.5 for amine modified CNT for 1 hour and 4 hours, whereas the pristine CNT showed only 53.9 mg/g. The increase was attributed to the availability of additional amino functional groups on the surface of the CNTs that adsorbs CO_2 via formation of carbamate ions. The breakthrough analysis concluded that the adsorption process is fairly rapid with approximately 70-80 % of the total adsorption taking place within the first five minutes [62].

Gui et al. (2013) found that amino functionalized CNT showed highest adsorption capacity 75.4 mg/g that is much higher than the pristine one. The CO_2 adsorption capacity was found greatly improved after amine functionalization with APTES, suggesting a better adsorption of CO_2 with the presence of amine groups on the MWCNTs surface, which enabled the adsorption of CO_2 via physisorption (by the MWCNTs) and chemisorption (by the amine groups). Reflux time of 5 hours was fund to be appropriate because excessive refluxing may although increase the extent of ammonification but also may reduce the adsorption capacity due to entangling of the functional groups leading to unavailability of some of them and reduced surface area [56].

Khalili et al. (2013) compared the CO_2 adsorption capacities of CNT with carbon in the temperatures range of 289-318 K and pressures up to 40 bar. The adsorption capacity increased with a decrease in temperature and increase in pressure. The maximum adsorption capacity of MWCNT and carbon at 298 K and 40 bar were 15 and 7.93 mm/g. The increased adsorption capacity was attributed to the large pore volume, surface area, hollowness and light mass [63].

In addition to the experimental investigation of the adsorption of CO_2 on CNT, several theoretical analysis has been done to understand the adsorption phenomena. Rahimi et al. (2013) conducted Monte Carlo simulations and adsorption experiments to study the effect of intertube distance and tube diameter on the mechanism of CO_2 adsorption onto the bundles of double-walled carbon nanotubes of 5 nm diameter at 303 K. It was found that the intertube spacing has a strong effect on the heat adsorption and the adsorption capacity because it influence the mechanism of adsorption. For low pressure (P≤ 14 bar), maximum adsorption was found to be at d=0.5 nm whereas at 14<P<40 bar, the peak in the absorption was found at d=1 nm. It was also observed that the intertube distance also influences the mechanism and the sequence of adsorption of CO_2 in various position in the CNT bundles [64]. Recently Bareberio et al. (2015) studied the mechanism of adsorption on CO_2, in addition to other gas, on to silver coated MWCNTs using thermal desorption spectroscopy. They found that at lower temperature and pressure the processes are a first-order desorption mechanism without formation of chemical bonds between substrate molecules and adsorbate. This clearly specifies the physisorption nature of adsorption with the formation of weak van der Waals interactions between gases and composites [65].

It is well understood that the mechanism of adsorption in a disordered network of nanotubes can be described as a sequence of gas adsorption in different sites. The adsorption starts at the grooves between adjacent tubes on the outside surface of the bundles, passes at the accessible interstitial channels in the interior of the bundles, and proceeds on the external surface of bundles. In our case, there is a further step of adsorption on NP–NP and NP–tube pores, which leads to a higher desorption energy [66].

Long et al. (2001) studied the adsorption of nitrous oxide (NO) on carbon nanotubes in presence of oxygen. It was shown that CNTs are good and reversible adsorbent with the uptake capacity of 78 mg/g of NO when applied to a mixture of 1000 ppm NO in presence of 5% O_2 at room temperature. The adsorbed NO was removed at a temperature lower than 300 °C. The authors claim that this is the highest value of adsorption capacity reported so far. The good capacity for NOx might be related to the unique structure, electronic properties, and surface functional groups (e.g., CdO on graphitic edges and defects) of carbon nanotubes[67].

Vasylenko et al. (2015) studied the adsorption process of He and NO on SWCNT using ab initio computational technique. The effect of sizes, chirality and the presence of vacancies on SWCNT on the adsorption were studied. The observed change of the band gap of zigzag SWCNTs under adsorption of NO enables us to conclude that a semiconducting type of SWCNT responds in a more pronounced way to the adsorption of NO and therefore can be a preferable type of SWCNT as a basis for nanosensors. It was found that the presence of vacancies on SWCNTs not only prompts a surface reconstruction but is also a reason for chemisorption of NO gas molecules and functionalization of CNTs [68]. Dai et al. (2009) studied the adsorption of NO_X (x=1,2,3) molecules on the SWCNT using first-principle calculations using DFT. The curvature is found to have a sizable effect on the interactions of NOx molecules with SWCNT surfaces [55].

Sulphur dioxide has been adsorbed on the surface of CNT. In a latest comparative study between different types of activated carbon, and SWCNT and MWCNT are carried out by Sun et al. (2013) [69]. The SO_2 adsorption capacities for coal-based activated carbon, coconut shell activated carbon, SWCNT and MWCNT are 12.21, 21.21, 3.51, & 1.04 mg/g, respectively, for 1000 ppm SO_2 in balance nitrogen at 303 K. Carrying out adsorption process carefully at different temperature revealed that the mechanism of adsorption is temperature dependent; at temperatures between 303-333 K, physical adsorption was found to be the dominant mode whereas, at elevated temperature (333-393 K), chemisorption is found to be the dominant mode. According to above analysis, it is obvious that physical adsorption is the main adsorption type during SO_2 adsorption process on CBAC and CSAC. The adsorption capacities of activated carbon were found to be higher than the CNTs. The higher-density p–p* in CNTs might be the active sites for SO_2 chemical adsorption, the micropores smaller than 0.7 nm were the best SO_2 adsorption place for both activated carbons and CNTs. CNTs have a smaller number of pores with a diameter less than 0.7 nm. Wider pore size distribution is not conducive to the creation of adsorption potential energy field and accordingly is not conducive to SO_2 adsorption. On top of above analysis, it can be concluded that the micropores whose diameter are smaller than 0.7 nm are the best pore structure for the SO_2 adsorption by carbonaceous materials.

Rahimi et al. (2015) studied the Grand canonical Monte Carlo simulations and adsorption experiments are combined to find optimized carbon nanotube (CNT) arrays for gas adsorption at low pressures and 303 K. The interstitial region and grooves are important adsorption sites. Bundles of 3D aligned double-walled carbon nanotube (DWCNT) with an inner diameter of 8 nm and different intertube distances were made experimentally. The intertube distance leads to increase in adsorption at p = 1 bar. Molecular simulation investigations were performed on double-walled CNTs with inner diameters 1-8 nm and intertube distances of 0-15 nm. It was found that at lower pressure (P<0.5 bar), inter-tube distance of 0.5 nm shows the highest adsorption capacity whereas at higher pressures (0.5<P<3 bar), the highest adoption capacity is shown by the inter-tube distance of 1 nm. Furthermore, it was found that the intertube distance has a much larger effect on gas adsorption than the tube diameter. Having high adsorption is not the only main goal since it is important to have low qst for economically reusable adsorbent for CO_2. Hence, clearly, there is a trade-off in the amount of adsorption and energy required for reusability [64, 66]. Similar theoretical studies have been performed to gain insight into the adsorption process of SO_2 on Pt/CNT and Au/CNT [70].

4.2 Ordered mesoporous carbon

Conventional carbon materials such as activated carbon and CNTs have a majority of their pores as disordered micropores with pore diameter less than 2 nm. These pores are usually highly interconnected and have large non-uniform pore size distribution. These features of the conventional materials are suited for gas adsorption materials. The presence of predominantly large percentage of microspores leads to slower gas diffusion rates owing to the large mass transfer resistance to the adsorbing molecules [71]. The presence of larger mesopores (2-50 nm) can better facilitate the gas transport and diffusion into microspores by reducing the resistance to mass transfer and pathway distance. Therefore, periodic pore arrangement, uniform micro and mesoporous size, high surface area, and large pore volume would be ideal for good gas adsorption. Ordered mesoporous carbon (OMCs) materials are a new class of materials that have been developed with tailored pore structure and uniform pore size distribution [72]. Their large specific surface area and pore volume, well-defined mesostructure, tunable pore size and open framework make mesoporous materials useful in many valuable applications [73].

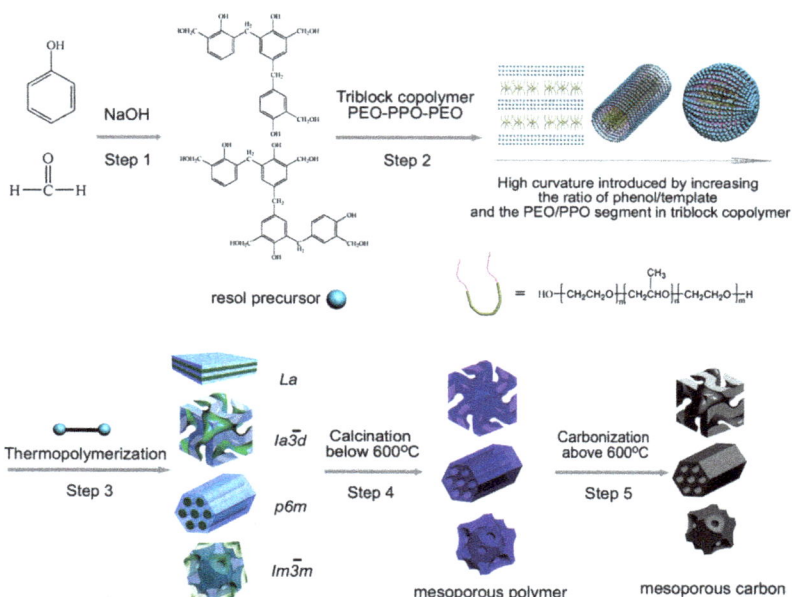

Figure 1: Preparation of the ordered mesoporous carbon frameworks with resol as carbon source precursor [74]. Reprinted with permission from (ref. 74). Copyright (2006) American Chemical Society.

Among different adsorbents, mesoporous carbons such as CMK-1, CMK-3 and CMK-5, because of their large specific surface area, highly ordered mesoporous structure and high thermal stability, have attracted much attention as adsorbents for different compounds [75].

OMCs are mostly prepared by the hard or soft templating method. In the hard template method, a mesoporous silica template, such as MCM-41, SBA-15 and a carbon source called as precursor, e.g., sucrose, furfuryl alcohol, naphthalene, C_2H_2, polyacrylonitrile, and phenolic resin, can be utilized. This process is a multistep and complex process as shown in Figure 1, and time-consuming. This is in contrast to the soft-template method to prepare ordered mesoporous carbons from self-assembly of amphiphilic block copolymers and phenolic resins. Details of these processes can be found somewhere else [72-78].

Ordered mesoporous carbon materials are expected to be excellent gas adsorbent. Several important types of research have been done in the last decade to study the adsorption of CO_2 on OMCs and its various functionalized versions. Yuan et al. studied the CO_2 adsorption characteristics of OMC prepared via a soft template method. It showed a very large surface area 2255 m^2/g and large pore volume 2.17 cm^3/g. The maximum CO_2 uptake at 100 kPa and 278 and 298 K are 3.0 and 2 mmol/g, respectively [79]. Lakhi et al. synthesized OMC with rod-shaped morphology with high surface area (1205 m^2/g) and large pore volume (1.46 cm^3/g) by using mesoporous SBA-15 prepared by the combined microwave and static method as a template. The resulting carbon was of rod-shaped morphology with a uniform size in length and width, which was replicated from the templates and was used as adsorbents for the capture of CO_2 molecules[80]. It is well known that the materials with long-range and well-defined structural order, uniform and regular particle morphology, high specific surface and large pore volume have higher CO_2 adsorption capacities compared to those with irregular particle morphology and lower specific surface and pore volume. The adsorption capacity of OMC synthesized by this method was found to be 24.4 mmol/g at 0 °C and 30 bar pressure. It is also found that the adsorption capacity decreased with increasing temperature and decreasing pressure. They compared the adsorption capacity of OMC with activated carbon, CNT, and mesoporous carbon nitride (MCN). It is surprising that even MCN, which have a basic functional group on their surface, showed lower adsorption capacity for CO_2 compared with OMC. It is believed that the excellent textural parameters with the controlled morphology of the later dictate the total adsorption capacity for CO_2 [81]. Goel et al. synthesized mesoporous carbon adsorbents, having high nitrogen content, *via* nano casting technique with melamine–formaldehyde resin as a precursor and mesoporous silica as a template. The maximum uptake of CO_2 was found to 0.83 mmol/g at 30 °C in 12.5% CO_2 rest N_2 atmosphere. This low value of CO_2 uptake could be a very low surface area of the adsorbent [82]. Nelson et al. used tannin, the naturally occurring poly phenolic biomass, tannin, as carbon source precursor as an alternative to providing a much greener and cost effective method to prepare OMC. The tannin-based porous carbon was then exposed to ammonia activation at high temperature to increase the surface area and micropore volume while incorporating nitrogen functionality into the framework. These activated carbons showed high CO_2 adsorption up to 3.44 mmol/g at 0

°C at 1 bar[76]. Since CO_2 is slightly acidic, it is obvious that adding basic nitrogen group to the surface of the OMC is expected to enhance the CO_2-uptake. Huang et al. modified the CMK-3 OMC with chitosan to infiltrate nitrogen containing functional groups into the structure. The adsorption capacity was lower than 3.6 mmol/g at 30 °C and 0.1 MPa [83]. However, functionalization of the OMC with basic functional groups reduces the surface area and pore volume which negatively impacts the adsorption capacity of the material.

Another type of mesoporous material is mesoporous carbon nitride (MCN). In these materials, the nitrogen atom is incorporated during the synthesis, which enhances the basic sites available on the surface. These materials have excellent properties such as strong basicity, inherent semiconducting nature, large surface area, high pore volume, controllable pore size and excellent thermal and mechanical stability. mesoporous carbon nitride based hybrids (MCN) are particularly attractive for capturing acidic CO_2 molecules because the free NH_2 groups on its surface can selectively adsorb the CO_2 molecules through acid–base interaction. Venu et al. conducted series of experimental work on the adsorption of CO_2 on MCN. Apart from other effects, the morphology of the MCN has a great impact on their performance as an adsorbent. In one study they reported that MCN with large pores and 3D porous structures showed adsorption capacity of 13.5 mm/g at 30 bar and 0 °C. In another study with 2D MCN with well-ordered particle, morphology could adsorb 16.5 mm/g at 30 bar and 0 °C. It should be pointed out that adsorption capacity is expected to be enhanced with an increase in specific surface area and pore volume. But, it is difficult to synthesize MCN materials with a very high surface area and large pore volume. This limits their application as adsorbent [81].

Chen 2014 & Cao et al. studied the NO adsorption on the OMCs and cerium oxide modified OMC. They synthesized the OMC by using the evaporation-induced self-assembly method. The resulting OMC and cerium oxide modified OMC had specific surface areas as 1444 and 1110 m^2/g, respectively. The adsorption studies were carried out in a 500 ppm NO with a trace amount of oxygen and balance argon. The NO adsorption capacity of OMC was 19.40 mg/g, which was more than twice of that of AC. With the introduction of 2 wt% cerium into OMC, the NO adsorption capacity was further improved to 22.00 mg/g. In absence of oxygen the corresponding, NO uptake was 7.62 mg/g. The improved performance of OMC over AC carbon can be attributed to the large surface and pore volume, which is twice and 3.5 times that of AC. Moreover, presence of large percentage (60%) microspores in AC also hinders effective transport of NO whereas in OMC most of the pores are mesoporous. The NO adsorption process involves the formation of monodentate nitrito (C-O-N=O) which is formed as a result of adsorption of NO on the oxidized active sites (CO) on the OMC. Cerium oxide is a well-known oxygen source for many applications. With the introduction of cerium, excess adsorbed oxygen atoms were provided from the redox cycle between Ce^{4+} and Ce^{3+}. These adsorbed oxygen atoms could react with active sites and generate C(O) groups to adsorb NO[84-85].

4.3 Graphene

Graphene is a sp2 hybridized carbon-based material with a hexagonal (benzene ring) monolayer network. Graphene is an allotrope of carbon, which is a strictly 2D material with the exceptionally large specific surface area (theoretical value of 2630 m2 g-1) [86]. In comparison with other carbon allotropes, graphene offers the greatest intrinsic carrier mobility at room temperature, with perfect atomic lattice, high mechanical strength, chemical and thermal stability [87-88]. Because of these extraordinary properties graphene is finding application in several applications, including the role as an adsorbent for the harmful gases such as CO_2, NO_X and SO_X[17, 89]. There are several members of the graphene family; graphene, graphene oxide and reduced graphene oxide. Furthermore, due to the planar geometry and presence of OH and COOH groups on the surface, graphene can be tailored for any application [90]. Therefore, graphene with its amazingly attractive properties, graphene certainly has the potential to be a selective and efficient adsorbent for CO_2 capture interestingly, by strategically removing atoms from its hexagonal lattice to create pores with radii large enough to admit CO_2, adsorbents with concomitantly high CO_2 selectivity and high CO2 capacity can be developed. These characteristics of graphene render it even more attractive compared to the conventional CO_2 adsorbents. In addition, graphene's planar geometry makes it amenable for modification or functionalization, providing essentially infinite possibilities to fabricate adsorbents with properties that are precisely tuned for the desired capture setting. The exceptional mechanical and thermal stability and strength make graphene a perfect adsorbent to be used in TPA or PSA. Therefore, an enormous amount of researches have been devoted to the explore graphene, graphene oxide and their functionalized forms as an adsorbent for CO_2, NOX and SOX both theoretically and experimentally [17].

One of the earliest works on graphene as adsorbent came from Rao et al. They measured the volumetric CO_2 isotherm of graphene samples synthesized from different carbon precursors. At -78 °C and 1 bar, graphene obtained from exfoliation of GO adsorbed up to 7.8 mmol g^{-1} whereas those from thermal conversion of nanodiamonds up to 8.6 mmol g^{-1}. Regardless of the precursor used, CO_2 adsorption on graphene was completely reversible and no significant hysteresis was observed on the desorption branches of the isotherms, suggesting weak intermolecular forces were involved. Moreover, based on density functional theory calculations the binding energy of a CO_2 molecule was estimated to be -59.1 kJ mol^{-1}, indicating the occurrence of physical adsorption of CO_2 onto graphene[91].

One of the major drawbacks of graphene is the formation of agglomerates which reduces the surface area and therefore the adsorption capacity is also hampered [92-93]. The introduction of nanopores into graphene sheets has been identified as one of the most effective methods for improving the adsorption performance of graphene materials. The introduction of nanopores into the graphene can improve the surface area and adsorption capacity. Meng and Park thermally exfoliated GO sheets in a vacuum producing graphene nanoplates (GNPs) with broad pore size distribution (including super micropores with widths 1.2 nm, mesopores between 20 and 50 nm, and macropores around 90nm) and appreciable porosity (1.7 cm^3 g^{-1}) [94-96]. The GNPs proved to be extremely suitable for separating CO_2 from flue gases at high pressures and room temperature: 56.4 mmol g^{-1} at 25 °C and 30 bar. This is much higher than the conventional adsorbents such as zeolite and carbon. Several methods to improve the adsorption capacity of graphene have been reported in the literature. This includes graphene nanomesh (GNM) [97], doping with a heteroatom (S or N) [98-99], non-covalent functionalization with poly (ionic liquid) [100], and incorporation of bulk materials into organic and inorganic matrices [101]. The introduction of any basic group or nitrogen atom into the graphene sheets increases the surface area and porosity in addition to creating more basic sites for adsorption of acidic gases resulting in increased CO_2 adsorption capacity [100].

Graphene oxide has also been receiving great attention. It is the more oxidized form of graphene. It is particularly attractive because of their simple and inexpensive production, tunable porosity and high chemical, thermal stability ease of regeneration. Due to the presence of an oxygen-containing functional groups, a wide variety of organic and inorganic materials. There are several excellent review papers which have been published recently and the interested readers are directed for further details [102].

Apart from CO_2 adsorption, the first study on the adsorption of SO_2 on graphene oxide has been reported very recently by Babu et al. [103]. Pure SO_2 adsorption experiments were carried out at 35 °C and up to the saturation temperature. The amount of pure SO_2 adsorbed on GO is much higher when compared to pure CO_2 adsorption under similar conditions of temperature and pressure. This was attributed to the stronger van der Waals interaction resulting from the larger dipole moment of SO_2 compared to CO_2 under these adsorption conditions. Near-ambient pressure, GO has an SO_2 adsorption capacity of 156 mg g^{-1}. With an increase in pressure, the adsorption capacity increases to 257 mgg^{-1} at 2.6 bar. GO with its moderate specific surface area of 268 m^{2}g^{-1} exhibits an adsorption capacity at par with other carbon materials like activated carbon, carbon molecular sieves or activated carbon nano fibres, which typically have a very large specific surface areas. Some theoretical studies have been done for the adsorption of NO_X on graphene and graphene oxide. Still not many experimental studies exist on NO_X adsorption over graphene [104-105].

5.   Metal-organic framework

A new class of hybrid materials called Metal-organic frameworks (MOFs) have gained remarkable popularity in recent years. These hybrid organic-inorganic materials are constructed by joining metal-containing with organic ligands, using strong bonds (reticular synthesis) to create open crystalline frameworks with permanent porosity [106-107]. The flexibility with which the metal and organic ligands can be varied has led to the possibility of preparing an infinite number of MOFs. MOFs can be designed for the desired properties by proper selection of the metal ion and the ligands. Up to this point, more than 6000 MOFs have been registered in the Cambridge Structural Database [107]. Besides the pre-design in synthesis, post-synthetic modifications have also been successfully used in tuning the pore properties of MOFs. Apart from many applications, MOFs are generally believed to be a unique adsorbent material for gases and liquid adsorption [108]. MOFs hold several records in porous materials including highest surface areas (7000 m^2/g). It also shows excellent high porosity and pore volume [109]. Compared to other porous materials, such as zeolites and activated carbon, due to very large pore volume and surface area, most MOFs have shown record CO_2 uptakes. MIL-53, HKUST-1, MOF-74, and DMOF-1 have shown excellent CO_2 adsorption capacity [110-111].

Benzenetricarboxylate MOFs with copper as the central cation and $BaCl_2$ as a second component (Ba/Cu-BTC) were evaluated in SO_2 adsorption experiments. The SO_2 uptake was shown to be high at elevated temperatures and exceeded the stoichiometric capacity, which is the theoretical maximum or total capacity, as a result of the contribution of chemical bonding between SO_2 and the metal cations of the MOF. At low temperatures, Cu- BTC (also known as HKUST-1) behaved as a good host material for highly dispersed barium salts [112]. Fernandez et al. studied the performance of FMOF-2 for the adsorption of SO_2.The relatively large uptake was attributed to the breathing effect of FMOF-2 during gas inclusion. It also showed excellent adsorption capacity[113]. Song et al. predicted the adsorption for SO_2 on ten MOFs using the Monte-Carlo simulation. In this study, we predicted the adsorption of SO_2 in ten metal–organic frameworks (MOFs) and investigated the effects of the heat of adsorption, free volume, and surface area of sulphur dioxide uptake using grand canonical Monte Carlo simulations over a wide range of pressures [114].

The NO adsorption and storage capability of CPO-27 (MOF-74) using different metals (Co, Ni) was studied by McKinlay et al., who reported an extremely high adsorption capacity of about ~7.0 mmol of NO/g at 25 °C over the pressure range of 0.1−1 bar [115]. Moreover, the studied MOFs exhibited relatively good storage stability (i.e., the recovery of the stored NO). Adsorption studies of hydrogen sulphide have been done and results are found to be satisfactory [116-117].

6. Adsorption equipment design

Adsorption process requires intimate contact between the adsorbent and adsorbate so that the adsorbate are effectively recovered by the adsorbent. Adsorption equipment comes in a variety of sizes and types. It varies from a simple container where adsorbent and adsorbate are mixed to very large continuous fluidized beds of several meters in diameters. The adsorbents usually come in the form of pellets by extrusion or pressing or in granular form by crushing of large masses or in the spherical or globular forms by the precipitation method [118]. All these adsorption equipment becomes saturated with adsorbate over a period of use and looses its capacity to further adsorb. For continuous operation, the exhausted adsorbent should be removed and replaced periodically. However, for a economical operation of the adsorption process, this exhausted adsorbent cannot be discarded but should be regenerated to achieve its original conditions as far as possible for further utilization [119]. For this reason, when continuous operation is necessary, at least two adsorbers are required, one on adsorption and the other on regeneration and cooling. Processes with three adsorption columns are also common when the breakthrough is not conducive. There are several ways in which the lost adsorption capacity of the adsorbent can be recovered. The exact way in which the adsorption and regeneration processes are carried out depends largely on the phases involved and the type of fluid-solid contacting pattern employed [118]. There are mainly three ways in which the fluid- solid contacting is carried out:

1. Fixed bed adsorption

2. Fluidized bed adsorption

3. Moving bed adsorption

4. Rotatory bed adsorbed

6.1 Fixed bed

Fixed bed adsorbers are the most common type of adsorption equipment used in the industry. The adsorbent in the form of pellets is placed in a bed supported on a perforated support to result in a randomly packed arrangement. The bed has the void volume to allow the fluid to pass through it. For most of the commercial large-scale processes, the adsorbent particle size varies from 0.06 to 6 mm, It is essential that the fixed bed will have a narrow particle size distribution. Down flow of the gases are preferred because for up-flow at high flow rate may fluidize the bed, which will cause loss of fine particle and attrition. Two beds are commonly employed, one for the adsorption while the other one for the regeneration. When the first bed is saturated, the valves open automatically to direct the gases to the other bed to start the regeneration process. The absorbers must be designed with consideration for pressure drop and must contain a means of supporting the adsorbent and a means of assuring that the incoming fluid is evenly distributed to the face of the bed. The pressure drop in a packed bed depends on the diameter to bed depth ratio and particle size. The bed depth dictates adsorption cycle time. High bed depth results in a high adsorption cycle but will lead to higher pressure drop and the initial cost of construction. While vertical beds with height 45 ft and 8-10 ft dia are in use, but the most common height is 1-4 ft are generally recommended to avoid excessive pressure drop and capital cost of construction. Bed diameters are selected to provide a superficial velocity of 0.5 to 1.5 ft/sec [32].

6.1.1 Regeneration

As pointed out earlier that for an economical run of the adsorption equipment the adsorbent has to be regenerated to or near its pre-adsorbed capacity. The following regeneration techniques are usually employed include (1) Pressure-swing adsorption (PSA) (2) Vacuum Pressure-swing adsorption (VPSA) (3) Temperature swing adsorption (TSA) (4) electric- swing adsorption (ESA).

Separation of a component from a gaseous mixture can be achieved by passing the gas through a fixed bed adsorption column containing high surface area adsorbents at pressure and temperature that favours the adsorption equilibrium. The component that needs to be separated selectively interacts with the adsorbents and are retained in the column while the rest of the gas elutes unaltered as they show weak interaction with the adsorbent. The adsorbent plays a key role in the adsorption processes. The adsorbent should show high adsorption capacity and selectivity for the target species, regenerationability, fast adsorption and desorption kinetics, stability in the operating conditions and, of course, should have low cost. Apart from the requirement of the adsorbent, choosing the operating conditions like temperature and pressure is also very important. It is very well established that the adsorption capacity of an adsorbent increases with increase in pressure and decrease in temperature. **Pressure swing adsorption (PSA)** is a mature process where adsorption is carried out at elevated pressures, and when the adsorbent is exhausted, the adsorption capacity of the adsorbent is restored by regenerating it at low pressure[45, 120]. The feed containing CO_2 needs to be compressed. This way adsorbent can be reused for many such adsorption/regeneration cycles. PSA has been tried as an option for separating CO_2 from the flue gas[121]. Excellent reviews are available on the application of PSA for many gas separation applications. It has shown some promising results because of its ease of applicability over a wide range of temperature and pressure conditions, its low energy requirements, and its low capital investment and generation of a pure CO_2 stream[122]. Another variant for the PSA is the **vacuum Pressure-swing adsorption (VPSA)** where the vacuum is used for the regeneration. Generally, the CO_2 is present in the flue gas from conventional coal-fired power plants only in the range 4-8% and is at atmospheric pressure. For this reason, a highly selective adsorbent is required [123]. Such adsorbents have a very steep isotherm profile in the low-pressure zone. Consequently, very low vacuum pressures are required for the regeneration, which has high associated cost. As a result, it is not possible to achieve the target of 90+% recovery and 95+% purity of CO_2 is not feasible by PSA or VPSA. In **Temperature Swing Adsorption (TSA)** the adsorption of the target adsorbate is assisted by using low temperature followed by heating the adsorbent rapidly to a high temperature by hot gas (steam, air nitrogen etc.) for the desorption step. The TSA is particularly attractive because it can utilize cheaper, low-grade thermal energy resources available in the conventional power plants for the regeneration of the adsorbents [124]. This has the potential to reduce the operating cost substantially. However, there are several challenges in this process. Firstly, a lot of hot gases is required in order to heat the adsorbent bed for regeneration because of the low heat capacity of the gases[125]. This leads to the desorption of the CO_2 in the stream of the gases resulting in the significant dilution of the CO_2. It is well established that the CO_2 must be recovered as pure as possible for storage and transport. Some have suggested to first heat the bed indirectly using heating jacket, coils, or electric heating tapes wrapped around the bed, and heat exchanger. Then hot gases in small amount is used to sweep the adsorbed CO_2 resulting in the more concentrated CO_2 stream. However, their scale up can be an issue and still use sweep gas will lead to significant dilution. Secondly, it takes a long time for the heating and specifically the cooling cycle. Therefore, the productivity is lower compared to other adsorption technologies. These difficulties can be avoided by using the concept of **Electric- swing adsorption (ESA)** where the heating required for the desorption for regeneration of the adsorbent in the packed bed is provided by the electrical heating of the bed by the Joule-effect effect. Electric current is supplied to the bed externally[126]. The adsorbent, however, needs to be a good conductor of electricity. Activated carbon, CNT, graphene and other carbon material be suited for the adsorbent for ESA. Several combinations the above-mentioned techniques are also been explored such as PTSA.

6.2 Fluidized bed

Continuous fluidized bed adsorption equipment has been used for the gas separation using solid adsorbent. They offer excellent heat and mass transfer characteristics due to intimate contact between adsorbate gas and very high surface area adsorbent particles[127]. Fluidized bed adsorption equipment prevents the formation of the hot spot by near isothermal operation due to the even and fast circulation of the particle. However, they suffer from the attrition of the expensive adsorbent particles and the loss of efficiency due to the axial mixing. Moving bed adsorbers also have been proposed where the adsorbent is moved between the adsorbers vessel and the regenerator patterned after the famous fluidized catalytic cracker (FCC) in an oil refinery. Pennline et al. proposed a moving bed adsorption system with the cross flow[128]. These systems offer low-pressure drop, but the contact time between the gas stream and the adsorbent is usually very low [129-130].

References

[1]   J.C. Abanades, B. Arias, A. Lyngfelt, T. Mattisson, D.E. Wiley, H. Li, M.T. Ho, E. Mangano, S. Brandani, Emerging CO2 capture systems, International Journal of Greenhouse Gas Control, 40 (2015) 126-166.

[2]   D. U.S. Energy Information Administration: Washington, Annual Energy Outlook, 2016.

[3]   D. Toporov, Combustion of pulversied coal in a Mixture of Oxygen and Recycled Flue Gas, Elsevier Ltd., London, UK, 2014.

[4]   P.J.Reddy, Clean Coal Technologies for Power Generation, CRC Press, Florida, USA, 2013.

[5]   J. Gale, H. Herzog, J. Braitsch, U.E. Aronu, H.F. Svendsen, K.A. Hoff, O. Juliussen, Greenhouse Gas Control Technologies 9Solvent selection for carbon dioxide absorption, Energy Procedia, 1 (2009) 1051-1057.

[6]   IEA, Energy Technology Perspectives 2010: Scenarios and Strategies to 2050, OECD Publishing Paris, 2009.

[7]   W.F.J. Burgers, P.S. Northrop, H.S. Kheshgi, J.A. Valencia, Worldwide development potential for sour gas, Energy Procedia, 4 (2011) 2178-2184.

[8]   A.E.O. 2016, U.S. Energy Information Administration: Washington, DC.

[9] C.A. Scholes, K.H. Smith, S.E. Kentish, G.W. Stevens, CO_2 capture from pre-combustion processesâ€"Strategies for membrane gas separation, International Journal of Greenhouse Gas Control, 4 (2010) 739-755.

[10] AL-Othman ZA, Inamuddin, Naushad M (2011) Adsorption thermodynamics of trichloroacetic acid herbicide on polypyrrole Th(IV) phosphate composite cation-exchanger. Chem Eng J 169:38–42.

[11] Al-Othman ZA, Inamuddin, Naushad M (2011) Determination of ion-exchange kinetic parameters for the poly-o-methoxyaniline Zr(IV) molybdate composite cation-exchanger. Chem Eng J 166:639–645.

[12] Y. Liu, T.M. Bisson, H. Yang, Z. Xu, Recent developments in novel sorbents for flue gas clean up, Fuel Processing Technology, 91 (2010) 1175-1197.

[13] M. Wang, A. Lawal, P. Stephenson, J. Sidders, C. Ramshaw, Post-combustion CO2 capture with chemical absorption: A state-of-the-art review, Chemical Engineering Research and Design, 89 (2011) 1609-1624.

[14] R.S. Haszeldine, Carbon Capture and Storage: How Green Can Black Be?, Science, 325 (2009) 1647-1652.

[15] S. Choi, J.H. Drese, C.W. Jones, Adsorbent Materials for Carbon Dioxide Capture from Large Anthropogenic Point Sources, ChemSusChem, 2 (2009) 796-854.

[16] F. Rezaei, A.A. Rownaghi, S. Monjezi, R.P. Lively, C.W. Jones, SOx/NOx Removal from Flue Gas Streams by Solid Adsorbents: A Review of Current Challenges and Future Directions, Energy & Fuels, 29 (2015) 5467-5486.

[17] R. Balasubramanian, S. Chowdhury, Recent advances and progress in the development of graphene-based adsorbents for CO2 capture, Journal of Materials Chemistry A, 3 (2015) 21968-21989.

[18] A.L. Chaffee, G.P. Knowles, Z. Liang, J. Zhang, P. Xiao, P.A. Webley, CO2 capture by adsorption: Materials and process development, International Journal of Greenhouse Gas Control, 1 (2007) 11-18.

[19] I. Kabalan, B. Lebeau, H. Nouali, J. Toufaily, T. Hamieh, B. Koubaissy, J.-P. Bellat, T.J. Daou, New Generation of Zeolite Materials for Environmental Applications, The Journal of Physical Chemistry C, 120 (2016) 2688-2697.

[20] Y. Kamimura, M. Shimomura, A. Endo, CO2 adsorptionâ€"desorption properties of zeolite beta prepared from OSDA-free synthesis, Microporous and Mesoporous Materials, 219 (2016) 125-133.

[21]   B.M. Weckhuysen, J. Yu, Recent advances in zeolite chemistry and catalysis, Chemical Society Reviews, 44 (2015) 7022-7024.

[22]   E.J. GarcÃa, J. PÃ©rez-Pellitero, G.D. Pirngruber, C. Jallut, M. Palomino, F. Rey, S. Valencia, Tuning the Adsorption Properties of Zeolites as Adsorbents for CO2 Separation: Best Compromise between the Working Capacity and Selectivity, Industrial & Engineering Chemistry Research, 53 (2016) 9860-9874.

[23]   S. Kesraoui-Ouki, C.R. Cheeseman, R. Perry, Natural zeolite utilisation in pollution control: A review of applications to metals' effluents, Journal of Chemical Technology & Biotechnology, 59 (1994) 121-126.

[24]   D. Barthomeuf, Conjugate acid-base pairs in zeolites, The Journal of Physical Chemistry, 88 (1984) 42-45.

[25]   J. Pawlesa, A.t. Zukal, J.ɜ .ΙŒejka, Synthesis and adsorption investigations of zeolites MCM-22 andΙ MCM-49 modified by alkali metal cations, Adsorption, 13 (2007) 257-265.

[26]   P.J.E. Harlick, F.H. Tezel, An experimental adsorbent screening study for CO2 removal from N2, Microporous and Mesoporous Materials, 76 (2004) 71-79.

[27]   S. Cavenati, C.A. Grande, A.r.E. Rodrigues, Adsorption Equilibrium of Methane, Carbon Dioxide, and Nitrogen on Zeolite 13X at High Pressures, Journal of Chemical & Engineering Data, 49 (2004) 1095-1101.

[28]   R.V. Siriwardane, M.-S. Shen, E.P. Fisher, J. Losch, Adsorption of CO2 on Zeolites at Moderate Temperatures, Energy & Fuels, 19 (2005) 1153-1159.

[29]   S.U. Rege, R.T. Yang, A novel FTIR method for studying mixed gas adsorption at low concentrations: H2O and CO2 on NaX zeolite and ³ɀ-alumina, Chemical Engineering Science, 56 (2001) 3781-3796.

[30]   F. Branddani, D.M. Ruthven, The effect of water on the adsosprtion of CO2 and C3H8 on the type X zeolites, Industrial & Engineering Chemistry Research, 43 (2004) 8339-8344.

[31]   F. Gholipour, M. Mofarahi, Adsorption equilibrium of methane and carbon dioxide on zeolite 13X: Experimental and thermodynamic modeling, The Journal of Supercritical Fluids, 111 (2016) 47-54.

[32]   L. Hauchhum, P. Mahanta, Carbon dioxide adsorption on zeolites and activated carbon by pressure swing adsorption in a fixed bed, International Journal of Energy and Environmental Engineering, 5 (2014) 349-356.

[33]   A. Demirbas, Adsorption of Sulfur Dioxide from Coal Combustion Gases on Natural Zeolite, Energy Sources, Part A: Recovery, Utilization, and Environmental Effects, 28 (2006) 1329-1335.

[34]   M. Sakizci, B. ErdoÄŸan Alver, E. YÃ¶rÃ¼koÄŸullari, Influence of the exchangeable cations on SO2 adsorption capacities of clinoptilolite-rich natural zeolite, Adsorption, 17 (2011) 739.

[35]   A.K. Gupta, S. Ibrahim, A. Al Shoaibi, Advances in sulfur chemistry for treatment of acid gases, Progress in Energy and Combustion Science, 54 (2016) 65-92.

[36]   T. Kopaqâ—, E. Kaymakgi, M. Kopac, DYNAMIC ADSORPTION OF S02 ON ZEOLITE MOLECULAR SIEVESâ€ Chemical Engineering Communications, 164 (1998) 99-109.

[37]   I.-C. Marcu, I. Sandulescu, Study of sulfur dioxide adsosprtion on Y zeolite, Journal of Serbian Chemical Society, 67 (2004) 563-569.

[38]   A. Srinivasan, M.W. Grutzeck, The Adsorption of SO2 by Zeolites Synthesized from Fly Ash, Environmental Science & Technology, 33 (1999) 1464-1469.

[39]   K. Skalska, J.S. Miller, S. Ledakowicz, Trends in NOx abatement: A review, Science of The Total Environment, 408 (2010) 3976-3989.

[40]   H. Yahiro, M. Iwamoto, Copper ion-exchanged zeolite catalysts in deNOx reaction, Applied Catalysis A: General, 222 (2001) 163-181.

[41]   S.C. Ma, J. Yao, X. Ma, L. Gao, M. Guo, Removal of SO2 and NOX Using Microwave Swing Adsorption over Activated Carbon Carried Catalyst, Chemical Engineering & Technology, 36 (2013) 1217-1224.

[42]   Y. Belmabkhout, A. Sayari, Effect of pore expansion and amine functionalization of mesoporous silica on CO2 adsorption over a wide range of conditions, Adsorption, 15 (2009) 318-328.

[43]   R. Serna-Guerrero, A. Sayari, Applications of Pore-Expanded Mesoporous Silica. 7. Adsorption of Volatile Organic Compounds, Environmental Science & Technology, 41 (2007) 4761-4766.

[44]   Y. Wang, M.D. LeVan, Adsorption Equilibrium of Carbon Dioxide and Water Vapor on Zeolites 5A and 13X and Silica Gel: Pure Components, Journal of Chemical & Engineering Data, 54 (2009) 2839-2844.

[45] N. Casas, J. Schell, R. Pini, M. Mazzotti, Fixed bed adsorption of CO2/H2 mixtures on activated carbon: experiments and modeling, Adsorption, 18 (2012) 143-161.

[46] F.-Y. Chang, K.-J. Chao, H.-H. Cheng, C.-S. Tan, Adsorption of CO2 onto amine-grafted mesoporous silicas, Separation and Purification Technology, 70 (2009) 87-95.

[47] S.-w. Choi, H.-K. Bae, Adsorption of CO2 on amine-impregnated mesoporous MCM41 silica, KSCE Journal of Civil Engineering, 18 (2014) 1977-1983.

[48] M.R. Mello, D. Phanon, G.Q. Silveira, P.L. Llewellyn, C.l.M. Ronconi, Amine-modified MCM-41 mesoporous silica for carbon dioxide capture, Microporous and Mesoporous Materials, 143 (2011) 174-179.

[49] V. Zele⟨ˠ⟩k, M. Badaniᵹ₃ov⟨ⁱ⟩, D. Halamov⟨ⁱ⟩, J. ₃Œejka, A. Zukal, N. Murafa, G. Goerigk, Amine-modified ordered mesoporous silica: Effect of pore size on carbon dioxide capture, Chemical Engineering Journal, 144 (2008) 336-342.

[50] B. Guo, L. Chang, K. Xie, Adsorption of Carbon Dioxide on Activated Carbon, Journal of Natural Gas Chemistry, 15 (2006) 223-229.

[51] N.P. Wickramaratne, M. Jaroniec, Activated Carbon Spheres for CO2 Adsorption, ACS Applied Materials & Interfaces, 5 (2013) 1849-1855.

[52] R. Saxena, V.K. Singh, E.A. Kumar, Carbon Dioxide Capture and Sequestration by Adsorption on Activated Carbon, Energy Procedia, 54 (2013) 320-329.

[53] S.B. Sinnott, R. Andrews, Carbon Nanotubes: Synthesis, Properties, and Applications, Critical Reviews in Solid State and Materials Sciences, 26 (2001) 145-249.

[54] Y. Yan, J. Miao, Z. Yang, F.-X. Xiao, H.B. Yang, B. Liu, Y. Yang, Carbon nanotube catalysts: recent advances in synthesis, characterization and applications, Chemical Society Reviews, 44 (2015) 3295-3346.

[55] J. Dai, P. Giannozzi, J. Yuan, Adsorption of pairs of NOx molecules on single-walled carbon nanotubes and formation of NO + NO3 from NO2, Surface Science, 603 (2009) 3234-3238.

[56] M.M. Gui, Y.X. Yap, S.-P. Chai, A.R. Mohamed, Multi-walled carbon nanotubes modified with (3-aminopropyl)triethoxysilane for effective carbon dioxide adsorption, International Journal of Greenhouse Gas Control, 14 (2013) 65-73.

[57]   M. Rahimi, J.K. Singh, D.J. Babu, J.r.J. Schneider, F. Müller-Plathe, Understanding Carbon Dioxide Adsorption in Carbon Nanotube Arrays: Molecular Simulation and Adsorption Measurements, The Journal of Physical Chemistry C, 117 (2013) 13492-13501.

[58]   C.W. Tan, K.H. Tan, Y.T. Ong, A.R. Mohamed, S.H.S. Zein, S.H. Tan, Energy and environmental applications of carbon nanotubes, Environmental Chemistry Letters, 10 (2015) 265-273.

[59]   Y.T. Ong, A.L. Ahmad, S.H.S. Zein, S.H. Tan, A review on carbon nanotubes in an environmental protection and green engineering perspective, Brazilian Journal of Chemical Engineering, 27 227-242.

[60]   M. Cinke, J. Li, C.W. Bauschlicher Jr, A. Ricca, M. Meyyappan, CO2 adsorption in single-walled carbon nanotubes, Chemical Physics Letters, 376 (2003) 761-766.

[61]   F. Su, C. Lu, W. Cnen, H. Bai, J.F. Hwang, Capture of CO2 from flue gas via multiwalled carbon nanotubes, Science of The Total Environment, 407 (2009) 3017-3023.

[62]   N. Omidfar, A. Mohamadalizadeh, S.H.C.A.P.J.R. Mousavi, Carbon dioxide adsorption by modified carbon nanotubes, Asia-Pacific Journal of Chemical Engineering, 10 (2015) 885-892.

[63]   S. Khalili, A.A. Ghoreyshi, M.Jahanshahi, K.Pirzadeh, Clean Soil Air Water. 10/2013, CLEAN – Soil, Air, Water, 41 (2013) 935-938.

[64]   M. Rahimi, D.J. Babu, J.K. Singh, Y.-B. Yang, J.r.J. Schneider, F. Müller-Plathe, Double-walled carbon nanotube array for CO2 and SO2 adsorption, The Journal of Chemical Physics, 143 (2015) 124701.

[65]   M. Barberio, P. Barone, A. Imbrogno, F. Xu, CO2 adsorption on silver nanoparticle/carbon nanotube nanocomposites: A study of adsorption characteristics, physica status solidi (b), 252 (2015) 1955-1959.

[66]   M. Rahimi, J.K. Singh, F. Müller-Plathe, CO2 Adsorption on Charged Carbon Nanotube Arrays: A Possible Functional Material for Electric Swing Adsorption, The Journal of Physical Chemistry C, 119 (2015) 15232-15239.

[67]   R.Q. Long, R.T. Yang, Carbon Nanotubes as a Superior Sorbent for Nitrogen Oxides, Industrial & Engineering Chemistry Research, 40 (2001) 4288-4291.

[68]   A.I. Vasylenko, M.V. Tokarchuk, S. Jurga, Effect of a Vacancy in Single-Walled Carbon Nanotubes on He and NO Adsorption, The Journal of Physical Chemistry C, 119 (2015) 5113-5116.

[69] F. Sun, J. Gao, Y. Zhu, G. Chen, S. Wu, Y. Qin, Adsorption of SO2 by typical carbonaceous material: a comparative study of carbon nanotubes and activated carbons, Adsorption, 19 (2013) 959-966.

[70] M. Yoosefian, M. Zahedi, A. Mola, S. Naserian, A DFT comparative study of single and double SO2 adsorption on Pt-doped and Au-doped single-walled carbon nanotube, Applied Surface Science, 349 (2015) 864-869.

[71] M. Enterría, J.L. Figueiredo, Nanostructured mesoporous carbons: Tuning texture and surface chemistry, Carbon, 108 (2016) 79-102.

[72] C. Liang, Z. Li, S. Dai, Mesoporous Carbon Materials: Synthesis and Modification, Angewandte Chemie International Edition, 47 (2008) 3696-3717.

[73] H. Wang, F.L.Y. Lam, X. Hu, K.M. Ng, Ordered Mesoporous Carbon as an Efficient and Reversible Adsorbent for the Adsorption of Fullerenes, Langmuir, 22 (2006) 4583-4588.

[74] Y. Meng, D. Gu, F. Zhang, Y. Shi, L. Cheng, D. Feng, Z. Wu, Z. Chen, Y. Wan, A. Stein, D. Zhao, A Family of Highly Ordered Mesoporous Polymer Resin and Carbon Structures from Organic−Organic Self-Assembly, Chemistry of Materials, 18 (2006) 4447-4464.

[75] T.-Y. Ma, L. Liu, Z.-Y. Yuan, Direct synthesis of ordered mesoporous carbons, Chemical Society Reviews, 42 (2013) 3977-4003.

[76] C. Liu, M. Yu, Y. Li, J. Li, J. Wang, C. Yu, L. Wang, Synthesis of mesoporous carbon nanoparticles with large and tunable pore sizes, Nanoscale, 7 (2015) 11580-11590.

[77] K.M. Nelson, S.M. Mahurin, R.T. Mayes, B. Williamson, C.M. Teague, A.J. Binder, L. Baggetto, G.M. Veith, S. Dai, Preparation and CO2 adsorption properties of soft-templated mesoporous carbons derived from chestnut tannin precursors, Microporous and Mesoporous Materials, 222 (2016) 94-103.

[78] M. Wang, L. Yao, J. Wang, Z. Zhang, W. Qiao, D. Long, L. Ling, Adsorption and regeneration study of polyethylenimine-impregnated millimeter-sized mesoporous carbon spheres for post-combustion CO2 capture, Applied Energy, 168 (2016) 282-290.

[79] B. Yuan, X. Wu, Y. Chen, J. Huang, H. Luo, S. Deng, Adsorption of CO2, CH4, and N2 on Ordered Mesoporous Carbon: Approach for Greenhouse Gases Capture and Biogas Upgrading, Environmental Science & Technology, 47 (2013) 5474-5480.

[80] K.S. Lakhi, W.S. Cha, J.-H. Choy, M. Al-Ejji, A.M. Abdullah, A.M. Al-Enizi, A. Vinu, Synthesis of mesoporous carbons with controlled morphology and pore diameters from SBA-15 prepared through the microwave-assisted process and their CO_2 adsorption capacity, Microporous and Mesoporous Materials, 233 (2016) 44-52.

[81] K.S. Lakhi, W.S. Cha, S. Joseph, B.J. Wood, S.S. Aldeyab, G. Lawrence, J.-H. Choy, A. Vinu, Cage type mesoporous carbon nitride with large mesopores for CO_2 capture, Catalysis Today, 243 (2015) 209-217.

[82] C. Goel, H. Bhunia, P.K. Bajpai, Mesoporous carbon adsorbents from melamineâ€"formaldehyde resin using nanocasting technique for CO_2 adsorption, Journal of Environmental Sciences, 32 (2015) 238-248.

[83] C.-C. Huang, S.-C. Shen, Adsorption of CO_2 on chitosan modified CMK-3 at ambient temperature, Journal of the Taiwan Institute of Chemical Engineers, 44 (2013) 89-94.

[84] F. Cao, J. Chen, M. Ni, H. Song, G. Xiao, W. Wu, X. Gao, K. Cen, Adsorption of NO on ordered mesoporous carbon and its improvement by cerium, RSC Advances, 4 (2014) 16281-16289.

[85] J. Chen, F. Cao, S. Chen, M. Ni, X. Gao, K. Cen, Adsorption kinetics of NO on ordered mesoporous carbon (OMC) and cerium-containing OMC (Ce-OMC), Applied Surface Science, 317 (2014) 26-34.

[86] S. Gadipelli, Z.X. Guo, Graphene-based materials: Synthesis and gas sorption, storage and separation, Progress in Materials Science, 69 (2015) 1-60.

[87] M.J. Allen, V.C. Tung, R.B. Kaner, Honeycomb Carbon: A Review of Graphene, Chemical Reviews, 110 (2010) 132-145.

[88] W. Choi, I. Lahiri, R. Seelaboyina, Y.S. Kang, Synthesis of Graphene and Its Applications: A Review, Critical Reviews in Solid State and Materials Sciences, 35 (2010) 52-71.

[89] K.C. Kemp, H. Seema, M. Saleh, N.H. Le, K. Mahesh, V. Chandra, K.S. Kim, Environmental applications using graphene composites: water remediation and gas adsorption, Nanoscale, 5 (2013) 3149-3171.

[90] K.S. Novoselov, V.I. Falko, L. Colombo, P.R. Gellert, M.G. Schwab, K. Kim, A roadmap for graphene, Nature, 490 (2012) 192-200.

[91] A. Ghosh, K.S. Subrahmanyam, K.S. Krishna, S. Datta, A. Govindaraj, S.K. Pati, C.N.R. Rao, Uptake of H2 and CO2 by Graphene, The Journal of Physical Chemistry C, 112 (2008) 15704-15707.

[92] R. Kumar, V.M. Suresh, T.K. Maji, C.N.R. Rao, Porous graphene frameworks pillared by organic linkers with tunable surface area and gas storage properties, Chemical Communications, 50 (2014) 2015-2017.

[93] W. Huang, X. Zhou, Q. Xia, J. Peng, H. Wang, Z. Li, Preparation and Adsorption Performance of GrO@Cu-BTC for Separation of CO2/CH4, Industrial & Engineering Chemistry Research, 53 (2014) 11176-11184.

[94] V. Presser, J. McDonough, S.-H. Yeon, Y. Gogotsi, Effect of pore size on carbon dioxide sorption by carbide derived carbon, Energy & Environmental Science, 4 (2011) 3059-3066.

[95] M.E. Casco, M. MartÃnez-Escandell, J. Silvestre-Albero, F. RodrÃguez-Reinoso, Effect of the porous structure in carbon materials for CO2 capture at atmospheric and high-pressure, Carbon, 67 (2014) 230-235.

[96] L.-Y. Meng, S.-J. Park, Effect of exfoliation temperature on carbon dioxide capture of graphene nanoplates, Journal of Colloid and Interface Science, 386 (2012) 285-290.

[97] G. Ning, C. Xu, L. Mu, G. Chen, G. Wang, J. Gao, Z. Fan, W. Qian, F. Wei, High capacity gas storage in corrugated porous graphene with a specific surface area-lossless tightly stacking manner, Chemical Communications, 48 (2012) 6815-6817.

[98] J. Li, M. Hou, Y. Chen, W. Cen, Y. Chu, S. Yin, Enhanced CO2 capture on graphene via N, S dual-doping, Applied Surface Science, (2014).

[99] W. Xing, C. Liu, Z. Zhou, L. Zhang, J. Zhou, S. Zhuo, Z. Yan, H. Gao, G. Wang, S.Z. Qiao, Superior CO2 uptake of N-doped activated carbon through hydrogen-bonding interaction, Energy & Environmental Science, 5 (2012) 7323-7327.

[100] P. Tamilarasan, S. Ramaprabhu, Integration of polymerized ionic liquid with graphene for enhanced CO2 adsorption, Journal of Materials Chemistry A, 3 (2015) 101-108.

[101] D. Zhou, Q.-Y. Cheng, Y. Cui, T. Wang, X. Li, B.-H. Han, Grapheneâ€"terpyridine complex hybrid porous material for carbon dioxide adsorption, Carbon, 66 (2014) 592-598.

[102] Y. Zhao, H. Ding, Q. Zhong, Preparation and characterization of aminated graphite oxide for CO2 capture, Applied Surface Science, 258 (2012) 4301-4307.

[103] D.J. Babu, F.G. Kuhl, S. Yadav, D. Markert, M. Bruns, M.J. Hampe, J.J. Schneider, Adsorption of pure SO2 on nanoscaled graphene oxide, RSC Advances, 6 (2016) 36834-36839.

[104] S. Tang, Z. Cao, Adsorption of nitrogen oxides on graphene and graphene oxides: Insights from density functional calculations, The Journal of Chemical Physics, 134 (2011) 044710.

[105] A.S. Rad, E. Abedini, Chemisorption of NO on Pt-decorated graphene as modified nanostructure media: A first principles study, Applied Surface Science, 360, Part B (2016) 1041-1046.

[106] A.H. Chughtai, N. Ahmad, H.A. Younus, A. Laypkov, F. Verpoort, Metal-organic frameworks: versatile heterogeneous catalysts for efficient catalytic organic transformations, Chemical Society Reviews, 44 (2015) 6804-6849.

[107] H.-C. Zhou, J.R. Long, O.M. Yaghi, Introduction to Metal–Organic Frameworks, Chemical Reviews, 112 (2012) 673-674.

[108] D. Farrusseng, S. Aguado, C. Pinel, Metal–Organic Frameworks: Opportunities for Catalysis, Angewandte Chemie International Edition, 48 (2009) 7502-7513.

[109] O.K. Farha, I. Eryazici, N.C. Jeong, B.G. Hauser, C.E. Wilmer, A.A. Sarjeant, R.Q. Snurr, S.T. Nguyen, A.Ã.z.r. YazaydÄ±n, J.T. Hupp, Metal–Organic Framework Materials with Ultrahigh Surface Areas: Is the Sky the Limit?, Journal of the American Chemical Society, 134 (2012) 15016-15021.

[110] B.r. Arstad, H. Fjellv¥g, K.O. Kongshaug, O. Swang, R. Blom, Amine functionalised metal organic frameworks (MOFs) as¹ adsorbents for carbon dioxide, Adsorption, 14 (2008) 755-762.

[111] J.-R. Li, Y. Ma, M.C. McCarthy, J. Sculley, J. Yu, H.-K. Jeong, P.B. Balbuena, H.-C. Zhou, Carbon dioxide capture-related gas adsorption and separation in metal-organic frameworks, Coordination Chemistry Reviews, 255 (2011) 1791-1823.

[112] H. Dathe, Peringer E, R. V, J. A, L. JA, Metal organic frameworks based on Cu2þ and benzene-1,3,5-tricarboxylate as host for SO2 trapping agents, C R Chemie, 8 (2005) 753-763.

[113] C.A. Fernandez, P.K. Thallapally, R.K. Motkuri, S.K. Nune, J.C. Sumrak, J. Tian, J. Liu, Gas-Induced Expansion and Contraction of a Fluorinated Metalâ€"Organic Framework, Crystal Growth & Design, 10 (2010) 1037-1039.

[114] X.-D. Song, S. Wang, C. Hao, J.-S. Qiu, Investigation of SO2 gas adsorption in metalâ€"organic frameworks by molecular simulation, Inorganic Chemistry Communications, 46 (2014) 277-281.

[115] A.C. McKinlay, B. Xiao, D.S. Wragg, P.S. Wheatley, I.L. Megson, R.E. Morris, Exceptional Behavior over the Whole Adsorptionâ€"Storageâ€"Delivery Cycle for NO in Porous Metal Organic Frameworks, Journal of the American Chemical Society, 130 (2008) 10440-10444.

[116] G. Nickerl, M. Leistner, S. Helten, V. Bon, I. Senkovska, S. Kaskel, Integration of accessible secondary metal sites into MOFs for H2S removal, Inorganic Chemistry Frontiers, 1 (2014) 325-330.

[117] C. Petit, B. Mendoza, T.J. Bandosz, Hydrogen Sulfide Adsorption on MOFs and MOF/Graphite Oxide Composites, ChemPhysChem, 11 (2010) 3678-3684.

[118] H. Brauer, Y.B.G. Varma, Design and Operation of Adsorption Equipment, Air Pollution Control Equipment, Springer Berlin Heidelberg, Berlin, Heidelberg, 1981, pp. 307-333.

[119] J.A. Delgado, M.a.A. Uguina, J.L. Sotelo, B. Ruíz, J.M. Gálmez, Fixed-bed adsorption of carbon dioxide/methane mixtures on silicalite pellets, Adsorption, 12 (2006) 5-18.

[120] C.A. Grande, Advances in Pressure Swing Adsorption for Gas Separation, ISRN Chemical Engineering, 2012 (2012) 13.

[121] P.R. Mhaskar, A.S. Moharir, Heuristics for synthesis and design of pressure-swing adsorption processes, Adsorption, 18 (2012) 275-295.

[122] F.G. Wiessner, Basics and industrial applications of pressure swing adsorption (PSA), the modern way to separate gas, Gas Separation & Purification, 2 (1988) 115-119.

[123] A. Andersen, S. Divekar, S. Dasgupta, J.H. Cavka, Aarti, A. Nanoti, A. Spjelkavik, A.N. Goswami, M.O. Garg, R. Blom, On the development of Vacuum Swing adsorption (VSA) technology for post-combustion CO2 capture, Energy Procedia, 37 (2013) 33-39.

[124] M.-W. Yang, N.-c. Chen, C.-h. Huang, Y.-t. Shen, H.-s. Yang, C.-t. Chou, Temperature Swing Adsorption Process for CO2 Capture Using Polyaniline Solid Sorbent, Energy Procedia, 63 (2014) 2351-2358.

[125] A. Ntiamoah, J. Ling, P. Xiao, P.A. Webley, Y. Zhai, CO2 Capture by Temperature Swing Adsorption: Use of Hot CO2-Rich Gas for Regeneration, Industrial & Engineering Chemistry Research, 55 (2016) 703-713.

[126] R.P.P.L. Ribeiro, C.A. Grande, A.E. Rodrigues, Electric Swing Adsorption for Gas Separation and Purification: A Review, Separation Science and Technology, 49 (2014) 1985-2002.

[127] B.-C. Chiang, M.-Y. Wey, C.-L. Yeh, Control of acid gases using a fluidized bed adsorber, Journal of Hazardous Materials, 101 (2003) 259-272.

[128] C.R. Mohanty, S. Adapala, B.C. Meikap, Removal of hazardous gaseous pollutants from industrial flue gases by a novel multi-stage fluidized bed desulfurizer, Journal of Hazardous Materials, 165 (2009) 427-434.

[129] S. Onitsuka, M. Ichiki, T. Watanabe, Method of removing NOx by adsorption, NOx adsorbent and apparatus for purifying NOx-containing gas, Google Patents, 1992.

[130] W.G. Matthews, H.C. Shaw, Selective adsorption of NOx from gas streams, Google Patents, 1979.

Chapter 9

Removal of arsenic from water through adsorption onto metal oxide-coated material

Sharf Ilahi Siddiqui and Saif Ali Chaudhry*

Department of Chemistry, Jamia Millia Islamia, New Delhi-110025, India

*saifchaudhary09@gmail.com

Abstract

Arsenic, a metalloid having a terrible impact on human health, is threatening the world continuously. It has become a curse for more than 20 countries and can be gauged from the fact that over 100 million peoples of Bangladesh and Bengal province of India are consuming arsenic contaminated ground water having concentration more than the WHO permissible limit. Traditional techniques used for removal of both forms of arsenic were not suitable for the As(III) form. Adsorption is a preferable method and different types of solid materials particles being used are fine powders that are difficult to separate from water. Moreover, fine powders cannot be used in column applications because of their low hydraulic conductivity. That is why; researchers have proposed metal oxide-coated media for use in adsorption technique. Recently, researchers have developed metal oxide coated adsorbents including, sand, natural rock, ceramic materials, activated carbon, perlite, zeolite, and some organic polymers being used as surface materials. In this review, most of the valuable literature available on arsenic remediation by adsorption using coated adsorbents and existing purification methods for drinking water; wastewater; industrial effluents, and technological solutions for arsenic have been listed. Herein, arsenic sorption by different coated materials surveyed and their sorption efficiencies have been compared. Arsenic adsorption behaviour in presence of other impurities and their separation/regeneration techniques has also been discussed.

Keywords

Heavy Metal, Arsenic, Arsenic Effect, Arsenic Remediation, Adsorption, Iron Oxide

Contents

1. Introduction

1.1   Arsenic: forms, exposure and effects

In 8th century, an Arabian, Geber discovered arsenic when he heated orpiment, As_2S_3, and its name was derived from the Greek word arsenikon meaning potent [1]. Arsenic is a metalloid having atomic number 33, mass 74.9216 $gmol^{-1}$, melting point 817 °C, boiling point 613 °C [2]. It belongs to the VA group with 5 valence electrons and electronic configuration of $3d^{10} 4s^2 4P^3$. Arsenic is 20th most abundant element in the earth crust and over 200 minerals contain it [3]. Mostly, in these minerals arsenic is found in association with various transition metals. Arsenic chemistry resembles with sulphide chemistry and the concentration of arsenic in sulphide minerals is high because it substitutes sulphur in various minerals. The principal minerals are pyrite, nicolite, cobaltite, tennatite and enargite [4]. Selective arsenic minerals and their composition are given in **Table 1**. The concentration of arsenic in most rocks ranges from 0.5 to 2.5 mg/kg and varies in different parts of the earth (**Table 2, 3**). In the terrestrial environment, concentration ranges from 1.5-3.0 mg/kg and is higher in finer grained argillaceous sediments and phosphorites [5]. Arsenic is mobilised into water through weathering reaction, biological activities, volcanic emissions, and other anthropogenic activities. However, the environmental arsenic problem is due to mobilisation under natural condition. In few places, arsenic concentration is extra high because of natural activities like mining, smelting, fossil fuel combustion and arsenic-containing pesticide and herbicides etc. Soil erosion alone contributes 612×10^8 g/year and leaching contributes 2380×10^8 g/year of arsenic in dissolved and suspended forms in the oceans [6]. About 40 % arsenic released in the atmosphere by anthropogenic activities come from wood preservation, pesticides, herbicides, steel production, lead and zinc smelting and incineration [7]. Arsenic formulated pesticides put directly into the soil through runoff contaminate water also. Generally, arsenic exists in four oxidation states; -3, 0, +3 and +5 that are represented as arsenide (As^{3-}), elemental arsenic (As^0), arsenite (As^{III}) and arsenate (As^V), respectively [8].

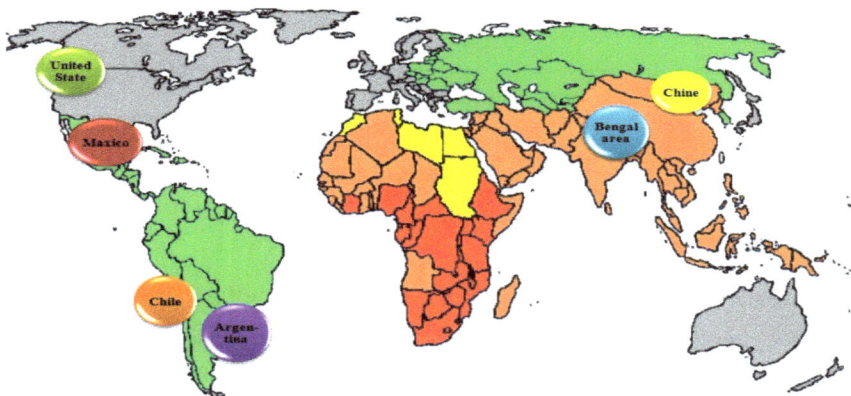

Figure 1: Selective countries having high arsenic concentration [158-164].

Arsenic after dissolution in water exists as H_3AsO_3 and H_3AsO_4 acids and under deprotonation get converted to different conjugate bases depending on pH conditions [9].

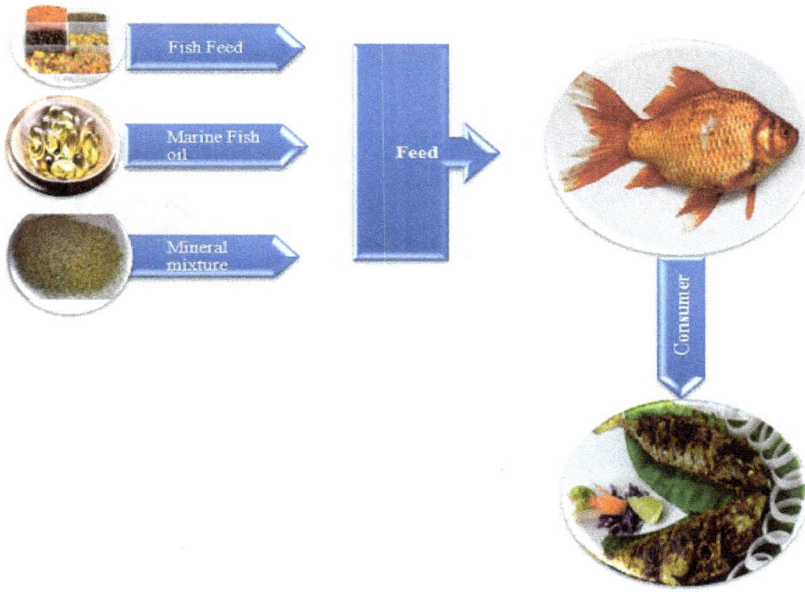

Figure 2: Possible routes of arsenic exposure to marine feeding stuff.

As(III) acts as hard acids and have a tendency to bind with oxygen and nitrogen atoms and conversely, As(V) behaves like a soft acid, form compounds with sulphur and phosphorus [10]. Arsenic compounds with sulphur, oxygen, nitrogen and metals are known as inorganic arsenic and those with carbon and hydrogen are referred to as organic arsenic. The form of arsenic in water depends upon pH and redox potential. At high or moderate redox potential, arsenic becomes stable as pentavalent oxyanions, H_3AsO_4, $H_2AsO_4^-$, $HAsO_4^{-2}$ and AsO_4^{-3}, whereas under reducing condition and low redox potential, the trivalent form, H_3AsO_3, is predominant species [8]. Trivalent arsenic, AsO_3^{3-}, and pentavalent arsenic, AsO_4^{3-}, are most preferable inorganic forms present in water, are referred to as As(III) and As(V), respectively. Some of the organic arsenic species found in water in fewer amounts are arsenobetain (AsB), monomethylarsonic acid (MMA), dimethylarsinic acid (DMA), tetramethylarsonium ion (TMAI), trimethylarsine oxide (TMAO), and arsenochloline (AsC) [11].

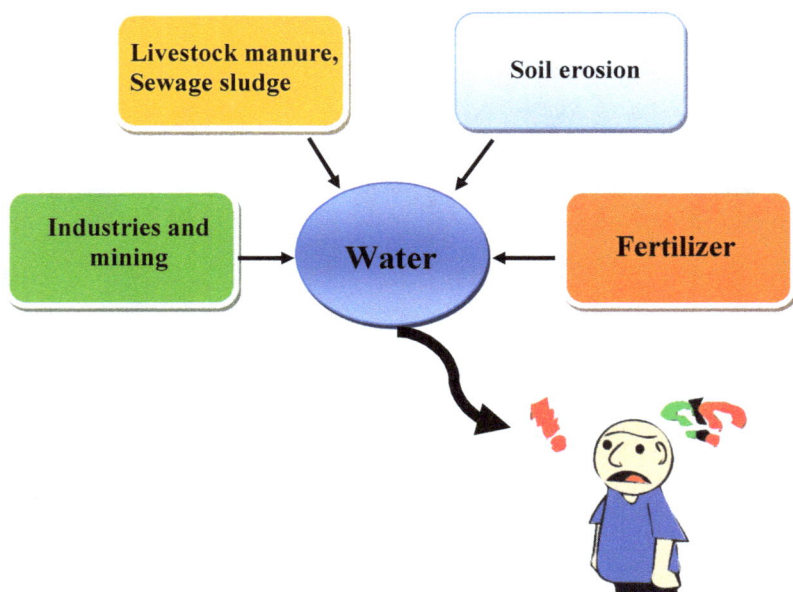

Figure 3: Possible routes of exposure to arsenic.

Different species of arsenic in water are given in **Table 4**. Unidentified arsenic has also been investigated in seawater and freshwater [12]. Generally, arsenic is found at low concentration in natural water, its level range between 1-2 mg/L, and may be up to 12 mg/L in areas containing natural sources of arsenic such as geothermal fields, uranium and gold mining areas [15, 16]. The occurrence of much higher arsenic concentration in groundwater has been reported from many parts of the world and is given in **Table 7**.

Groundwater of Argentina, Chile, Mexico, USA, Hungary, Vietnam, China, Bangladesh and adjoining Bengal province of India, etc. have been reported having elevated arsenic concentration. Approximately, 150 million people are affected globally and newly affected areas are continuously discovered [17]. **Figure 1** is showing the most affected areas having arsenic contaminated water. **Figure 2 and 3** show the possible routes of arsenic exposure for marine feeding stuff and human being [18].

Arsenic is a well-known carcinogen and considered one of the world's most hazardous chemicals. Exposure to excessive arsenic for 5-10 years through drinking water and food result arsenicosis, a name used for arsenic-related health problems. Diseases caused by arsenic include changes of skin pigmentation, gastrointestinal disturbances, bladder, kidney, and lung cancers, muscular weakness and neurological changes in humans. Arsenic toxicity has no effective medicine for treatment, and only arsenic-free water can reduce the sufferings of people. The harmful effects of arsenic in the human body are given in **Table 8** and **Figure 4, 5**. The inorganic form of arsenic is about 100 times more toxic than organic arsenic. Among inorganic arsenic, As(III) is about 60 times more toxic than As(V) [13]. As(III) have greater cellular uptake and ability to bind to sulfhydryl sites of proteins and enzymes that divert the functioning of protein and enzymes which are responsible for the hazardous effect on human body [14].

Currently, in Bangladesh and six districts of the Indian province of Bengal, about 100 million people are affected by arsenic poisoning caused due to drinking underground water [19]. About 21.3% of deaths in Bangladesh had been attributed to arsenic poisoning in drinking water [20]. To avoid these dangerous effects towards animals, plants and humans, efforts are being made to minimise these hazardous effects. For this purpose, it is necessary to document the levels of arsenic in drinking water, and its chemical speciation for establishing regulatory standards and guidelines. **Table 9** shows different methods developed for the measurement of concentration and speciation of arsenic and arsenic compounds in various water samples.

Figure 4: Possible effect of arsenic towards human body.

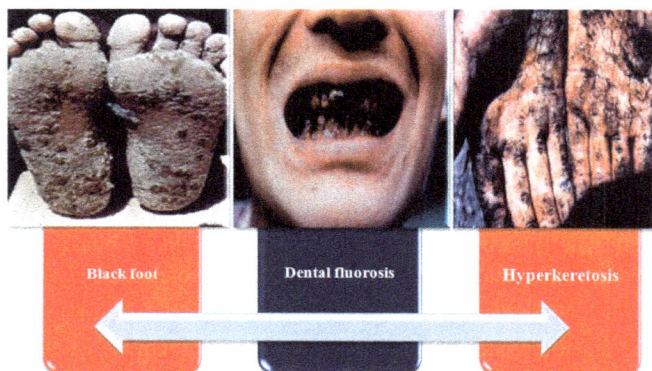

Figure 5: Diseases caused by arsenic exposure.

Various environmental agencies have established guidelines for drinking water quality to ensure public health and safety as given in **Table 5**. Most of the countries have adopted WHO provisional guideline of 0.01 mg/L standard and few still have retained the earlier WHO guideline of 0.05 mg/L as their national standard as given in **Table 6**.

1.2   Arsenic remediation from water

The removal of arsenic from water is the best way to save millions of people across the world. Treatment methods such as ion exchange, coagulation, precipitation, membrane separation, reverse osmosis etc. shown in **Figure 6,** are usually applied for arsenic removal. The detail of these treatment methods is given in **Table 10**. The estimation of arsenic levels has been discussed in the literature [21]. After the determination of concentration and its chemical speciation, it is essential to remove this toxic element for making water suitable for drinking purpose. Various removal techniques discussed above, however, some of these are effective only for As(V) and not for As(III). As(III) is about 60 times more toxic than As(V), so it is very necessary to remove As(III) effectively [22]. However, adsorption has been proved as an efficient technique for both As(V) and As(III) removal. Adsorption technique has significant advantages such as cost-effectiveness, effective at low arsenic concentration, highly selective, high efficiency, ease of handling, minimum chemical or biological sludge regeneration [23].

Figure 7 depicts the advantages of adsorption over other treatment methods. Sorption behaviour largely depends on pH of water and concentration, and adsorption technique is better capable of removing arsenic from a wide range of pH and effective even at low concentration [24]. On the basis of this reason, here in this review adsorption process is discussed.

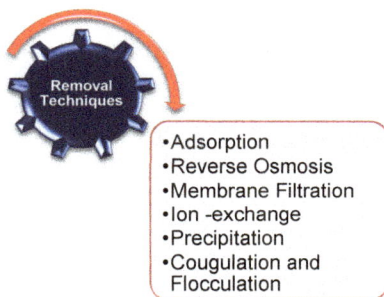

Figure 6: Water treatment techniques.

2.   Arsenic remediation through adsorption

Adsorption is a process that occurs when a liquid solute (arsenic) accumulates on a solid adsorbent, forming a molecular or atomic film around the surface. In this process, an arsenic ion from water interacts with the surface of a solid having energy-rich sites. The electronic and spatial property makes the adsorbent surface energetically heterogeneous and suitable for adsorption [25, 26].

Figure 7: Advantages of adsorption process in arsenic water treatment.

2.1 Adsorption analysis

Generally, adsorption processes are analysed in two ways, batch adsorption analysis and column adsorption analysis.

2.1.1 Batch adsorption analysis

To determine the capacity of an adsorbent for a solute (arsenic) adsorption, mostly batch experimental method is used. The batch experiments are carried out with a series of Erlenmeyer flasks containing a solute solution (arsenic) and a definite amount of an adsorbent. These flasks are agitated mechanically in a thermostatic water bath at fixed temperature for a fixed interval of time. The solutions are then centrifuged and the supernatants analysed for unadsorbed arsenic in the solution. The adsorption data obtained from batch experiments is used to calculate the equilibrium uptake capacity and percentage adsorption by following relationships:

$$Q_e = (C_o - C_e)\frac{V}{m}$$

Percentage adsorption $= \left(\frac{C_o - C_e}{C_o}\right)100$

Where Q_e is the amount of arsenic adsorbed onto m weight of an adsorbent at equilibrium. V is volume; C_o and C_e are the initial and equilibrium concentrations of arsenic solution.

Arsenic concentration, pH of water, a dose of the adsorbent, contact time and temperature affect the adsorption and are optimised for maximum uptake. Then experimental data, at different temperatures, is fitted to various isotherm and kinetic models to apply the process on a practical level. Adsorption data is also fitted to thermodynamic relationships to judge spontaneity and mechanism of the process.

2.1.1.1 Adsorption isotherm modelling

The fitting of adsorption data to an isotherm is important for both theoretical and practical applications. The correlation of equilibrium data is necessary for optimisation of adsorption system. Langmuir, Freundlich and Dubinin-Radushkevich isotherm equations are generally used to test the experimental sorption data. The parameters obtained from these models at particular temperatures, provide important information related to the adsorption mechanisms, surface properties and affinities of the adsorbent for adsorbate.

2.1.1.1.1   Langmuir adsorption isotherm

Langmuir adsorption isotherm is the relationship between the number of active sites on adsorbent surface and concentrations of adsorbate (arsenic) species in the water at equilibrium [27]. Arsenic ion chemically binds to the sites, available on the surface of adsorbent, which is energetically equivalent. Once adsorbed, then the ion does not interact with other ions and thus only homogeneous monolayer is formed on the surface [28]. The linearized mathematical form of Langmuir isotherm is given below:

$$\frac{C_e}{Q_e} = \frac{C_e}{Q_o} + \frac{1}{Q_o b}$$

Where Q_o (mg/g) is maximum monolayer adsorption capacity of an adsorbent that represents practical adsorption capacity and is a useful quantity to compare the performance of different adsorbents. b is a Langmuir constant that is related to the energy of adsorption. Both Q_o and b are evaluated from the slope and the intercept, respectively, of the isotherm plot. Langmuir constant is used to calculate an essential dimensionless constant called separation factor, R_L, given as follows [28]:

$$R_L = \frac{1}{(1 + b C_e)}$$

R_L value describes the feasibility of adsorption process. If $R_L > 1$, the adsorption process would be unfavourable; for $R_L = 0$, the adsorption would be irreversible, whereas $0 < R_L < 1$ indicates the energetically favourable adsorption process.

2.1.1.1.2   Freundlich isotherm

Freundlich isotherm is applied to an adsorption system when the number of adsorption sites exceeds the number of contaminant molecules or ions (arsenic). According to this isotherm, each site has specific binding energy and site with higher binding energy is occupied first and adsorption energy decreases exponentially upon the completion of the adsorption process. The isotherm, thus, describes multilayer physical adsorption over the heterogeneous surface [28]. The linearized form of Freundlich isotherm is given by the following equation:

$$\log Q_e = \log k_F + \frac{1}{n} \log C_e$$

Where, k_F [(mg/g)(L/mg)$^{1/n}$] and n are Freundlich constants, which represent the quantity of adsorbed arsenic from a unit concentration, a relative adsorption capacity of an adsorbent, and the exponent which measure adsorption intensity or surface heterogeneity, respectively [28]. For any favourable adsorption process, the numerical values of n lie in the range 1-10. The intercept and the slope of $logQ_e$ versus $logC_e$ plot give values of $logk_F$ and $1/n$ from which k_F and n can be obtained.

2.1.1.1.3   Temkin isotherm

For heterogeneous adsorbent and non-ideal liquid adsorbate (arsenic), Temkin isotherm is applied which is based on the assumption that adsorbate-adsorbent interaction decreases the heat of adsorption with the coverage of surface [29]. The linearized form of isotherm is:

$$Q_e = \frac{RT}{b_T} \ln A_T + \frac{RT}{b_T} \ln C_e$$

Where, A_T (L/g) and b_T (kJ/mol) are constants which are related to the maximum binding energy and heat of adsorption, respectively. These constants are calculated from the slope and intercept of the plot Q_e vs lnC_e, respectively.

2.1.1.1.4   Dubinin-Radushkevich isotherm

Dubinin-Radushkevich (D-R) isotherm can be utilised for knowing the type of adsorption. The linearized form of the D-R isotherm is:

$$\ln Q_e = \ln Q_{D-R} - \beta \varepsilon^2$$

Where, Q_{D-R} (mg/g) is Dubinin-Radushkevich theoretical monolayer saturation capacity, and β (mol^2/J^2) is a constant which is related to the mean free energy of adsorption. Polyani potential, ε, which is related to the equilibrium concentration, C_e, is given by the following relationship: represents

$$\varepsilon = RT \ln\left(1 + \frac{1}{C_e}\right)$$

Where, R (8.314 J/K/mol) and T are universal gas constant and temperature in Kelvin scale, respectively. The value of β and Q_{D-R} can be obtained from the slope and intercept of the straight line plot between lnQ_e vs ε^2, respectively.

2.1.1.2   Thermodynamics of adsorption

Langmuir constant, b, is used for calculation of thermodynamic parameters [30]. Most important of the thermodynamic parameter, Gibbs free energy change, $\Delta G°$ (kJ/mol), of adsorption is related to Langmuir constant:

$$\Delta G^{\circ} = -RT \ln b$$

A negative value of free energy change, ΔG°, indicates the spontaneity of adsorbate (arsenic) adsorption onto adsorbent at the test temperature. When the magnitude of ΔG° does not change with the change in temperature then the adsorption spontaneity would be independent of temperature. Low magnitude of energy may be due to the formation of bidentate complexes by arsenic with the adsorbent surface, which is a slower process. ΔG° in the range -20 - 0 kJ/mol indicates adsorptions of arsenic onto adsorbent as a physical process. ΔG° is also related to change in entropy, ΔS°, and change in enthalpy, ΔH°, as shown by the following relationships:

$$\Delta G^{\circ} = \Delta H^{\circ} - T\Delta S^{\circ}$$

Intercept and slope of ΔG° versus T plot for arsenic give the values of ΔH° and ΔS°, respectively. Negative and positive values of ΔH° indicate exothermic and endothermic nature of adsorption process, respectively. ΔS° value indicates the entropy change depending upon the sign and magnitude.

2.1.1.3 Adsorption kinetic studies

Kinetics study determines the efficiency and mechanism of the adsorption process. The kinetic parameter, rate constant etc., values depend upon the physical and chemical characteristics of adsorbent and adsorbate (arsenic). Adsorbate (arsenic) gets transferred from aqueous phase to solid phase through several steps. One or combination of these steps can control the rate of the whole process. In order to investigate the mechanism of arsenic adsorption onto adsorbent, the experimental kinetic data is fitted to various kinetic models including pseudo-first order and pseudo-second order models etc.

2.1.1.3.1 Pseudo-first order kinetic model

The first kinetic model used for the study of sorption in the liquid-solid system was proposed by Lagergren and later on modified by Ho is called pseudo-first order model [31]. The linearized mathematical form of Lagergren model is given by the following equation:

$$\log(Q_e - Q_t) = \log Q_e - \frac{k_1}{2.303}t$$

Where k_1(min^{-1}) is pseudo-first order rate constant. Q_t is the amount of arsenic adsorbed onto m weight of an adsorbent at any time t. These parameters can be calculated from the straight line plot between $log(Q_e-Q_t)$ and t at any specific temperature. On the basis of discrepancies in the theoretical and experimental Q_e values, pseudo-first order kinetic model is accepted or rejected.

2.1.1.3.2   Pseudo-second order kinetic model

The pseudo-second-order kinetics suggests that the number of adsorption sites on the adsorbent surface and the number of arsenic ions in the liquid phase both determine the rate of adsorption [32]. The chemical bond formation between adsorbate (arsenic) and appropriate sites on the adsorbent surface is the rate-limiting step. Linearized form of pseudo-second order kinetic model is represented by the following expression:

$$\frac{t}{Q_t} = \frac{1}{k_2 Q_e^2} + \frac{t}{Q_e}$$

or

$$\frac{t}{Q_t} = \frac{1}{h} + \frac{t}{Q_e}$$

Where $h=k_2 Q_e^2$ is initial rate constant and k_2 the overall rate constant for pseudo-second order adsorption reaction. Various constants related to pseudo-second order model for arsenic adsorption can be calculated from the plot between t/Q_t vs t.

2.1.1.3.3   Elovich model

Zeldowitsch [33] developed a kinetic model called Elovich kinetic model which assume that actual solid adsorbent surfaces are energetically heterogeneous and no lateral interaction takes place between the adsorbed solute. Linearized form of Elovich model is represented by the following expression:

$$Qt = 1/\beta \ln(\alpha\beta) + 1/\beta \ln t$$

Where, α (µg/g/min) and β (g/µg/min) are Elovich coefficients that represent the initial adsorption rate and rate of desorption coefficient, respectively. β is related to the extent of surface coverage and activation energy for chemisorption. The plot between the Q_t and $ln(t)$ gives the value of constants, α and β, from slope and intercept of the plot, respectively.

2.1.1.4   Mechanism of adsorption

Generally, the rate of liquid-solid adsorption process depends on intra-particle diffusion, film diffusion, and mass action. However, in a physical process, adsorption does not depend on the mass action step due to high speed associated.

2.1.1.4.1   Intra-particle diffusion

In intra-particle diffusion process a molecule or ion of solute (arsenic) is adsorbed on the solid surface, get diffused into the interior pores on the adsorbent surface and form a chemical or physical bond. If it is a slow step or rate determining step then solute uptake varies proportionally with $t^{0.5}$ [34]. The transportation of solute molecule (arsenic) from the solution to the surface of the adsorbent can be analysed by Weber and Morris intraparticle diffusion model [35]. The mathematical form of intraparticle diffusion model is given by the following equation:

$$Q_t = k_{ipd} t^{0.5} + C$$

If intra-particle diffusion is the rate determining step then plot between Q_t vs $t^{0.5}$ yield a straight line that passes through the origin.

2.1.1.4.2   Film diffusion

In this case, the liquid film is formed on the adsorbent surface and plays an important role during the adsorption process [36]. Linearized form of film diffusion model is represented by the following expression:

$$\ln(1 - F) = -k_{fd} t$$

Where $F = Q_t/Q_e$ is the fractional attainment of equilibrium and k_{fd} is film diffusion rate constant. The plot between the $ln(1-F)$ vs t determines the film diffusion step and give a straight line which passes through the origin.

2.1.2   Fixed-bed column adsorption study

Application of an adsorbent is well examined via column tests. The column efficiency depends on the characteristics of adsorbate (arsenic), adsorbent, the length of the column, adsorbent particle size, adsorbate (arsenic) concentration, flow rate and pH of the solution [37]. Column adsorption experiments may be carried out to judge the effect of flow rate on the adsorption of solute onto adsorbents. The column experiment may be performed using a burette type column with different inner diameter and height. The bottom of the column is filled with a specific amount of adsorbent on a glass wool support to yield a definite bed height. A solute (arsenic) solution of the given concentration is fed in a down-flow manner in column and effluents are collected at particular flow rates (mL/min). The concentration of solute in the affluent solution is analysed. The breakthrough curves can be obtained by plotting C_t/C_o vs t.

2.1.2.1   Thomas's kinetic model

Thomas kinetic model is derived from Langmuir isotherm and second order reversible reaction kinetics. The model is suitable where external and internal diffusion limitations are absent and is applied to predict the dynamic behaviour of column [38]. The mathematical form of Thomas model is given below:

$$\frac{C_t}{C_o} = \frac{1}{1 + \exp[(k_{Th}Q_{Th}w/z) - k_{Th}C_o t]}$$

Where, k_{Th} (mL/min/mg) and Q_{Th} (mg/g) are Thomas rate constant and adsorption capacity of adsorbent, respectively. C_o and C_t (mg/L) are inlet and outlet concentrations at any time t, respectively. w is the weight of adsorbent (g) and z is flow rate (mL/min). The linearized form of Thomas model is expressed as:

$$\ln\left(\frac{C_o}{C_t} - 1\right) = \frac{k_{Th}Q_{Th}w}{z} - k_{Th}C_o t$$

The slope and the intercept of the straight line plot between $ln(C_o/C_t - 1)$ versus t give k_{Th} and Q_{Th} values [39].

2. Nature of the adsorbent for removal of arsenic

Adsorption capacity generally depends on the nature of the solid surface of the adsorbent and nature of solute. Generally, adsorption is a surface phenomenon and high surface area shows higher adsorption capacity. The characteristics of an adsorbent required for significant removal of arsenic are given in **Figure 8**.

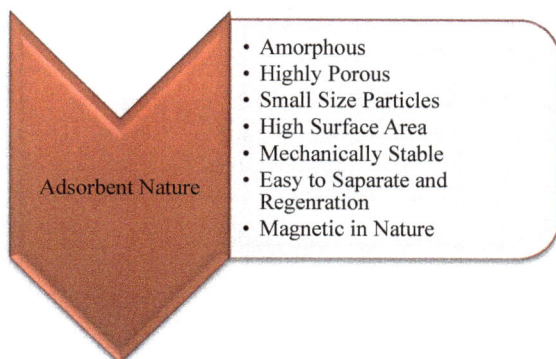

Adsorbent Nature

- Amorphous
- Highly Porous
- Small Size Particles
- High Surface Area
- Mechanically Stable
- Easy to Saparate and Regenration
- Magnetic in Nature

Figure 8: Suitable Adsorbent for efficient removal of arsenic.

Different types of adsorbents are being used since ancient times. Activated carbon is the most widely used adsorbent which is a highly porous and amorphous solid consisting of microcrystallites with a graphite lattice [40]. Activated carbon is usually prepared in small pellets or a powder form but its generation is difficult. In addition to carbon, some other widely used adsorbents used for arsenic adsorption are clay minerals, biomaterials, industrial solid waste, and zeolites etc. [41]. But now a day, metal oxide micro and nanoparticles are being frequently used for water treatment due to their large surface area and porous structures. Both As(III) and As(V) can strongly adsorb on solid metal oxide [42]. The high surface area to mass ratio, high surface reactivity, and unique catalytic activities are the most important properties of a metal oxide micro and nanoparticles which increase efficiency as an adsorbent as compare to the macro sized material. A number of reports showed that micro and nano-sized metal oxides are more effective in comparison to their macro-sized particles [43]. Nanoparticles of metal oxides such as Iron oxides, Titanium oxide, Aluminium oxide, Copper oxide and Zirconium oxide have been used in water treatment processes [44, 45]. Mayo et al. **[46]** and Tuu et al. **[47]** have used magnetite Fe_3O_4 nanoparticles for the removal of arsenic from water with a highest level of adsorption capacity for both As(III) and As(V). However, the separation of exhausted metal oxide particles after adsorption is a difficult job and requires costly instrumentation. In the meantime, the novel invention in adsorption technology is the use of magnetic nano-sized metal oxide as an adsorbent.

3. Support coated adsorbent

Although, adsorption capacity increases on enlarging surface area but the separation of smaller size particles is not easy, therefore, another way of increasing surface area is coating these smaller particles on supporting material. Moreover, the metal oxides are usually fine powders and not suitable for use in column applications because of their low hydraulic conductivity [48]. That is why; researchers have proposed metal oxide-coated media as an emerging technology for arsenic removal at small scale water facilities. Different metal oxide coated adsorbents with advantages in terms of hydraulic conductivity, cost and applications sustainability have been developed. Iron-oxide-coated sand [49], iron-manganese oxide coated kaolinite [50], nanoscale iron-manganese oxide-loaded zeolite [51] and zirconium oxide-coated sand [52] have been studied.

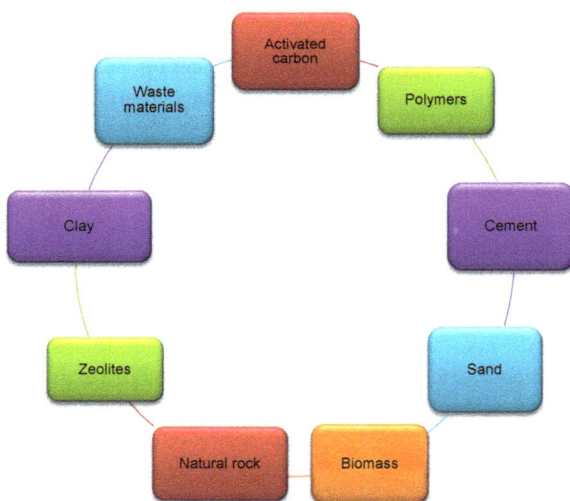

Figure 9: Supporting materials used for coating with metal oxide.

Moreover, these days, researchers are searching for more cost-effective, porous and easily available materials where the small sized particles (nanomaterials) could be applied. Activated carbon, Sand, ceramic materials, zeolite and some organic polymers have recently been used as surface materials for coating are given in **Figure 9.** Metal oxides, preferably, iron oxide micro or nano sized particles have been coated on these surface materials and their adsorption capacities were illustrated. Under these circumstances, nanomaterials loaded or embedded or coated materials would be promising adsorbents in reducing environmental problems and arsenic removal application as well.

3. Metal oxide coated materials for arsenic removal

The removal performance and mechanism of metal oxide coated materials for both ionic and molecular forms of arsenic were described by Jovanovic et al. [53]. The authors revealed that adsorption was a multistage process involving both macropore and intraparticle diffusions. Iron oxide coated materials have exhibited higher affinity towards As(III) as shown through maximum adsorption capacity and Gibbs free energy, and the process was spontaneous and favourable. Langmuir and Freundlich's isotherms indicated physical and chemical nature of adsorption bonding, whereas, Dubinin-Radushkevich isotherm revealed physical process. The literature available on coated materials used for the removal of both forms of arsenic was surveyed, and detail studies are given herein.

5.1 Metal oxide coated sand

The metal oxide coated sand has been used and showed a good result for As(III) and As(V) removal [49, 54]. **Figures 10** and **11** are showing the schematic adsorption of arsenic onto the metal oxide coated sand. **Figure 13** is showing the schematic process of arsenic removal from water with iron oxide coated sand. Among various metal oxides, iron oxides are most frequently coated on the sand surfaces due to their characteristics and low cost. Adsorption efficiency of iron oxide coated sand with uncoated sand has been compared and revealed that maximum adsorption of As(III) for coated sand was found to be much higher (0.0285 mg/g) than that for uncoated sand (0.00563 mg/g) at pH 7.5 in 2 h contact time [55]. The maximum As(III) removal efficiency achieved was 99 % for coated sand with an adsorbent dose of 20 g/L and solution having an initial concentration of 100 µg/L in a batch experiment. The adsorption capacity of iron oxide coated sands, IOCS, in batch mode was performed at pH 7.0 ± 0.4, for As(V) adsorption onto IOCS surface [56]. The removal efficiencies of total arsenic were in order, As(V)>As(V) + As(III)>As(III). Moreover, adsorption isotherms for As(V) and As(III) fitted well to Langmuir model satisfactorily for the four different initial pH conditions and initial arsenic concentrations. Fe-oxyhydroxide coated sand was used for investigating the efficiency of adsorption of As(III) and As(V) [57]. The Fe-oxy hydroxide coating was confirmed to be ferrihydrite with EXAFS spectroscopy.

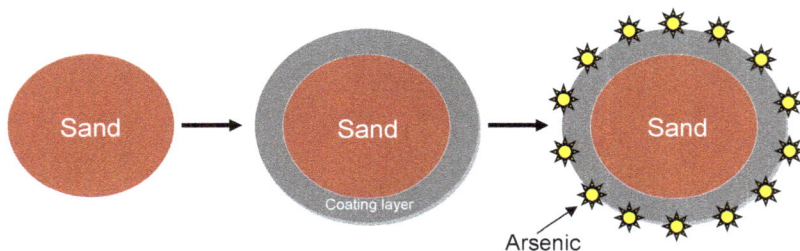

Figure 10: Schematic diagram of iron oxide coated sand and adsorption of metal ions onto coated sand.

This adsorbent exhibited extremely fast adsorption of As(III) and As(V), and As(V) adsorption was affected by changes in pH. At slightly acidic pH, substantially higher amounts of As(V) was adsorbed, but As(III) adsorption did not depend on pH. The Fe-content of the coated sand also affected the arsenic adsorption, higher Fe content on sand grains, causes higher adsorption of arsenic. Rahman et al. [58] proposed a new technique for the treatment of spent iron-oxide coated sand (IOCS) from filters used in arsenic removal. Chelant-washing of the arsenic-loaded IOCS was combined with the solid phase extraction treatment for the regeneration. The process was cost-effective including recycling of the solvent and decontamination of the spent arsenic-rich sludge. Similarly, comparative adsorption studies of iron oxide-coated sand (IOCS), manganese oxide-coated sand (MOCS), and iron-manganese oxide coated sand (IMOCS) was carried out.

Figure 11: Possible binding of arsenic (III) and (V) onto the iron oxide coated materials.

Huang et al. [59] revealed that all these adsorbent removed As(III), significantly, however, capacity of IMOCS due to oxidation of As(III) to As(V) was much better than

IOCS and MOCS. Kinetics followed pseudo-second-order rate and adsorption data fitted well to Langmuir as well as Freundlich isotherms for all these adsorbents.

5.2   Metal oxide coated natural rock

Natural rocks were used as surface material for metal oxide coating and showed better result for arsenic removal. The adsorption activity of iron oxide coated natural rock, IOCNR, towards arsenic was investigated in batch experiment at room temperature [60]. 75 % of total arsenic removal was observed from sample with 5 g/L dose of IOCNR, when the initial arsenic concentration and pH were 0.04 mg/L and 7.5, respectively. The solution was agitated for contact time of 6 h with speed of 180 rpm. Adsorption process followed Langmuir isotherm and maximum adsorption capacity of IOCNR was found 0.36 mg/g. The adsorption process followed the first-order reaction kinetics and film diffusion was the rate-limiting step. The fixed bed column adsorption efficiency of IOCNR for As(III) removal was investigated [61] with a column having diameter of 2 cm and varying bed depths of 10, 15 and 20 cm. The columns were designed using the logit method to evaluate the adsorption rate constant k and column efficiency N. This study proposed an in situ two-step mechanism, firstly oxidation of As(III) to As(V), and subsequently adsorption on IOCNR. Column performance of real arsenic-bearing groundwater was evaluated and results indicated that the treated effluent water quality was suitable for domestic use. In 2013, Maji et al. [62] examined the As(III) adsorption efficiency of iron-oxide-coated natural rock (IONR) using the batch experiment. This showed that in 6 h of contact time, about 99% As(III) was removed from water sample at dosage of 13 g/L, when the initial concentration was 0.6 mg/L at room temperature. The As(III) adsorptive capacity of IOCNR was found to be 1.647 mg/g. The effects of various parameters for As(III) adsorption, including contact time, dose, pH, initial As(III) concentrations, the presence of common cations and anions, and organic contaminants were evaluated. The thermodynamic study showed negative Gibbs free energy change which were evidence of spontaneity and favourable adsorption of As(III) on IOCNR.

5.3   Coated cement

Iron oxide treated cement was also used for the removal of As(III) and As(V) [63]. Kundu et al. [64] investigated the equilibrium sorption of As(III) and As(V) from aqueous environment onto iron oxide-coated cement (IOCC). The adsorption data were evaluated using six error analysis method such as the linear coefficient of determination, the sum of the squares of errors, the sum of absolute errors, the average relative error, the hybrid fractional error function and the Marquardt's percent standard. The error values indicated that Freundlich isotherm showed best fitting for all the experimental data. Isotherms parameters indicated that the adsorption of arsenic onto IOCC was a temperature dependent phenomenon. Further, batch experiments were conducted to study adsorption capacity of iron oxide coated cement (IOCC) for As(III) [65]. This study was examined at ambient temperature as a function of adsorbent dose, pH, contact time, initial arsenic concentration and temperature. Kinetics revealed that the uptake of As(III) ion was very fast and most of fixation occurred within the first 20 min of contact time. Langmuir, Freundlich, Redlich-Peterson, R-P, and Dubinin-Radushkevich, D-R models were used to describe the adsorption isotherms at different As(III) concentrations and 30 g/L adsorbent dose. The maximum adsorption capacity of IOCC for As(III) determined from the Langmuir isotherm was 0.69 mg/g. The mean free energy of adsorption, E, calculated from the D-R isotherm was found to be 2.86 kJ/mol, suggested the adsorption as physical process. Thermodynamic parameters indicated exothermic nature of adsorption and a spontaneous and favourable process.

5.4   Coated activated carbon

For the adsorption of arsenic from water, Rodriguez and Mendez [66] used different activated carbons modified with iron hydro oxide nanoparticles. The surface area of the modified activated carbons ranged from 632 to 1101 m^2/g. Adsorption isotherms were determined at different pH, temperature and As(V) concentrations less than 1 ppm. The maximum adsorption capacity for As(V) varied from 0.37 to 1.25 mg/g and temperature did not affected adsorption. However, arsenic adsorption decreased by 32% when the solution pH was increased from 6 to 8. In the presence of competitive ions adsorption capacity of adsorbent also decreased. The kinetics of adsorption followed the pseudo-second-order model with regression coefficient range 0.98-0.99. This result indicated that iron modified activated carbons were efficient adsorbents for arsenic at concentrations lower than 0.30 mg/L. Similarly, Iron oxide nanoparticles also were deposited on activated carbon (AC) with microwave hydrothermal (MH) treatment technique and used for arsenic removal [67]. BET and porous texture analyses showed some pore filling in AC, but pore volume increased on depositing iron oxide particles. With the MH technique, porous iron oxide was obtained with a high loading value of 20.27% in just 9 minute contact time. As(V) adsorption onto iron oxide deposited AC obeyed Langmuir and pseudo-second order models. Batch adsorption experiments revealed a high efficiency of As(V) removal. Maximum adsorption capacity was 27.78 mg/g, and for a loading of 0.75 g/L, 99.90% uptake was reached within just 5 minute at pH range of 6-8. The results suggested MH synthesized iron oxide particles as promising materials for water treatment. Hydrous iron oxide loaded granule activated carbon showed good adsorption capacity against arsenic. The maximum adsorption capacity of iron loaded granule activated carbon was evaluated as 26 mg/g, when the influent contained 0.3 mg/L arsenic [68]. Similarly, Iron oxide coated modified activated carbon have also been used for arsenic removal. Zhang et al. [69] comparatively studied the removal efficiency of iron oxide, coal-based activated carbon and iron oxide coated activated carbon using X-ray diffraction, nitrogen adsorption, and scanning electronic microscope. Batch experiments were performed for the determination of adsorption capacity and pH effects was also observed. At pH range 4.0-5.5 and 3.0-7.0 iron oxide and iron oxide coated modified activated carbon showed maximum adsorption capacity, and virgin activated carbon has the lowest adsorption capacity in the entire pH range. Iron oxide coated activated carbon and virgin activated carbon adsorption data fitted well to Langmuir isotherm and iron oxide adsorption data fitted well with Freundlich adsorption isotherm.

5.5 Coated polymeric materials

Polymeric materials discovered as surface materials for metal oxide coating, as Jiang et al. [70] coated iron oxide onto the surface of polystyrene. This PS-Fe_3O_4 was then used as adsorbent for the removal of arsenic which showed excellent results. The maximum adsorption capacity of PS-Fe_3O_4 was 139.3 mg/g, which was 77.7% greater than that of bulk Fe_3O_4. The spent adsorbent was readily separated from water under a low magnetic field <0.035T. The continuous adsorption-desorption cyclic results demonstrated that As(V)-loaded PS-Fe_3O_4 could be effectively regenerated by NaOH solution and regenerated composite beads could be employed for repeated use without significant capacity loss. In 2002, for the removal of arsenic from aqueous solution, Katsoyiannis and Zoubolis [71] modified the polystyrene and high internal phase emulsions HIPEs polymer using iron hydroxide and studied the adsorption activity. This experiment showed that both modified media were capable of removing arsenic from the aqueous stream and leading to residual concentration below 0.01 mg/L. Cellulose modified with iron nanoparticle was also used as adsorbent and considered as an exceptional adsorbent material due to its magnetic properties, high surface area and a good adsorption capacity. Hokkanen et al. [72] evaluated adsorption activity of magnetic nanoparticle activated micro-fibrillated cellulose towards As(V) and effects of contact time, pH, initial As(V) concentration, and regeneration were also investigated. The equilibrium adsorption data were fitted to Langmuir, Freundlich, and Sips adsorption models. The monolayer adsorption capacity of the adsorbent, evaluated using Langmuir isotherm, was 2.460 mmol/g. The kinetics of adsorption process was described by pseudo-second-order model. On the basis of the results from EXAFS spectra, Guo et al. [73] described the reaction mechanism of As(III) and As(V) on cellulose bead loaded with iron oxyhydroxide. Prepared bead cellulose loaded with iron oxyhydroxide (BCF) containing 47% Fe was applied for the elimination of arsenic from aqueous solution. Both As(III) and As(V) were strongly and specifically adsorbed by akageneite adsorptive centres on BCF by an inner-sphere mechanism. No change in oxidation state was observed following interactions between the arsenic species and the BCF surface. The arsenic species adsorbed on akageneite dominantly formed bidentate binuclear complex through corner-sharing 2C between As(V) tetrahedra or As(III) pyramids and adjacent edge-sharing FeO_6 octahedra. Sigdel et al. [74] developed iron oxide impregnated alginate beads for arsenic removal. Both As(III) and As(V) were removed from water using iron oxide impregnated alginate beads at different pH. Neutral pH was suitable for As(III)adsorption and As(V)was significantly removed at pH range 3-6. However, the adsorption capacity of iron oxide impregnated alganite beads for As(III) and As(V) were slightly affected by iron loading as well as the presence of phosphate ion. On increasing iron loading, the removal efficiency for As(III) increased, while As(V) removal slightly decreased. The arsenic uptake process followed pseudo-second-order kinetics and intra-particle diffusion was rate limiting step. Zouboulis et al. [75] developed iron oxide coated alginate biopolymer and employed for the removal of arsenic from wastewater. Alginate beads loaded with hydrous ferric oxides were placed in a fixed bed column and experiment was performed to evaluate adsorption capacity. The composition of iron oxide loaded alginate was 3.9 mg of Fe/g of wet alginate bead and 230-bed volumes of 0.05 mg/L As(V) solution were treated till the breakthrough point.

5.6   Coated nanotubes

Multiwalled boron nitride nanotubes (BNNTs) functionalized with Fe_3O_4 nanoparticles NPs have been used for arsenic removal from aqueous solution. Batch experiments were performed by the Chen et al. [76] for evaluation of adsorption activity of iron oxide coated boron nitride nanotubes towards arsenic. Langmuir, Freundlich, and Dubinin-Radushkevich adsorption isotherms were applied for 1-40 mg/L range of As(V) concentrations under the same conditions. Nanoscale iron oxide particles synthesized by Sabbatini et al. [77] were deposited on porous alumina tubes to develop tubular ceramic absorbers for the removal of arsenic. The support tubes and membrane were characterized by surface area and porosity measurements, permeability tests and scanning electron microscopy SEM imaging. The concentrations of arsenic were determined by inductively coupled plasma-optical emission spectroscopy ICP-OES. In the same way, ethylenediamine functionalized multiwall carbon nanotubes, e-MWCNT, were loaded with iron oxide by Velickovic et al. [78] by oxidation of Fe(II) using base and precipitation of Fe(III). The obtained e-MWCNT/Fe^{3+} and e-MWCNT/Fe^{2+} adsorbents followed pseudo-second-order kinetic model. The maximum adsorption capacities obtained from Langmuir model for As(V) on e-MWCNT/Fe^{2+} and e-MWCNT/Fe^{3+} were 23.47 and 13.74 mg/g at 25°C, respectively. Thermodynamic parameters showed that the adsorption of As(V) was spontaneous and endothermic for both e-MWCNT/Fe^{2+} and e-MWCNT/Fe^{3+}. Iron oxide coated multi-walled carbon nanotubes, IOMCNTs, were also used as an adsorbent for As(III) removal [79]. Comparative study showed that modified multi-walled carbon nanotubes have higher As(III) removal efficiency than non-modified. More than 75% As(III) could be removed from water using IOMCNTs. Iron oxide decorated graphene-carbon nanotubes showed excellent adsorption activity for arsenic. **Figure12** shows coating of iron oxide onto the carbon nanotubes. Vadahanambi et al. [80] synthesized this 3D-iron oxide nanoparticles incorporated graphene-carbon nanotubes with high surface-to- volume ratio and open pore network.

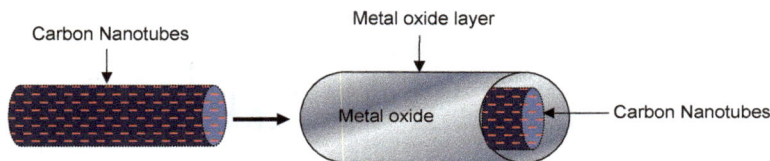

Figure: 12- Metal oxide coated carbon nanotubes.

5.7   Coated soil constituent

Xie et al. [81, 82] developed the aquifer iron coating method via four step injection of oxidant, iron salt and oxygen free water and used in situ for arsenic adsorption. Arsenic removal efficiency was investigated in various quartz sand filled columns under anoxic conditions. As(V) reagent and 0.005 mol/L iron sulphate (FeSO$_4$), and 0.0025 mol/L of sodium hypochlorite (NaClO) were continuously injected in quartz filled column resulting uniform goethite layer coating on the surface of quartz sand and As(V) adsorption taking place simultaneously. Characterization study revealed that arsenic formed bidentate binuclear complexes with fine goethite particles through adsorption or co-precipitation. This study revealed that in situ aquifer iron coating method enhanced the adsorption capacity of goethite particles due to increase in the specific surface area of goethite particles.

Glocheux et al. [83] optimised 3D organised mesoporous silica, OMS, coated with both iron and aluminium oxides for the optimal removal of As(III) and As(V) from the synthetic water. Batch experiments were performed for this purpose and the effect of initial arsenic concentration and pH, kinetics and diffusion mechanisms were studied. The effect of total Al to Fe oxides coating on the selective removal of As(III) and As(V) was also studied and it was shown that 8 % metal coating was the optimal configuration for the coated OMS materials in removing arsenic. The advantage of an organised material over an unstructured adsorbent was very limited in terms of kinetic and diffusion under the experimental conditions. The physisorption was the main adsorption process involved in arsenic removal by the coated OMS. Maximum adsorption capacity for As(V) was 55 mg/g at pH 5.0 for material coated with 8 % Al oxides while 35 mg As(V)/g was removed at pH 4.0 for equivalent material coated with Fe oxides.

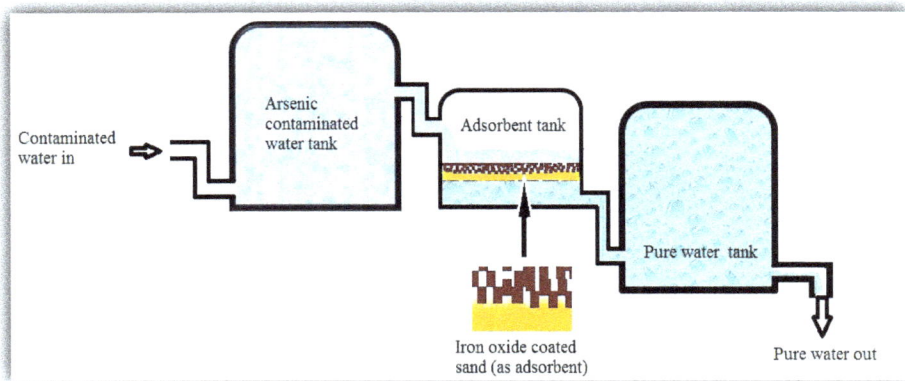

Figure: 13- Schematic diagram of purification of arsenic contaminated water using iron oxide coated sand.

Jeon et al. [84] coated iron oxide onto zeolite (IOCZ) and then performed batch and column experiments to examine sorption activity for As(V). Langmuir isotherm model was suitable to explain the adsorption characteristics of As(V) onto IOCZ. The optimum dose of ICOZ was 33.3 g/L at a concentration of 20.12 mg/L and the adsorption capacity of ICOZ was 0.68 mg/g and As(V) was completely removed within 30 minutes in a concentration of 2.0 mg/L, with a 100 g/L dosage of IOCZ. They also investigated the effect of solution pH and presence of competitive ion. The pH effect was negligible while sulphate ions inhibited sorption of As(V). Similarly, granule iron oxide coated diatomite (IOCD) was employed for As(V) removal from aqueous solution and compared to raw diatomic [85]. IOCD had a great affinity to adsorb As(V) as compared to raw diatomite. This improvement was attributed to increases in both surface area and affinity, and enhanced complexation between the negatively charged As(V) ions and partial positive surface charge on the surface of the iron oxide. Therefore, as pH was increased from 3.5 to 9.5 adsorption rate decreased due to increased negative charge on the surface. Freundlich adsorption isotherm well described the process and followed the pseudo-second-order kinetics. Nano-sized iron oxide-coated perlite [86] was also used for As(V) removal which showed a tremendous result. It was observed that a 100% As(V) adsorption was achieved at pH 4-8 from the initial concentration containing 1.0 mg As(V)/L and the adsorption percentage depended on the initial concentration. Kinetics experimental data fitted well with the pseudo-second-order model. Langmuir isotherm adequately fitted adsorption data than Freundlich isotherm in the simulation of adsorption of As(V). The adsorption rate constant and the maximum adsorption capacity were 44.84 l/mg and 0.39 mg/g, respectively.

5.8   Coated waste material

Waste materials were also used for coating with iron oxide. Coal bottom ash and fly ash were preferable. Coal bottom ash is the waste material of coal-fired power plant, was coated with iron oxide and used as an adsorbent for arsenic removal by Mathieu et al. [87]. This low-cost adsorbent removed 90 % arsenic within 60 minute contact time. The adsorption process followed pseudo-second-order kinetics. The maximum monolayer adsorption capacity of iron oxide coated coal bottom ash was 0.20 mg/g. The impact of groundwater arsenic concentration on removal capacity was observed. When the groundwater arsenic concentration was decreased from 560 to 170µg/g, adsorption rate increased from 2.4×10^5 to 7.2×10^5 g/mol/min. Yadav et al. [88] used iron oxide coated bagasse fly ash with a surface area of 168 m²/g for arsenic removal from water. This iron oxide coated bagasse fly ash showed maximum adsorption capacities for As(III) and As(V) as 0.039 and 0.025 mg/g, respectively. This adsorption process followed the pseudo-second order kinetics.

5.9   Coated biomass

Pokhrel and Viranghavan [89] used iron oxide-coated A. Niger biomass for the removal of Dimethylarsinic acid (DMA) and also examined the parameters that influenced DMA removal from aqueous solution using a two-level seven-factor fractional factorial design analysis. Iron coated honeycomb briquette cinders (Fe-HBC) removed a high amount of As(V) from water [90]. Optimisation study showed that within 14 h, 0.96 mg/g As(V) was removed from the water at pH 7.5. Adsorption data fitted well to Langmuir isotherm with correlation coefficient 0.999. Competitive ions effect was observed in the order $PO_4^{3-} > HCO_3^- > F^- > Cl^-$. Nguyen et al. [91] used iron oxide coated sponge (IOCSp) for removal of arsenic and studied the adsorption efficiency of IOCSp. The maximum adsorption capacity of IOCSp for As(III) and As(V) were 4.2 and 4.6 mg/g, respectively. A filter packed with a small amount of 25 g IOCSp maintained a consistent arsenic removal efficiency of 95 % from a synthetic solution containing an arsenic concentration of 1.0 mg/L.

These were the recently used coated adsorbents which showed excellent results towards the arsenic removal from water. But the main challenge of the coated adsorbents is to make a permanent coating on the surface of support. Therefore, a detailed study was needed to develop techniques in preparation of coated adsorbents in order to get higher adsorption efficiency and regenerability that's why we tried to cover up sufficient data on metal oxide coated materials.

4. Comparative study

Metal oxide coated adsorbent such as iron oxide doped/coated activated carbon and soil constituents like sand, clay, perlite and zeolites have been used for the removal of arsenic from water. Their adsorption capacities are available in the literature, however, the comparative study of these adsorbents is not easy. The adsorption capacity of an adsorbent depends on the various factors including pH, temperature, concentration, adsorbent dosage, arsenite/arsenate ratio, separation and regeneration. The surface area of adsorbent is also responsible for evaluating sorption capacity. An adsorbent with the higher surface area have higher adsorption capacity but is not necessary all the time. Adsorption capacity for organics becomes higher with a higher surface area of adsorbent, while it is not true for inorganic adsorption. Therefore, comparing adsorbents by surface area alone is difficult. The adsorption capacity of adsorbents has been examined for a different sample like drinking water, ground water, aqueous solution and actual wastewater, etc. The different type of competitive ions and their concentrations in the water sample are also responsible for the adsorption capacity of adsorbent. Some adsorbent is not affected by the interfering ions, even in very high concentration, while some other are affected readily in a particular time. Instead of these, different experimental methods include batch experiments and column method, are also reported for the evaluation of adsorption capacities.

These both type of experimental method cannot be compared with each other. In batch experiments, Langmuir isotherm or the Freundlich isotherm were used for the evaluation of sorption capacities which also makes the comparison more difficult to pursue. Therefore, herein, we tried to compare different adsorbents on the basis of their removal capacity (**Table 11**). The 3D chart is also given in **Figure 14** for comparing the adsorption capacity of different metal oxide coated and non-coated adsorbents.

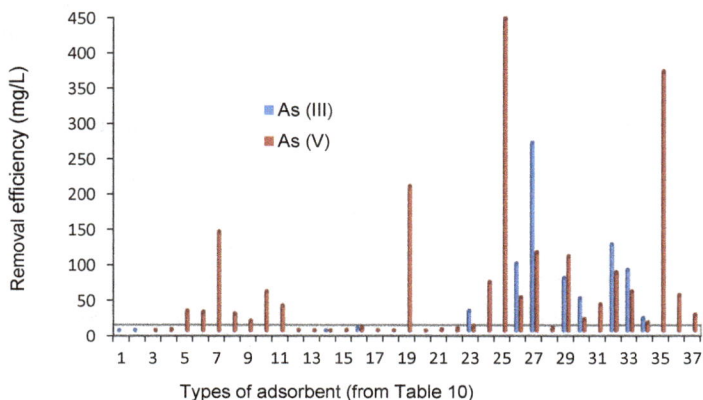

Figure: 14 Chart of comparative studies of different coated and uncoated adsorbents.

5. Competitive study

The interaction between the adsorbate and adsorbent become complicated in the multi-component system due to solute-solute, solute-arsenic ions competition. This competition occurs at the active sites at the surface of adsorbent which bring change in adsorption capacity for single metal ions and in the presence of competitive ions. To prevent this competition solute-solute, the single-ion system in deionized water is used frequently for adsorption purpose instead of multi-ion systems. However, it is well known that natural ground water or arsenic-contaminated water contains other types of ions too. These types of ions may interfere with adsorption efficiency for arsenic removal. For example, high concentrations of silicate, 6-28 mg Si/L, phosphates, 0.2-3.0 mg/L and bicarbonate, 50-670 mg/L ions were found in the groundwater of Bangladesh [92]. Obviously, more work out on multi-component systems and to produce that adsorbent which shows well adsorption behaviour in a multi-component system. In the presence of a common competitive ion in multi-component system, metal oxide coated materials showed better adsorption capacity for both As(III) and As(V) as compared to virgin metal oxide. However, metal doped metal oxide also showed good adsorption efficiency and do not affect greatly by competitive ions as compared to another adsorbent. **Figure 15** is showing the effect of competitive ions on the arsenic removal process. Maji et al. [62] examined the effect of the presence of some common positive/negative ion and organic contaminants towards the As(III) adsorption capacity of iron-oxide-coated natural rock (IONR). Pokhrel and Viranghavan [89] examined the presence of common ions such as Ca^{2+}, Fe^{2+}, SO_4^{2-} ions, and Cl^- ions that influenced the Dimethylarsinic acid DMA removal efficiency of iron oxide-coated A. Niger biomass. The magnitude of the influence of ions considered on DMA removal was observed in the order $Ca^{2+}>$ $SO_4^{2-}>Fe^{2+}>Cl^-$.Effect of F^-, Cl^-,PO_4^{3-}, HCO_3^- on As (V) removal efficiency of iron coated honeycomb briquette cinders Fe-HBC observed as in order $PO_4^{3-}> HCO_3^-> F^-> Cl^-$ [90]. Nano-sized iron oxide-coated perlite [86] were also used for As(V) removal which showed promising results. It was also observed that the adsorption efficiency did not affect by competitive ion such as phosphate ion. This study showed the influence of various ions on the adsorption of arsenic in multi-component systems, thus lots of work is needed to established mechanistic guidelines for arsenic sorption in multi-component systems and it is necessary to develop more adsorbent for a multi-component system which does not affect by others ions.

6. Separation and desorption

Regeneration is very important to make adsorbent cost effective. After the establishment of equilibrium during adsorption process of solute, the adsorbent gets exhausted. For reuse of this adsorbent, regeneration is required. But before the regeneration and after the establishment of equilibrium during adsorption process, it is necessary to separate solute-loaded adsorbent from water/aqueous solution, for the purpose of desorption/regeneration. Several methods of separation are reported based on adsorbent nature such as centrifugation [93, 94], filtration [95, 96] and magnetic separation [97, 98].

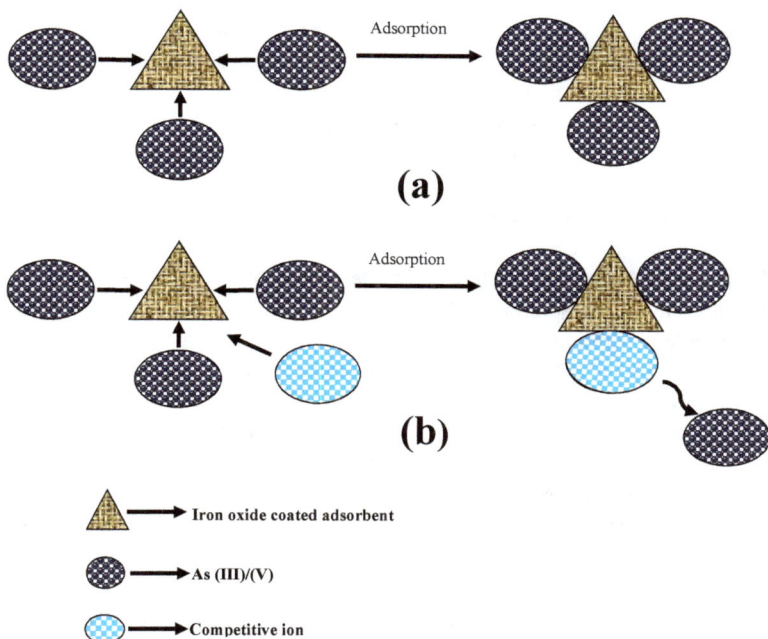

(a)

(b)

Iron oxide coated adsorbent

As (III)/(V)

Competitive ion

Figure: 15- Adsorption of As (III)/As (V) (a) In the absence of competitive ion, (b) in the presence of competitive ion.

The non-magnetic adsorbent can be separated by centrifugation while filtration method is used for the membrane and size based adsorbent. Many researchers have used filtration process for the separation purpose [95]. The magnetic separation method is used for the separation of magnetic particles, magnetite and other metal oxide and has been proved more effective than other methods due to easy handling and high efficiency [97, 98]. Magnetic separation depends on magnetic field gradients, magnetic properties of adsorbent and size of adsorbent. An adsorbent with magnetic in nature and nano-size can easily separate by this process. Generally, the magnetic adsorbent is incorporated by magnetic elements such as nickel, cobalt and iron. Iron containing adsorbents like magnetite, maghemite and others iron oxides are frequently being used as adsorbent due to their magnetic nature. This adsorbent has very good magnetic properties that is why; they can be easily separated from water by applying low gradient magnetic field [97, 98]. Magnetic field column separator consisting of a stainless steel may be used for this purpose [99, 100]. A distinct advantage of these adsorbents is that adsorbents can readily be isolated from water by application of an external magnetic field (**Figure 16**). Moreover, coagulation method is also being used for the filtration of micron-scale particles and proved as cost effective separation method [87]. After separation, contaminant loaded adsorbent is ready to regeneration. Regeneration of exhausted adsorbent makes the adsorption process economically effective. The main aim of desorption/regeneration is to regenerate adsorption capacity of an adsorbent and reuse for the adsorption process. pH is responsible for desorption/regeneration because the adsorption of cations is negligible in the acidic solution while anion adsorption is negligible in basic solution [101]. Therefore, by the adjustment of solution pH, desorption/regeneration may be made to occur. For this purpose, a desorbing agent like base or acids could be used. Sodium hydroxide [102] and HCl are most common desorbing agents used for desorption of arsenic [103]. Nano-sized iron-oxide-coated quartz (IOCQ) was regenerated using HCl [103]. This modified adsorbent was also regenerated in alkali solutions. Iron oxyhydroxide loaded bead cellulose adsorbent was regenerated with 2M NaOH solution [108] and was used through four cycles. The continuous adsorption-desorption cyclic results demonstrated that As(V)-loaded PS-Fe_3O_4 could be effectively regenerated by NaOH solution and the regenerated composite beads could be employed for repeated use without significant capacity loss [105]. Similarly, hydrated Fe(III) oxide, HFO, dispersed on a polymeric exchanger [102] and iron oxide coated cement IOCC [98] were regenerated using 10 % NaOH. The adsorbents were subsequently reused for more than three cycles. Alginate beads [74] were regenerated using NaOH solution and successfully reused for multiple cycles.

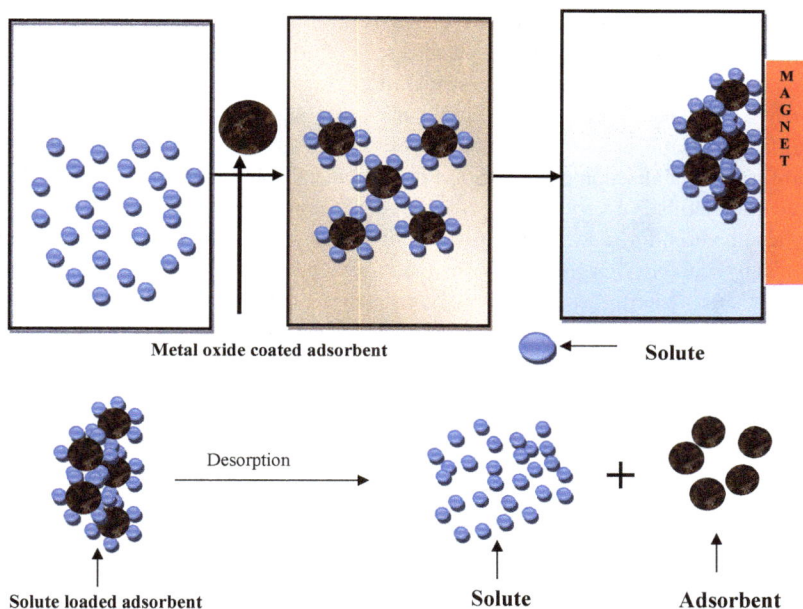

Figure: 16- Separation and regeneration of metal oxide coated adsorbent/desorption of arsenic.

7. Disposal of desorbed arsenic

Of course, regeneration of exhausted adsorbent enhances the adsorption capacity of adsorbent but the disposal of desorbed arsenic is not easy even very difficult task. It is very difficult to dispose of the concentrated arsenic and stop this to recycle in water source again. The most common way of disposal of arsenic waste is to dump in sea or landfills. For this purpose, Kyzas and Matis [104] and Leist et al. [105] reviewed various disposal techniques. Kyzas and Matis [104] reviewed that solidification/stabilisation and verification techniques which may prove a useful way to the disposal of a concentrated arsenic contaminant into landfills. Solidification/stabilisation is the encapsulation of contaminant into the physically stable mass that chemically reduces the hazard potential of waste by converting them into less soluble, mobile or toxic form. Pozzolanic materials such as cement and lime are most frequently used binders for solidification of arsenic. Mohan et al.[106] reviewed that Portland cement, Portland cement and iron(II), Portland cement and iron(III), Portland cement and lime, Portland cement, iron and lime, Portland cement and fly ash, Portland cement and silicates good binders and frequently used for solidification/stabilisation of arsenic. Leist et al. [105] well described the solidification/stabilisation process. Solidified/stabilised contaminant followed by disposal in secure landfills is a very attractive way to dispose of arsenic waste. Vitrification [104] may also be used for reducing the mobility of metals by incorporating metal waste into a chemically durable, leach-resistant and vitreous mass. The technology requires large amounts of energy to achieve vitrification temperatures. Another method, pyrometallurgical recovery, may also be used for this purpose. Pyrometallurgical recovery is a process of conversion of contaminated waste feed into a product with a high concentration of a contaminant that can be reused [105]. These techniques are very expensive to operate, therefore, lots of work is needed in this area and it is necessary to find more ways to treat arsenic waste followed by disposal.

Conclusion

Metal oxides are important materials that have been used as an adsorbent by large numbers of researchers all over the world to decontaminate drinking water in general and arsenic in particular. However, most of the metal oxides are fine particles that are difficult to separate from the aqueous solution after equilibrium attained and make a suspension with an aqueous solution. Moreover, finely powdered adsorbent is not suitable for column sorption because of their low hydraulic conductivity. To overcome their separation and column applications, various metal oxides, especially iron oxides in different forms have been coated on various solid supports. Herein, the effect of different sized metal oxide coated adsorbent on arsenic removal reported, using different parameters like pH, dosage, kinetic studies, equilibrium studies and adsorption isotherm and their adsorption capacity have been compared. It was found that smaller sized metal oxide coated material have high adsorption capacity for arsenic and provide a better scope for future research to search more solid supports.

Acknowledgement

One of the authors is gratefully acknowledging the UGC for providing Non-NET fellowship and support by the Department of Chemistry, Jamia Millia Islamia, New Delhi, India, for carrying out the present work.

References

[1]   J. W. Mellor, A Comprehensive Treatise on Inorganic and Theoretical Chemistry. Longmans, Green, London, p 9 (1954).

[2]   B. K. Mandal, K. T. Suzuki, Arsenic round the world: a review, Talanta 58 (2002) 201-235. https://doi.org/10.1016/S0039-9140(02)00268-0

[3]   P. L. Smedley, D. G. Kinniburgh (2002) A review of the source, behaviour and distribution of arsenic in natural waters. Appl. Geochem. 17(5) (2002) 517-568. https://doi.org/10.1016/S0883-2927(02)00018-5

[4]   R. J. Bowell, C. N. Alpers, H. E. Jamieson, D. K. Nordstrom, J. Majzlan, Arsenic: Environmental Geochemistry, Mineralogy and Microbiology, Reviews in Mineralogy and Geochemistry, p 79. 2015. ISSN 1529-6466, ISBN 978-0-939950-94-2.

[5]   E. Lombi, W. W. Wenzel, D. C. Adriano, Arsenic-contaminated soils: II Remedial action. In: Wise DL, Trantolo DJ, Eichon EJ, Inyang HI, Stottmeister U (2000b) Remediation Engineering of Contaminated Soils. Marcel Dekker, New York, pp 739-758.

[6]   J. Matschullat, Arsenic in the geosphere - a review, Sci. Total Environ. 249(1-3) (2000) 297-312. https://doi.org/10.1016/S0048-9697(99)00524-0

[7]   E. T. Mackenzie, R. J. Lamtzy, V. Petorson, (1979), Global trace metals cycles and predictions. J. Int. Assoc. Math. Geol. 6 (1979) 99-142. https://doi.org/10.1007/BF01028961

[8]   S. Wang, C. N. Mulligan, Occurrence of arsenic contamination in Canada: sources, behavior and distribution, Sci. Total Environ. 366 (2006) 701-721. https://doi.org/10.1016/j.scitotenv.2005.09.005

[9]   Ringbom (1963) Complexation in Analytical Chemistry. Inter science-Wiley, New York.

[10]   I. Bodek, W. J. Lyman, W. F. Reehl, D. H. Rosenblatt, (1998) Environmental Inorganic Chemistry: Properties, Processes and Estimation Methods. Pergamon Press, USA.

[11]   H. Hasegawa, M. Matsui, S. Okamura, M. Hojo, N. Iwasaki, Y. Sohrin, Arsenic speciation including 'hidden' arsenic in natural waters, Appl. Organometal. Chem. 13 (1999) 113-119. https://doi.org/10.1002/(SICI)1099-0739(199902)13:2<113::AID-AOC837>3.0.CO;2-A

[12]   X. C. Le, (2002) Arsenic speciation in the environment and humans. In: Franken berger Jr WT (ed) Environmental Chemistry of Arsenic, Marcel Dekker, New York, p 95-116.

[13]   M. Styblo, L. M. D. Razo, L. Vega, D. R. Germolec, E. L. L. Cluyse ELL, G. A. Hamilton, W. Reed, C. Wang, W. R. Cullen, D. J. Thomas, (2000) Comparative toxicity of trivalent and pentavalent inorganic and methylated arsenicals in rat and human cells, Arch. Toxicol. 74 (2000) 289-299. https://doi.org/10.1007/s002040000134

[14]   K.T. Kitchin, K. Wallace, Arsenite binding to synthetic peptides based on the Zn finger region and the estrogen binding region of the human estrogen receptor-alpha, Toxicol. Appl. Pharmacol. 206 (2006) 66-72. https://doi.org/10.1016/j.taap.2004.12.010

[15]   WHO (2008) Guidelines for drinking-water quality [electronic resource]: incorporating 1st and 2nd addenda. Recommendations -3rd edn, p 1.

[16]   D. K. Nordstrom, An overview of arsenic mass-poisoning in Bangladesh and West Bengal, India In: Minor Elements Arsenic, Antimony, Selenium, Tellurium and Bismuth. ln: Young C (ed) Soc Mining Metallurgy and Exploration, p 21-30 (2000).

[17]   M. Rahman, M. Vahter, M. A. Wahed, N. Sohel, M. Yunus, P. K. Streatfield, (2006), Prevalence of arsenic exposure and skin lesions A population based survey in Matlab, Bangladesh J. Epidemiol. Community Health 60(3) (2006) 242-248. https://doi.org/10.1136/jech.2005.040212

[18]   R. V. Hedegaard, J. J. Sloth, Speciation of arsenic and mercury in feed: why and how?, Biotechnol. Agron. Soc. Environ. 15(1) (2011) 45-51.

[19]   M. Tondel, M. Rahman, A. Magnuson, I. A. Chowdhury, M. H. Faruquee, S. A. Ahmad, The relationship of arsenic levels in drinking water and the prevalence rate of skin lesions in Bangladesh, Environ. Health Perspect 107 (1999) 727-729. https://doi.org/10.1289/ehp.99107727

[20]   J. C. Ng, J. Wang, A. Shraim, Global health problems caused by arsenic from natural sources, Chemosphere 52 (2003) 1353-1359. https://doi.org/10.1016/S0045-6535(03)00470-3

[21]   R. Gürkan, U. Kır, N. Altunay, Development of a simple, sensitive and inexpensive ion-pairing cloud point extraction approach for the determination of trace inorganic arsenic species in spring water, beverage and rice samples by UV-

Vis spectrophotometry, Food Chem. 180(1) (2015) 32-41.
https://doi.org/10.1016/j.foodchem.2015.01.142

[22] S. A. Chaudhry, M. Ahmed, S. I. Siddiqui, S. Ahmed, Fe(III)-Sn(IV) mixed binary oxide-coated sand preparation and its use for the removal of As(III) and As(V) from water: Application of isotherm, kinetic and thermodynamics, J. Mol. Liq. 224 (2016) 431-441. https://doi.org/10.1016/j.molliq.2016.08.116

[23] T. S. Y. Choong, T. G. Chuah, Y. Robiah, F. L. G. Koay, I. Azni, Arsenic toxicity, health hazards and removal techniques from water: an overview, Desalination 217 (2007) 139-166; M. Horsfall Jnr, S. I. Ayebaemi, (2005) Effect of Temperature on the Sorption of Pb2+ and Cu2+ from Aqueous Solution by Caladium bicolor (Wild Cocoyam) Biomass, Elec. J. Biotech. 8(2) (2005) 162-169. https://doi.org/10.1016/j.desal.2007.01.015

[24] C. Han, H. Pu, H. Li, L. Deng, S. Huang, S. He, Y. Luo, The optimization of As (V) removal over mesoporous alumina by using response surface methodology and adsorption mechanism, J Hazard Mater 254-255 (2013) 301-309. https://doi.org/10.1016/j.jhazmat.2013.04.008

[25] C. Wang, H. Luo, Z. Zhang, Y. Wu, J. Zhang, S. Chen, Removal of As (III) and As (V) from aqueous solutions using nanoscale zero valent iron-reduced graphite oxide modified composites, J. Hazard. Mater. 268 (2014) 124-131. https://doi.org/10.1016/j.jhazmat.2014.01.009

[26] A. K. Darban, Y. Kianinia, E. T. Nassaj, Synthesis of nano- alumina powder from impure kaolin and its application for arsenite removal from aqueous solutions, J. Environ. Health. Sci. Eng. 11(19) (2013) 1-11.

[27] K. Y. Foo, B. H. Hameed, Insights into the modeling of adsorption isotherm systems, Chem. Eng. J. 156 (2010) 2-10. https://doi.org/10.1016/j.cej.2009.09.013

[28] AL-Othman ZA, Inamuddin, Naushad M (2011) Adsorption thermodynamics of trichloroacetic acid herbicide on polypyrrole Th(IV) phosphate composite cation-exchanger. Chem Eng J 169:38–42 https://doi.org/10.1016/j.cej.2011.02.046

[29] M. I. Temkin, V. Pyzhev, Kinetic of ammonia synthesis on promoted iron catalyst, Acta. Physiochim. USSR 12 (1940) 327-356.

[30] L. Zeng, Arsenic Adsorption from Aqueous Solutions on an Fe (III)-Si Binary Oxide Adsorbent, Water Qual Res J Canada 39(3) (2004) 267-275.

[31] J. Lin, L. Wang, Comparison between linear and non-linear forms of pseudo-first-order and pseudo-second-order adsorption kinetic models for the removal of

methylene blue by activated carbon, Frontiers of Environmental Science & Engineering in China 3(3) (2009) 320-32. https://doi.org/10.1007/s11783-009-0030-7

[32]   F. C. Wu, R. L. Tseng, S. C. Huang, R. S. Juang, Characteristics of pseudo-second-order kinetic model for liquid-phase adsorption: A mini-review, Chem. Eng. J 151(1-3) (2009) 1-9. https://doi.org/10.1016/j.cej.2009.02.024

[33]   Zeldowitsch, Adsorption site energy distribution, J. Acta. Physicochem. USSR 1 (1934) 364-449.

[34]   K. V. Kumar, Linear and non-linear regression analysis for the sorption kinetics of methylene blue onto activated carbon, J. Hazard Mater. 137 (2006) 1538-1544. https://doi.org/10.1016/j.jhazmat.2006.04.036

[35]   W. J. Weber, J. C. Morris, Kinetics of adsorption carbon from solutions, Sanit. J. (1963) Eng. Div. Am. Soc. Civ. Eng. 89 (1963) 31-59.

[36]   T. A. Khan, S. A. Chaudhry, I. Ali, Equilibrium uptake, isotherm and kinetic studies of Cd(II) adsorption onto iron oxide activated red mud from aqueous solution, J. Mol. Liq. 202 (2015) 165-175. https://doi.org/10.1016/j.molliq.2014.12.021

[37]   H. D. S. S. Karunarathne, B. M. W. P. K. Amarsinghe, (2013) Fixed Bed Adsorption Column Studies for the Removal of Aqueous Phenol from Activated Carbon Prepared from Sugarcane, Energy Procedia. 34 (2013) 83-90. https://doi.org/10.1016/j.egypro.2013.06.736

[38]   Z. Xu, J. G. Cai, B. C. Pan, Mathematically modeling fixed-bed adsorption in aqueous systems J. of Zhejiang University 14(3) (2013) 155-176.

[39]   D. W. Hand, S. Loper, M. Ari, J. C. Crittenden, Prediction of multicomponent adsorption equilibrium using ideal adsorbed solution theory, Environ. Sci. Technol. 19(11) (1985) 1037-1043. https://doi.org/10.1021/es00141a002

[40]   A. O. Ekpete, M. Horsfall Jnr, T. Tarawou, (2010) Potential of fluted and commercial activated carbons for phenol removal in aqueous systems, J. Eng. Appl. Sci. 5 (2010) 939-947.

[41]   E. Diamadopoulos, S. Loannidis, G. P. Sakellaropoulos, As (V) removal from aqueous solutions by fly ash, Water Res. 27(12) (1993) 1773-1777. https://doi.org/10.1016/0043-1354(93)90116-Y

[42]   L. Önnby, V. Pakade, B. Mattiasson, H. Kirsebom, Polymer composite adsorbents using particles of molecularly imprinted polymers or aluminium oxide

nanoparticles for treatment of arsenic contaminated waters, Water Res. 46(13) (2012) 4111-4120. https://doi.org/10.1016/j.watres.2012.05.028

[43   M. Pena, X. G. Meng, G. P. Korfiatis, C. Y. Jing, Adsorption mechanism of arsenic on nanocrystalline titanium dioxide. Environ Sci Technol 40 (2006) 1257-1262. https://doi.org/10.1021/es052040e

[44]   Y. M. Pajany, C. Hurel, C. Marmier, M. Romeo, Arsenic (V) adsorption from aqueous solution onto goethite, hematite, magnetite and zero-valent iron: effects of pH, concentration and reversibility, Desalination 281 (2011) 93-99. https://doi.org/10.1016/j.desal.2011.07.046

[45]   X. Guan, J. Du, X. Meng, Y. Sun, B. Sun, Q. Hu, Application of titanium dioxide in arsenic removal from water: A review. J Hazard Mater (215-216) (2012) 1-16. https://doi.org/10.1016/j.jhazmat.2012.02.069

[46]   J. T. Mayo, C. Yavuz, S. Yean, L. Cong, H. Shipley, W. Yu, J. Falkner, A. Kan, M. Tomson, V. L. Colvin, The effect of nanocrystalline magnetite size on arsenic removal, Sci. Tech. Adv. Mater. 8 (2007) 71-75. https://doi.org/10.1016/j.stam.2006.10.005

[47]   T. Tuutijärvi, J. Lu, M. Sillanpää, G. Chen, As (V) adsorption on maghemite nanoparticles, J. Hazard. Mater. 166(2-3) (2009) 1415-1420. https://doi.org/10.1016/j.jhazmat.2008.12.069

[48]   L. A. Zeng, A mothod for preparing silica containing iron(III) oxide adsorbents for arsenic removal, Water Res. 37 (2003) 4351-4358. https://doi.org/10.1016/S0043-1354(03)00402-0

[49]   T. Virghavan, O. S. Thirunavukkarasu, K. S. Suramanian, Removal of arsenic in drinking water by iron oxide coated sand and ferrihydrite - batch studies, Water Qual. Res. J. Can. 36 (2001) 55-70.

[50]   T. A. Khan, E.A. Khan, Shahjahan, Removal of basic dyes from aqueous solution by adsorption onto binary iron-manganese oxide coated kaolinite: Non-linear isotherm and kinetics modeling, Applied Clay. Sci. 107 (2015) 70-77. https://doi.org/10.1016/j.clay.2015.01.005

[51]   S. Kong, Y. Wang, H. Zhan, M. Liu, L. Liang, Q. Hu, Competitive adsorption of humic acid and arsenate on nanoscale iron-manganese binary oxide-loaded zeolite in groundwater, J. Geochemical Exploration 144 (2014) 220-225. https://doi.org/10.1016/j.gexplo.2014.02.005

[52] T. A. Khan, S. A. Chaudhry, I. Ali, Thermodynamic and kinetic studies of As (V) removal from water by zirconium oxide-coated marine sand, Environ. Sci. Pollut. Res. 20 (2013) 5425-5440. https://doi.org/10.1007/s11356-013-1543-y

[53] B. M. Jovanović, V. L. Pesić, L. V. Rajaković, Enhanced arsenic sorption by hydrated iron (III) oxide-coated materials--mechanism and performances, Water Environ. Res. 83(6) (2011) 498-506 https://doi.org/10.2175/106143010X12851009156484

[54] O. S. Thirunavukkarasu, T. Virghavan, K. S. Suramanian, Arsenic removal from drinking water using iron oxide-coated sand, Water Air Soil Pollut. 142 (2003) 95-111. https://doi.org/10.1023/A:1022073721853

[55] V. K. Gupta, V. K. Saini, N. Jain, Adsorption of As(III) from aqueous solutions by iron oxide-coated sand, J. Colloid Interface Sci. 288(1) (2005) 55-60. https://doi.org/10.1016/j.jcis.2005.02.054

[56] J. C. Hsu, C. J. Lin, C. H. Liao, S. T. Chen, Removal of As(V) and As(III) by reclaimed iron-oxide coated sands, J. Hazard Mater 153(1-2) (2008) 817-826. https://doi.org/10.1016/j.jhazmat.2007.09.031

[57] J. Mähler, I. Persson, Rapid adsorption of arsenic from aqueous solution by ferrihydrite-coated sand and granular ferric hydroxide, Appl Geochem 37 (2013) 179-189. https://doi.org/10.1016/j.apgeochem.2013.07.025

[58] I. M. M. Rahman, Z. A. Begum, H. Sawai, T. Maki, H. Hasegawa, Decontamination of spent iron-oxide coated sand from filters used in arsenic removal, Chemosphere 92 (2) (2013) 196-200. https://doi.org/10.1016/j.chemosphere.2013.03.024

[59] Y. Huang, J. K. Yang, A. A. Keller, Removal of arsenic and phosphate from aqueous solution by metal (Hydr-) oxide coated sand, ACS Sustainable Chem. Eng. 2(5) (2014) 1128-1138. https://doi.org/10.1021/sc400484s

[60] S. K. Maji, Y. H. Kao, C. W. Liu, Arsenic removal from real arsenic-bearing groundwater by adsorption on iron-oxide-coated natural rock (IOCNR), Desalination 280 (1-3) (2011) 72-79. https://doi.org/10.1016/j.desal.2011.06.048

[61] S. K. Maji, Y. H. Kao, C. J. Wang, G. S. Lu, J. J. Wu, C. W. Liu, Fixed bed adsorption of As (III) on iron-oxide-coated natural rock (IOCNR) and application to real arsenic-bearing groundwater, Chem. Eng. J. 203 (2012) 285-293. https://doi.org/10.1016/j.cej.2012.07.033

[62]   S. K. Maji, Y. H. Kao, P. Y. Liao, Y. J. Lin, C. W. Liu, Implementation of the adsorbent iron-oxide-coated natural rock (IOCNR) on synthetic As (III) and on real arsenic-bearing sample with filter, Appl. Surf. Sci. 284 (2013) 40-48. https://doi.org/10.1016/j.apsusc.2013.06.154

[63]   S. Kundu, A. K. Gupta, Adsorptive removal of As (III) from aqueous solution using iron oxide coated cement (IOCC): evaluation of kinetic, equilibrium and thermodynamic models, Sep. Purif. Technol. 52(2) (2006) 165-172. https://doi.org/10.1016/j.seppur.2006.01.007

[64]   S. Kundu, A. K. Gupta, Arsenic adsorption onto iron oxide-coated cement (IOCC): Regression analysis of equilibrium data with several isotherm models and their optimization, Chem. Eng. J. 122(1-2) (2006) 93-106. https://doi.org/10.1016/j.cej.2006.06.002

[65]   S. Kundu, A. K. Gupta, Adsorption characteristics of As (III) from aqueous solution on iron oxide coated cement (IOCC), J. Hazard Mater. 142(1-2) (2007) 97-104. https://doi.org/10.1016/j.jhazmat.2006.07.059

[66]   A. V. V. Rodriguez, J. R. R. Mendez, Arsenic removal by modified activated carbons with iron hydro(oxide) nanoparticles, J. Environ. Manage 114 (2013) 225-231. https://doi.org/10.1016/j.jenvman.2012.10.004

[67]   A. Yürüm, Z. O. K. Ataklı, M. Sezen, R. Semiat, Y. Yürüm, Fast deposition of porous iron oxide on activated carbon by microwave heating and arsenic (V) removal from water, Chem. Eng. J 242 (2014) 321-332. https://doi.org/10.1016/j.cej.2014.01.005

[68]   M. Jang, W. Chen, F. S. Cannon, Preloading hydrous ferric oxide into granular activated carbon for arsenic removal, Environ. Sci. Technol. 42(9) (2008) 3369-3374. https://doi.org/10.1021/es7025399

[69]   Q. L. Zhang, N. Y. Gao, Y. C. Lin, B. Xu, L. S. Le, Removal of arsenic(V) from aqueous solutions using iron-oxide-coated modified activated carbon, Water Environ. Res. 79(8) (2007) 931-936. https://doi.org/10.2175/106143007X156727

[70]   W. Jiang, X. Chen, Y. Niu, B. Pan, Spherical polystyrene-supported nano-Fe3O4 of high capacity and low-field separation for arsenate removal from water, J. Hazard Mater. 243 (2012) 319-325. https://doi.org/10.1016/j.jhazmat.2012.10.036

[71]   I. A. Katsoyiannis, A. I. Zouboulis, Removal of arsenic from contaminated water sources by sorption onto iron-oxide-coated polymeric materials, Water Res. 36(20) (2002) 5141-5155. https://doi.org/10.1016/S0043-1354(02)00236-1

[72]   S. Hokkanen, E. Repo, S. Lou, M. Sillanpä, Removal of arsenic (V) by magnetic nanoparticle activated microfibrillated cellulose, Chem. Eng. J. 260 (2015) 886-894. https://doi.org/10.1016/j.cej.2014.08.093

[73]   X. Guo, Y. Du, F. Chen, H. S. Park, Y. Xie, (2007) Mechanism of removal of arsenic by bead cellulose loaded with iron oxyhydroxide (β-FeOOH): EXAFS study, J. Colloid Interface Sci. 314 (2) (2007) 427-433. https://doi.org/10.1016/j.jcis.2007.05.071

[74]   A. Sigdel, J. Park, H. Kwak, P. K. Park, (2016) Arsenic removal from aqueous solutions by adsorption onto hydrous iron oxide-impregnated alginate beads, J. Ind. Eng. Chem. 35 (2016) 277-286. https://doi.org/10.1016/j.jiec.2016.01.005

[75]   A. I. Zouboulis, I. A. Katsoyiannis, Arsenic removal using iron oxide loaded Alginate, J. Ind. Eng. Chem. 41(24) (2002) 6149-6155. https://doi.org/10.1021/ie0203835

[76]   R. Chen, C. Zhi, H. Yang, Y. Bando, Z. Zhang, N. Sugiur, D. Golberg, Arsenic(V) adsorption on Fe3O4 nanoparticle-coated boron nitride nanotubes, J. Colloid Interface Sci. 359(1) (2011) 261-268. https://doi.org/10.1016/j.jcis.2011.02.071

[77]   P. Sabbatini, F. Yrazu, F. Rossi, G. Thern, A. Marajofsky, M. M. Fidalgo de Cortalezzi, Fabrication and characterization of iron oxide ceramic membranes for arsenic removal. Water Res 44(19) (2010) 5702-5712. https://doi.org/10.1016/j.watres.2010.05.059

[78]   Z. Veličković, G. D. Vuković, A. D. Marinković, M. S. Moldovan, A. A. Perić-Grujić, P. S. Uskoković, M. D. Ristić, Adsorption of arsenate on iron(III) oxide coated ethylenediamine functionalized multiwall carbon nanotubes, Chem. Eng. J. (181-182) (2012) 174-181. https://doi.org/10.1016/j.cej.2011.11.052

[79]   B. S. Tawabini, S. F. Al-Khaldi, M. M. Khaled, M. A. Atieh, Removal of arsenic from water by iron oxide nanoparticles impregnated on carbon nanotubes, J. Environ. Sci. Health A Tox. Hazard Subst. Environ. Eng. 46(3) (2011) 215-223. https://doi.org/10.1080/10934529.2011.535389

[80]   S. Vadahanambi, S. H. Lee, W. J. Kim, I. O. Kwon, Arsenic removal from contaminated water using three-dimensional graphene-carbon nanotube-Iron oxide nanostructures, Environ. Sci. Technol. 47(18) (2013) 10510-10517. https://doi.org/10.1021/es401389g

[81]   X. Xie, K. Pi, Y. Liu, C. Liu, J. Li, Y. Zhu, C. Su, T. Ma, Y. Wang, In-situ arsenic remediation by aquifer iron coating: Field trial in the Datong basin, China. J. Hazard. Mater. 302 (2016) 19-26. https://doi.org/10.1016/j.jhazmat.2015.09.055

[82]   X. Xie, Y. Wang, K. Pi, C. Liu, J. Li, Y. Liu, Z. Wang, M. Duan, (2015) In situ treatment of arsenic contaminated groundwater by aquifer iron coating: Experimental study, Sci. Total Environ. 527-528 (2015) 38-46. https://doi.org/10.1016/j.scitotenv.2015.05.002

[83]   Y. Glocheux, A. B. Albadarin, J. Galán, E. Oyedoh, C. Mangwandi, C. Gérente, S. J. Allen, G. M. Walker, Adsorption study using optimised 3D organised mesoporous silica coated with Fe and Al oxides for specific As(III) and As(V) removal from contaminated synthetic groundwater, Micropor. Mesopor. Mat. 198 (2014) 101-114. https://doi.org/10.1016/j.micromeso.2014.07.020

[84]   C. S. Jeon, K. Baek, J. K. Park, Y. K. Oh, S. D. Lee, Adsorption characteristics of As(V) on iron-coated zeolites, J. Hazard. Mater. 163(2-3) (2009) 804-808. https://doi.org/10.1016/j.jhazmat.2008.07.052

[85]   Y. F. Pan, C. T. Chiou, T. F. Lin, Adsorption of arsenic (V) by iron-oxide-coated diatomite (IOCD), Environ. Sci. Pollut. Res. Int. 17(8) (2010) 1401-1410. https://doi.org/10.1007/s11356-010-0325-z

[86]   M. G. Mostafa, Y. H. Chen, J. S. Jean, C. C. Liu, Y. C. Le, Kinetics and mechanism of arsenate removal by nanosized iron oxide-coated perlite, J. Hazard. Mater. 187(1-3) (2011) 89-95. https://doi.org/10.1016/j.jhazmat.2010.12.117

[87]   J. L. Matheu, A. J. Gadgil, S. E. Addy, K. Kowolik, Arsenic remediation of drinking water using iron-oxide coated coal bottom ash, J. Environ. Sci. Health. A Tox. Hazard. Subst. Environ. Eng. 45(11) (2010) 1446-1460. https://doi.org/10.1080/10934529.2010.500940

[88]   L. S. Yadav, B. K. Mishra, A. D. Kumar, K. K. Paul, (2014) Arsenic removal using bagasse fly ash- iron coated and sponge iron char, J. Environ. Chem. Eng. 2(3) (2014) 1467-1473. https://doi.org/10.1016/j.jece.2014.06.019

[89]   D. Pokhrel, T. Viraraghavan, Organic arsenic removal from an aqueous solution by iron oxide-coated fungal biomass: An analysis of factors influencing adsorption, Chem. Eng. J. 140(1-3) (2008) 165-172. https://doi.org/10.1016/j.cej.2007.09.038

[90]   T. Sheng, S. A. Baig, Y. Hu, X. Xue, X. Xu, Development, characterization and evaluation of iron-coated honeycomb briquette cinders for the removal of As(V)

from aqueous solutions, Arabian J. Chem. 7(1) (2014) 27-36.
https://doi.org/10.1016/j.arabjc.2013.05.032

[91] T. V. Nguyen, S. Vigneswaran, H. N. Ngo, J. Kandasamy, Arsenic removal by iron oxide coated sponge: Experimental performance and mathematical models, J. Hazard. Mater. 182(1-3) (2010) 723-729.
https://doi.org/10.1016/j.jhazmat.2010.06.094

[92] X. Meng, G. P. K. Ofiatis, S. Christodoulatos, S. Bang, (2001) Treatment of arsenic in Bangladesh wellwater using a household co-precipitation and filtration system, Water Res. 35 (2001) 2805-2810. https://doi.org/10.1016/S0043-1354(01)00007-0

[93] G. Chen, Y. Wang, L. H. M. Tan, X. Yang, L. S. Tan, Y. Chen, H. Y. J. Chen, (2010) Measuring ensemble-averaged surface-enhanced raman scattering in the hotspots of colloidal nanoparticle dimers and trimmers, J. Am. Chem. Soc. 132 (2010) 3644-3645. https://doi.org/10.1021/ja9090885

[94] L. Bai, X. J. Ma, J. F. Liu, X. M. Sun, D. Y. Zhao , D. G. Evans, (2010) Rapid separation and purification of nanoparticles in organic density gradients, J. Am. Chem. Soc. 132(7) (2010) 2333-2337. https://doi.org/10.1021/ja908971d

[95] E. A. Deliyanni, L. K. Nalbandian, A. Matis, (2006) Adsorptive removal of arsenites by a nanocrystalline hybrid surfactant-akaganeite sorbent, J. Colloid Interface Sci. 302 (2006) 458-466. https://doi.org/10.1016/j.jcis.2006.07.007

[96] E. A. Deliyanni, D. N. Bakoyannakis, A. I. Zouboulis, K. A. Matis, Sorption of As (V) ions by akaganeite-type nanocrystals, Chemosphere 50 (2003) 155-163.
https://doi.org/10.1016/S0045-6535(02)00351-X

[97] I. Akin, G. Arslan, A. Tor, M. Ersoz, Y. Cengeloglu, Arsenic(V) removal from underground water by magnetic nanoparticles synthesized from waste red mud, J. Hazard. Mater. (235-236) (2012) 62-68.
https://doi.org/10.1016/j.jhazmat.2012.06.024

[98] C. T. Yavuz, J. T. Oh, W. W. Yu, A. Prakash, J. C. Falkner, S. Yean, L. Cong, H. J. Shipley, A. Kan, M. Tomson, D. Natelson, V. L. Colvin, Low field magnetic separation of monodisperse Fe3O4 nanocrystals, Science 314(5801) (2006) 964-967. https://doi.org/10.1126/science.1131475

[99] V. Chandra, J. Park, Y. Chun, J. W. Lee, I. Hwang, K. S. Kim, Water dispersible magnetite reduced graphene oxide composites for arsenic removal, ACS Nano 4(7) (2010) 3979-3986. https://doi.org/10.1021/nn1008897

[100] A. Khodabakhshi, M. M. Amin, M. Mozaffari, Synthesis of magnetite nanoparticles and evaluation of its efficiency for arsenic removal from simulated industrial wastewater Iran, J. Environ Health Sci. Eng. 8(3) (2011) 189-200.

[101] I. Ali, New generation adsorbents for water treatment, Chem. Rev. 112(10) (2012) 5073-5091 https://doi.org/10.1021/cr300133d

[102] L. Cumbal, J. Greenleaf, D. Leun, A. K. S. Gupta, Polymer supported inorganic nanoparticles: characterization and environmental applications, React. Funct. Polym. 54(1-3) (2003) 167-180. https://doi.org/10.1016/S1381-5148(02)00192-X

[103] M. G. Mostafa, J. Hoinkis, Nanoparticle adsorbents for arsenic removal from drinking water: a review, Int. J. Environ. Sci. Manage. Eng. Res. 1 (2012) 20-31.

[104] G. Z. Kyzas, K. A. Matis, Methods of arsenic wastes recycling: Focus on flotation, J. Mol. Liq. 214 (2016) 37-45. https://doi.org/10.1016/j.molliq.2015.11.028

[105] M. Leist, R. J. Casey, D. Caridi, The management of arsenic wastes: problems and prospects, J. Hazard. Mater. 76(1) (2000) 125-138. https://doi.org/10.1016/S0304-3894(00)00188-6

[106] D. Mohan, J. Pittman, Arsenic removal from water/wastewater using adsorbents-a critical review, J. Hazard. Mater. 142 (2007) 1-53. https://doi.org/10.1016/j.jhazmat.2007.01.006

[107] M. Dannan, S. Dally, F. Conso, Arsenic induced encephalopathy, Neurology 34 (1984) 1529. https://doi.org/10.1212/wnl.34.11.1524

[108] G. L. Dekundt, A. Leonard, J. Arany, G. J. Dubuisson, E. Delavignett, In vivo studies in male mice on the mutagenesis effects of inorganic arsenic, Mutagenesis 1 (1986) 33-34. https://doi.org/10.1093/mutage/1.1.33

[109] O. Axelson, E. Dahlgren, C. D. Jansson, S. O. Rehnuland, Arsenic exposure and mortality, A case Ref study from a Swedish copper smelter, Br. J. Ind. Med. 35 (1978) 8-15.

[110] K. S. Squibb, B. A. Fowler, The toxicity of arsenic and its compounds. In: B. A. Fowler (Ed.), Biological and Environmental effects of arsenic, Elsevier, 1983, pp 233-269. https://doi.org/10.1016/B978-0-444-80513-3.50011-6

[111] H. H. Goebel, P. E. Schmidt, J. Bohl, B. Tettenborn, G. Kramer, L. Guttman, (1990) Poly neuropathy due to arsenic intoxication: Biopsy Studies, J. Neuropathol. Exp. Neurol. (1990), 137-149. https://doi.org/10.1097/00005072-199003000-00006

[112] A. Franzblau, R. Lilis, Acute arsenic intoxication from environmental arsenic exposure, Arch. Environ. Health 44 (1989) 385-390. https://doi.org/10.1080/00039896.1989.9935912

[113] L. K. Bickley, C. M. Papa, Chronic arsenicism with vitiligo, hyperthyroidism and cancer, N. J. Med. 86 (1989) 377-380.

[114] G. R. Hoffman, Genetic Toxicology. In: M. O. Amdur, J. Doull, C. D. Klassen, Toxicology Pergmon, 4th Edition, New York, 1991, pp 201-225.

[115] S. Nordstrom, L. Beckman, I. Nordenson, Occupational and Environmental Risks in and around a smelter in Northern Sweden, Spontaneous abortion among female employers and decreased birth weight in their offspring, Hereditas 90 (1979) 291-296. https://doi.org/10.1111/j.1601-5223.1979.tb01316.x

[116] M. Rahman, M. Tondel, S. A. Ahmad, O. Axelson, Diabetes mellitus associated with arsenic exposure in Bangladesh, Am. J. Epidemiol. 148 (1998) 198-203. https://doi.org/10.1093/oxfordjournals.aje.a009624

[117] IARC, Some drinking water disinfectants and contaminants, including arsenic IARC monographs on the evaluation of carcinogenic risks to humans. International Agency for research on Cancer, WHO, Lyon, 2004, France.

[118] G. R. Kingsley, R. R. Schaffert, Micro determination of arsenic and its application to biological material, Anal. Chem. 23 (1951) 914-919. https://doi.org/10.1021/ac60054a023

[119] W. Goessler, D. Kuehnelt, Analytical methods for the determination of arsenic and arsenic compounds in the environment. In: W. T. Frankenberger Jr, (Ed.) Environmental Chemistry of Arsenic. Marcel Dekker, New York, 2002, pp 27-50.

[120] US Environmental Protection Agency (1994b) Method 2008, Methods for the Determination of Metals in Environmental Samples, Washington DC.

[121] P. A. Gallagher, C. A. Schwegel, X. Wei, J. T. Creed, (2001) Speciation and preservation of inorganic arsenic in drinking water sources using EDTA with IC separation and ICP-MS detection, J. Environ. Monit. 3 (2001) 371-376. https://doi.org/10.1039/b101658j

[122] M. R. Karagas, C. X. Le, S. Morris, J. Blum, X. Lu, V. Spate, M. Carey, V. Stannard, B. Klaue, T. D. Tosteson, Markers of low level arsenic exposure for evaluating human cancer risks in a US population, Int. J. Occup. Med. Environ. Health 14 (2001) 171-175.

[123] D. J. Swart, J. B. Simeonsson, Development of an electrothermal atomization laser-excited atomic fluorescence spectrometry procedure for direct measurements of arsenic in diluted serum, Anal. Chem. 71 (1999) 4951-4955. https://doi.org/10.1021/ac9903508

[124] B. Agahian, J. S. Lee, J. H. Nelson, R. E. Johns, Arsenic levels in fingernails as a biological indicator of exposure to arsenic, Am. Ind. Hyg. Assoc. J. 51 (1990) 646-651. https://doi.org/10.1080/15298669091370293

[125] E. A. Crecelius, Modification of the arsenic speciation technique using hydride generation, Anal. Chem. 50 (1978) 826-827. https://doi.org/10.1021/ac50027a040

[126] Environmental Protection Agency Method (1996c) Inorganic Arsenic in Water by Hydride Generation Quartz Furnace Atomic Absorption. Washington, DC.

[127] K. J. Lamble, S. J. Hill, Arsenic speciation in biological samples by on-line high-performance liquid chromatography-microwave digestion-hydride generation-atomic absorption spectrometry, Anal. Chim. Acta. 344 (1996) 261-270. https://doi.org/10.1016/S0003-2670(96)00348-0

[128] X, C. Le, M. Ma, Speciation of arsenic compounds by using ion-pair chromatography with atomic spectrometry and mass spectrometry detection, J. Chromatogr. 764 (1997) 55-64. https://doi.org/10.1016/S0021-9673(96)00881-3

[129] S. Londesborough, J. Mattusch, R. Wennrich, Separation of organic and inorganic arsenic species by HPLC-ICP-MS, Fresenius, J. Anal. Chem. 363 (1999) 577-581. https://doi.org/10.1007/s002160051251

[130] W. H. Holl, Mechanisms of arsenic removal from water, Environ. Geochem. Health, 32 (2010) 287-290. https://doi.org/10.1007/s10653-010-9307-9

[131] A. M. Ingallinella, V. A. Pacini, R. G. Fernandez, R. M. Vidoni, G. Sanguinetti, Simultaneous removal of arsenic and fluoride from groundwater by coagulation-adsorption with polyaluminum chloride, J. Environ. Sci. Health A 46 (2011) 1288-1296. https://doi.org/10.1080/10934529.2011.598835

[132] B. An, Q. Liang, D. Zhao, Removal of arsenic (V) from spent ion exchange brine using a new class of starch-bridged magnetite nano particles, Water Res. 45-5 (2011) 1961-1972. https://doi.org/10.1016/j.watres.2011.01.004

[133] L. A. Richards, B. S. Richards, H. M. A. Rossiter, A. I. Schafer, Impact of speciation on fluoride, arsenic and magnesium retention by nanofiltration/reverse osmosis in remote Australian communities, Desalination 248 (2009) 177-183. https://doi.org/10.1016/j.desal.2008.05.054

[134] I. A. Katsoyiannis, A. I. Zouboulis, (2006) Comparative evaluation of conventional and alternative methods for the removal of arsenic from contaminated ground waters, Rev. Environ. Health 21(1) (2006) 25-41. https://doi.org/10.1515/REVEH.2006.21.1.25

[135] B. V. D. Bruggen, Advances in electro dialysis for water treatment, Advances in Membrane Technologies for Water Treatment, 2015, pp. 185-203.

[136] E. Smith, R. Naidu, A. M. Alston, Arsenic in the soil environment: A review, Adv. Agron. 64 (1998) 149-195. https://doi.org/10.1016/S0065-2113(08)60504-0

[137] M. E. O. Escobar, N. V. Hue, W. G. Cutler, Recent developments in arsenic: contamination and remediation. In: S. G. Pandalai (Ed.) Recent Research Developments in Bioenergetics, 2006, pp 4:1-6.

[138] US Environmental Protection Agency (US EPA) 2000 Office of research and development, Introduction to phytoremediation. EPA/600/R-99/107.

[139] L. C. Allen, Electronegativity is the average one-electron energy of the valence-shell electrons in ground-state atoms, J. Am. Chem. Soc. 111(25) (1998) 9003-9014. https://doi.org/10.1021/ja00207a003

[140] M. F. Hossain, Arsenic contamination in Bangladesh-An overview, Agri. Ecosys. Env. 113 (2006) 1-16. https://doi.org/10.1016/j.agee.2005.08.034

[141] G. Mu-iz, V. Fierro, A. Celzard, G. Furdin, G. G. Sánchez, M. L. Ballinas, Synthesis, characterization and performance in arsenic removal of iron-doped activated carbons prepared by impregnation with Fe (III) and Fe (II), J. Hazard. Mater. 165(1-3) (2009) 893-902. https://doi.org/10.1016/j.jhazmat.2008.10.074

[142] Z. Liu, F. S. Zhang, R. Sasai, Arsenate removal from water using Fe3O4-loaded activated carbon prepared from waste biomass, Chem. Eng. J. 160(1) (2010) 57-62. https://doi.org/10.1016/j.cej.2010.03.003

[143] A. O. A. Tuna, E. Özdemir, E. B. Şimşek, U. Beker, Removal of As (V) from aqueous solution by activated carbon-based hybrid adsorbents: Impact of experimental conditions, Chem. Eng. J. 223 (2013) 116-128. https://doi.org/10.1016/j.cej.2013.02.096

[144] V. Lenoble, O. Bouras, V. Deluchat, B. Serpaud, J. C. Bollinger, Arsenic adsorption onto pillared clays and iron oxides, .J Colloid Interf. Sci. 255(1) (2002) 52-58. https://doi.org/10.1006/jcis.2002.8646

[145] B. J. Lafferty, R. H. Loeppert, Methyl arsenic adsorption and desorption behavior on iron oxides, Environ. Sci. Technol. 39(7) (2005) 2120-2127. https://doi.org/10.1021/es048701+

[146] W. Tang, Q. Li, S. Gao, J. K. Shang, Arsenic (III, V) removal from aqueous solution by ultrafine α-Fe2O3 nanoparticles synthesized from solvent thermal method, J. Hazard. Mater. 192(1) (2011) 131-138. https://doi.org/10.1016/j.jhazmat.2011.04.111

[147] K. P. Raven, A. Jain, R. H. Loeppert, Arsenite and arsenate adsorption on ferrihydrite: kinetics, equilibrium, and adsorption envelopes, Environ. Sci. Technol. 32 (1998) 344-349. https://doi.org/10.1021/es970421p

[148] Z. Bujňáková, P. Baláž, A. Zorkovská, M. J. Sayagués, J. Kováč, M. Timko, Arsenic sorption by nanocrystalline magnetite: An example of environmentally promising interface with geosphere, J. Hazard. Mater. 262 (2013) 1204-1212. https://doi.org/10.1016/j.jhazmat.2013.03.007

[149] S. Lin, D. Lu, Z. Liu, Removal of arsenic contaminants with magnetic γ-Fe2O3 nanoparticles, Chem. Eng. J. 211-212 (2012) 46-52. https://doi.org/10.1016/j.cej.2012.09.018

[150] L. Feng, M. Cao, X. Ma, Y. Zhu, C. Hu, Superparamagnetic high-surface-area Fe3O4 nanoparticles as adsorbents for arsenic removal, J. Hazard. Mater. 217-218 (2012) 439-446. https://doi.org/10.1016/j.jhazmat.2012.03.073

[151] M. C. S. Faria, R. S. Rosemberg, C. A. Bomfeti, D. S. Monteiro, F. Barbosa, L. C. A. Oliveira, M. Rodriguez, M. C. Pereira, J. L. Rodrigue, Arsenic removal from contaminated water by ultrafine δ-FeOOH adsorbents, Chem. Eng. J. 237 (2014) 47-54. https://doi.org/10.1016/j.cej.2013.10.006

[152] G. Zhang, Z. Ren, X. Zhang, J. Chen, Nanostructured iron(III)-copper(II) binary oxide: A novel adsorbent for enhanced arsenic removal from aqueous solutions, Water Res 47(12) (2013) 4022-4031. https://doi.org/10.1016/j.watres.2012.11.059

[153] T. Basu, D. Nandi, P. Sen, U. C. Ghosh, Equilibrium modeling of As (III,V) sorption in the absence/presence of some groundwater occurring ions by iron(III)-cerium(IV) oxide nanoparticle agglomerates: A mechanistic approach of surface interaction, Chem. Eng. J. 228 (2013) 665-680. https://doi.org/10.1016/j.cej.2013.05.037

[154] Y. Zhang, X. Dou, B. Zhao, M. Yang, T. Takyama, S. Kato, (2010) Removal of arsenic by a granular Fe-Ce oxide adsorbent: Fabrication conditions and

performance, Chem. Eng. J. 162(1) (2010) 164-170.
https://doi.org/10.1016/j.cej.2010.05.021

[155] W. Zhang, J. Fu, G. Zhang, X. Zhang, (2014) Enhanced arsenate removal by novel Fe-La composite (hydr)oxides synthesized via coprecipitation, Chem. Eng. J. 251 (2014) 69-79. https://doi.org/10.1016/j.cej.2014.04.057

[156] A. Z. M. Badruddoza, Z. B. Z. Shawon, M. T. Rahman, K. W. Hao, K. Hidajat, M. Shahabuddin, Ionically modified magnetic nanomaterials for arsenic and chromium removal from water, Chem. Eng. J. 225 (2013) 607-615. https://doi.org/10.1016/j.cej.2013.03.114

[157] Y. Jin, F. Liu, M. Tong, Y. Hou, Removal of arsenate by cetyl tri methyl ammonium bromide modified magnetic nanoparticles, J. Hazard. Maters. (227-228) (2012) 461-468.

[158] M. A. Armienta, R. Rodriguez, A. Aguayao, N. Ceniceros, F. Juraez, O. Cruz, G. Villasenor, Point and regional sources of arsenic in the groundwater of Zimapan, Mexico, Acta Universitatis Carolinae Geologica 39 (1995) 285-290.

[159] W. Lianfang, H. Jianzhong, Chronic arsenicism from drinking water in some areas of Xinjiang, China. In: J. Nriagu (Ed), Arsenic in the Environment, Part II: Human Health and Ecosystem Effects. John Wiley Inc, New York, 1994, pp 159-172.

[160] Public Health Engineering Department (1993) First phase Action Plan Report on Arsenic Pollution in Groundwater in West Bengal. Arsenic Investigation Project, Government of West Bengal, p 46.

[161] P. Bagla, J. Kaiser, India's spreading health crisis draws global arsenic experts, Science 274 (1996) 174-175. https://doi.org/10.1126/science.274.5285.174

[162] A. H. Welch, D. B. W. John, D. R. Helsel, R. B. Wanty, Arsenic in ground water of the United States-occurrence and geochemistry, Ground Water 38(4) (2000) 589-604. https://doi.org/10.1111/j.1745-6584.2000.tb00251.x

[163] J. Bundschuh, B. Farias, R. Martin, A. Storniolo, P. Bhattacharya, J. Cortes, G. Bonorino, R. Albouy, Groundwater arsenic in the Chaco-Pampean Plain, Argentina Case study from Robles County, Santiago del Estero Province, Appl Geochem 19(2) (200) 231-243.

[164] J. M. Borogo, N. Vicent, H. Venturino, H. Infante, Arsenic in the drinking water of the city of Antofagasta: Epidemiological and clinical study before and after the installation of a treatment plant, Environ. Health. Perspect. 19 (1977) 103-105. https://doi.org/10.1289/ehp.7719103

Chapter 10

Adsorption and ion exchange: basic principles and their application in food processing

Suvardhan Kanchi[1*], Myalownkosi I Sabela[1], Krishna Bisetty[1*], Shakeel Ahmed[2]

[1]Department of Chemistry, Durban University of Technology, Durban, South Africa

[2]Department of Chemistry, Jamia Millia Islamia (A Central University), New Delhi-110025, India

ksuvardhan@gmail.com

Abstract

This chapter provides a critical review for a wide range of ion-exchange chromatography applications which is an integral part of our food processing industry. The characteristics of ion exchange chromatography are reviewed with particular emphasis on the basic principle and its column materials using state-of-the-art analytical instrumentation coupled with different detectors for analysis.

Keywords

Adsorption, Chromatography, Anion-Exchange Chromatography, Cation-Exchange Chromatography, Food Applications

Contents

1. Introduction

Ion-exchange chromatography is named on the basis of the exchangeable counterion. It is an important separation technique and is a part of ion chromatography for the determination of ionic compounds, together with ion-partition/interaction and ion-exclusion chromatography [1]. It is based on ionic or electrostatic interactions between ionic or polar analyses, ions present in the eluent and ionic functional groups present in the chromatographic support. When the stationary phase bears a positive charge and the exchangeable ion is an anion, the process is referred to as anion-exchange chromatography. If the opposite is true, it is referred to as cation-exchange chromatography.

1.1 Principle of ion-exchange chromatography

The procedure is limited to the purification of ionizable molecules because IEC separates molecules on the basis of their charged groups, which cause them to interact electrostatically with opposite charges on the stationary-phase matrix. Therefore, the stationary phase carries ionizable functional groups coupled to an inert matrix material. Because of the principles of electroneutrality, these immobilized charges are electrostatically associated with exchangeable counterions from solution. Charged molecules to be purified compete with these counterions for binding to the charged groups on the stationary phase and are thereby retained on the basis of their charge. Different types of molecules will bind to the matrix with affinities that depend on both the conditions used and a number of individual charged groups either anionic or cationic. These differences lead to the resolution of various types of molecule and bio-macromolecules by ion-exchange chromatography. For example, a typical protein purification can be achieved using ion exchange chromatography by applying onto the column, a mixture of many proteins derived from a bacterial cell. After molecules that do not bind are washed away, conditions can be gradually adjusted, such as by increasing the concentration of a simple counterion or by altering the pH, to release the molecule of interest from the stationary phase. Molecules with different charges will elute at specific points in the chromatography as adjustments are made. The ion-exchange chromatography works in five different steps:

Step-1:

Equilibration is the first step, in which the ion exchanger is brought to an initial state after conditioning in terms of pH and ionic strength, which allows the binding of the desired solute molecules. In this step, the groups present in the ion-changer interactions with the exchangeable counterions.

Step-2:

The solute molecules carry the appropriately charged counter-ions and interact reversibly to the gel through adsorption in which the loosely bound substances from the column bed can be washed out with buffer solution.

Step-3:

Desorption process takes place by adjusting the pH of the buffer solution which leads to the unfavourable conditions for ionic bonding of the solute molecules. The mechanism in this step involves desorption of solute molecules from the surface of the column bed by increasing salt concentration gradient. Therefore, the elution order depends on the order of their bonding strengths, the most weakly bound substances being eluted first.

Step-4&5:

The last two steps involve the cleaning of un-eluted substances in Step-3 and re-conditioning for Step-1. The mechanism is illustrated in Figure 1.

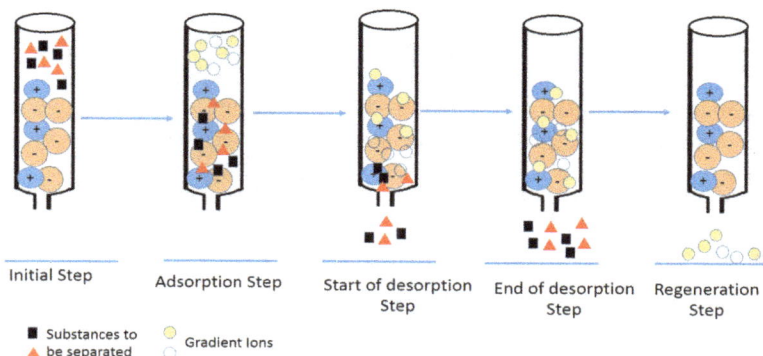

| Initial Step | Adsorption Step | Start of desorption Step | End of desorption Step | Regeneration Step |

■ Substances to ▲ be separated   ○ Gradient Ions

Fig. 1 Schematic illustration of Principle of ion exchange chromatography.

1.2   Classification of ion-exchange materials

Ion exchange materials are insoluble substances containing loosely held ions that are exchangeable with other ions in solutions as they come in contact with them. These exchanges take place without any physical alteration to the ion exchange material. Ion exchangers are insoluble acids or bases which salts are also insoluble, and this enables them to exchange either positively charged ions (cation exchangers) or negatively charged ones (anion exchangers). Many natural substances such as proteins, cellulose, living cells and soil particles exhibit ion-exchange properties which play an important role in the way they function in nature.

The materials in the ion-exchange chromatography is categorized as:

a) Synthetic inorganic ion-exchangers

Different researchers [2,3] have reported on the synthesis of inorganic materials, $Na_2Al_2Si_3O_{10}$ in which sodium ion is available for exchange. In a later stage, exchange enhanced "gel permutits" were synthesized [4]. Interestingly, this material has similar properties like zeolites [5]. Zeolites exhibit completely regular crystal structure because of their narrow, rigid and strictly uniform pore structure, they act as "molecular sieves" [6].

Major efforts have also been put forward to synthesise inorganic cation exchangers using tetra and trivalent elements by substituting the silicon. Fe_2O_3, Al_2O_3, Cr_2O_3, TiO_2, ZrO_2, ThO_2, SnO_2, MoO_2 and WO_3 are amphoteric in nature however, these materials are being used as cationic exchangers at pH above their isoelectric points [7,8].

b) Synthetic organic ion-exchangers

Synthetic organic ion-exchange resins are the most import class of ion exchangers and are also named as organic ion exchangers due to their synthesis from organic substances. These framework consists of an irregular macromolecular three-dimensional network of hydrocarbon chains and carries ionic group such as

$-SO_3^{2-}$, $-COO^-$, $-PO_3^{2-}$, $-ASO_4^{2-}$ in cation exchangers and $-NH_3^+$, $=NH_2^+$

in anion exchangers and these are crosslinked polyelectrolytes.

The synthetic organic ion-exchange resins are prepared by two types of polymerization reactions which include:

i) Condensation polymerization in anionic and cationic ion exchangers

Condensation polymers are formed by reacting polyfunctional molecules in such a way that new carbon-carbon, carbon-nitrogen or carbon-oxygen bonds are formed with elimination of water etc. *m*-phenylenediamine, aromatic amine is condensed with formaldehyde to form Anionic-exchange resins as shown in the below scheme:

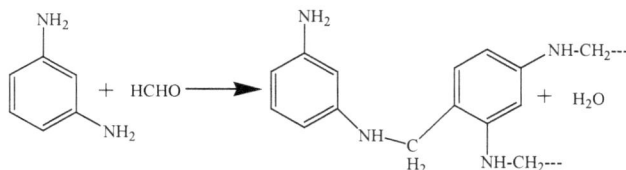

The amino substituents which are attached directly to the benzene ring are weakly basic and its strength can be enhanced by alkylation using dimethyl sulphate as an alkylating agent [9].

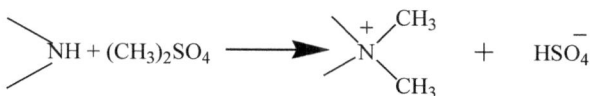

Anionic –exchange resins can also be synthesised using aliphatic polyamines which result in high base strength resins in contrast with aromatic amines. This is a condensation reaction with aldehydes and phenols or aromatic amines as shown in the below schematic illustration [10].

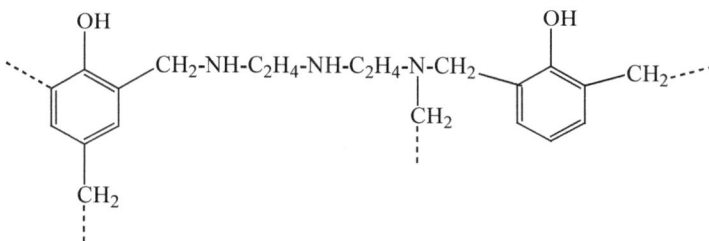

Anionic-exchange resins are also prepared using dichloroethane instead of an addition to an aldehyde. Aliphatic polyamines and epoxides such as epichlorohydrin are synthesised and the structure is shown below [11]:

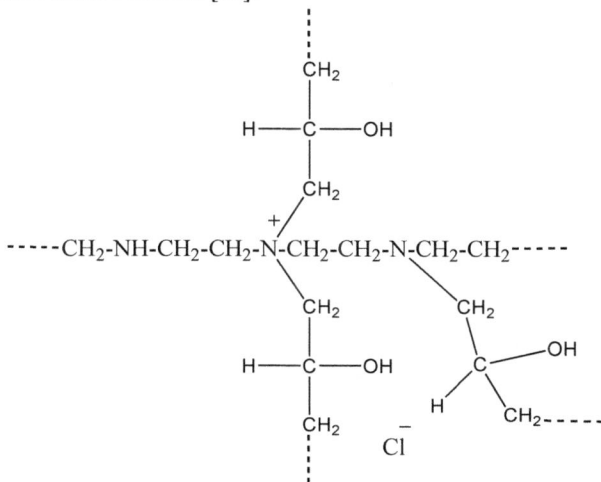

Anionic-exchange resins are synthesized by condensation process of phenol and formaldehyde in the presence of ammonium sulphate [12].

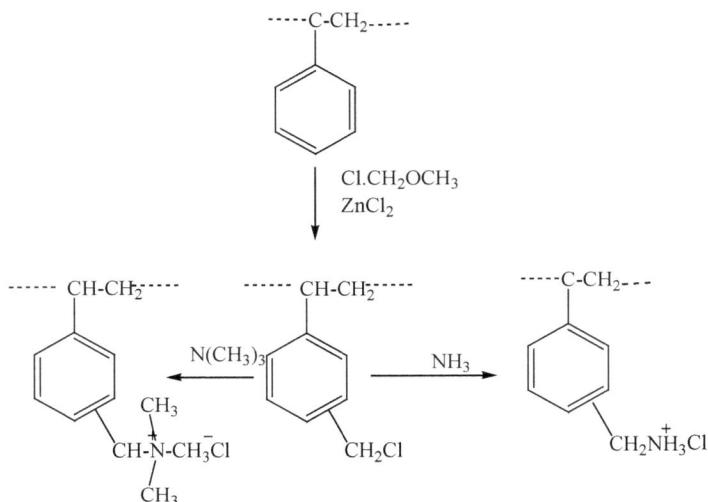

$$-----C-CH_2-----$$

Cl.CH$_2$OCH$_3$
ZnCl$_2$

N(CH$_3$)$_3$ NH$_3$

CH$_3$
CH-N-CH$_3$Cl   CH$_2$Cl   CH$_2$NH$_3$Cl
CH$_3$

Quaternary phosphonium groups based resins are synthesized by interacting the chlormethylated polymer with a tris-dialkylaminophosphate, as shown in the below reaction [13]. The resulted resin has strong basic in nature.

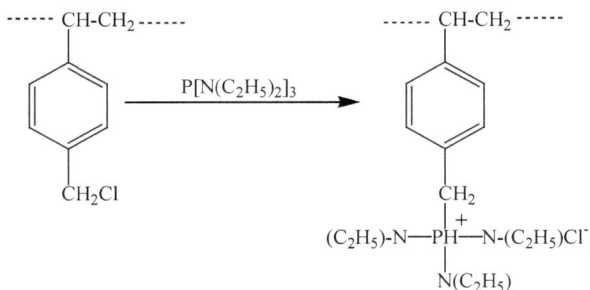

P[N(C$_2$H$_5$)$_2$]$_3$

CH$_2$Cl   CH$_2$
(C$_2$H$_5$)-N—PH—N-(C$_2$H$_5$)Cl
N(C$_2$H$_5$)

Strong basic monofunctional resins are also prepared by copolymerization of *p*-dimethylamidostyrene with a crosslinking formaldehyde. The typical structure of such resins is shown below:

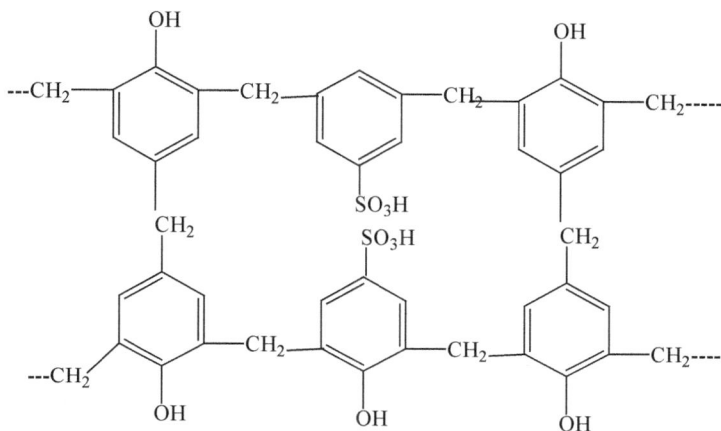

Cationic-exchangers are synthesized by phosphorylation of styrene furfuraldehyde copolymer and also by acetylation followed by phosphorylation of the same polymer [14].

ii) Addition polymerization in anionic and cationic ion-exchangers

In case of addition polymerization, olefinic compounds containing carbon-carbon bonds as the links will join as monomers. These polymers have the same empirical composition as the monomers used in their preparation and have a higher chemical and thermal stability than the condensation polymers. The degree of crosslinking and the particle size of the products are more readily controlled in this technique.

The commercially cationic exchange resins are prepared by addition of copolymers synthesized from vinyl monomers, which have high thermal and chemical stability when compared with the polymers prepared through a condensation process. Cross-linked polystyrenes with sulphonic acid groups are the important cationic exchange resins. The sulphonic acids groups are introduced after polymerization by treatment with concentrated sulphuric acid or chlorosulfonic acid as shown in below scheme:

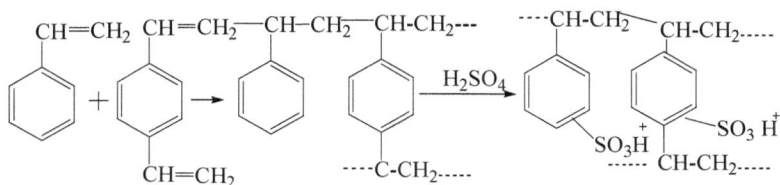

The most widely used weak acid cationic exchangers are prepared from acrylic acid and methacrylic acid as shown below:

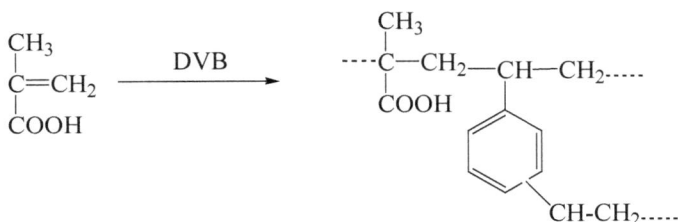

c) Coals or carbonaceous based ion-exchangers

When certain colas are treated with hot concentrated H_2SO_4, yielded products have the capability for ion-exchange. The so formed materials are physically and chemically stable towards acids and to some extent towards basic medium [15].

Sulphonation is the successful method for conversion of lignite and bituminous coals and anthracites to cationic exchangers using fuming sulphuric acid which results in the addition of sulphonic groups. Carboxylic groups are also added by the oxidation process. In certain aspects, sulfonated coals resembled organic ion-exchange resins [16,17].

d) Natural and modified organic ion-exchangers

Natural materials such as horns, hair, wool, leather and cellulose possess ion-exchange properties due to their proteinaceous nature also contain both basic and acidic residues, therefore these substances behave as amphoteric materials. Generally, cellulose contains aliphatic alcoholic groups which easily undergoes oxidation and results in the formation of carboxylic groups by sodium hypochlorite or nitrogen tetroxide and consequently, they exhibited carboxylic cation - exchange properties [18].

e) Mineral ion-exchangers

All mineral ion exchangers are natural materials in which aluminosilicates are crystalline in nature and exhibits cationic properties with open three-dimensional framework and aluminosilicate lattice. For example, Kaolinite (Figure 2) is an important clay mineral of minerals of $Al_4(Si_4O_{10})(OH)_8$ resulted from weathering of feldspar group of minerals, (K, Na)($AlSi_3O_8$), or Ca($Al_2Si_2O_8$).

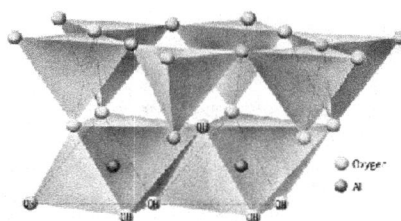

Fig. 2 Structure of kaolinite [19].

The two layers are shown above, the bottom octahedral layer represents the oxygen atoms and the 6-coordinated Al atoms or ions. Some of these oxygen atoms are shared with 4-co-ordionated silicon on the top layer, each tetrahedron represents a SiO_4 group.

On the other hand, very few aluminosilicates such as montmorillonite and feldspars which belonging to the sodalite and cancrinite exhibit anionic properties. However, the only mineral anion exchangers which have been used for practical purposes are apatite $[Ca_5(PO_4)_3]F$ and hydroxylapatite $[Ca_5(PO_4)_3]OH$ [20].

2. Applications

Together with the development of more sophisticated synthetic ion exchange and adsorbent resins, the number of potential applications on industrial scale rose as well. In recent decades, profound knowledge has been accumulated concerning the structure and properties of ion exchange and adsorbent resins.

Simultaneous separation and analysis of caffeine, theobromine and theophylline in Japanese foods were developed by Qing-chuan Chen and co-workers using two ion chromatographic methods [21]. These separations were based on isocratic elutions using UV detector at 274 nm. By considering the molecular structures of three alkaloids, these exist mainly as neutral molecules in neutral and weak acidic aqueous solutions and can be separated by reverse phase liquid chromatography (RPLC) with acetate or phosphate buffer solution containing an organic solvent. In principle, these alkaloids will be converted to cationic and anionic forms in strong acid and basic solutions, respectively, hence it is possible to separate them by cation-exchange or anion-exchange chromatography. In cation-exchange chromatography, HCl was used as a typical eluent rather than HNO_3 due to their oxidizing nature. The effects of the HCl concentration on the retention behaviour of the three alkaloids were also studied. The authors found that the retention time for theophylline reaches a maximum when eluent concentration was 20 mmol L^{-1} whereas for caffeine and theobromine, the retention times were maximum when the eluent concentrations were 100 mmol L^{-1} and decrease with further increase in the eluent concentration. In anion-exchange chromatography, Dionex IonPac AS4SC in hydrophobic medium and KOH as an eluent was used in this study. The obtained results showed that caffeine, theobromine and theophylline were sequentially eluted, and the retention times of theobromine and theophylline decreased when the KOH concentration increased from 2.5 to 100 mmol L^{-1}.

The obtained LOD's for the three compounds were all below sub-μg mL^{-1} level. These methods have been successfully applied to the determination of these three compounds in foods and pharmaceutical preparations with average recoveries ranging from 87% to 103%. The results obtained from this study revealed that ion-exchange chromatography is possibly a good alternative to conventional liquid chromatography for the separation and determination of some water-soluble, neutral organic molecules with certain hydrophobicity.

Novel high-performance anion-exchange chromatographic method coupled to pulse amperometric detector was developed for the characterization and determination of disaccharide lactose and three major bovine milk oligosaccharides (BMO) in dairy streams [22]. This method is having an advantages such as of minimal sample preparation and achieves good chromatographic separation of oligosaccharide isomers within 30 min due to the pulse amperometric detection. The linear dynamic range and LOD's were found to be 0.1 to 10 mg L^{-1} and 0.03 to 0.22 mg L^{-1}, respectively. The average recoveries of the BMO were excellent and ranged from 98.4 to 100.4%.

Separation protocol for the identification and quantification of artificial sweeteners (sodium saccharin, aspartame, acesulfame-K), preservatives (benzoic acid, sorbic acid), caffeine, theobromine and theophylline have been reported by Qing-Chuan Chen and Jing Wang [23]. The separation was performed within 45 min on an anion-exchange analytical column operated at 40 °C by an isocratic elution with 5 mM aqueous NaH_2PO_4 (pH 8.20) solution containing 4% (v/v) acetonitrile as eluent, while UV detection was employed. The LOD's for all sweeteners and preservatives were found to be below the sub-mg mL^{-1} level under optimal conditions. The method has been successfully applied for the analysis of various food and pharmaceutical preparations, and the average recoveries for real samples ranged from 85 to 104%.

Commercial anion-exchange resins were used to develop a preparative-scale ion-exchange chromatographic method for the separation of the proteins α-lactalbumin, β-lactoglobulin, bovine serum albumin, immunoglobulin G and lactose from a sweet dairy whey mixture. This chromatographic set-up was designed by the Chemical Technology Division at Oak Ridge National Laboratory, Oak Ridge, TN, USA. The apparatus consists of the preparative columns used, 20-l Nalgene polypropylene tanks for feed, buffers, cleaning solutions and fractions collection, a pumping system, associated piping, a diode array ultraviolet spectrophotometer and a pH meter, connected in series as illustrated in Figure 3.

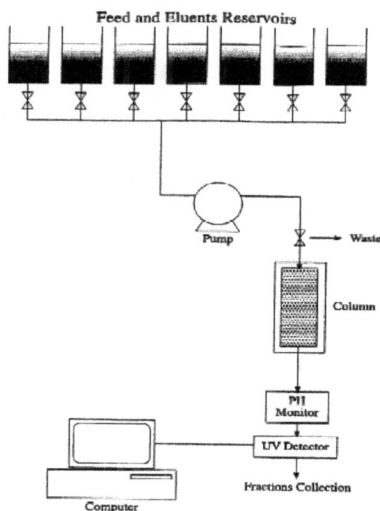

Fig. 3 Schematic diagram of chromatographic set-up [24].

In this experimental study, anion-exchange chromatography was chosen as the initial step for the separation of proteins from the whey due to their differences in the isoelectric points. The isoelectric point, pI, is the pH of the solution at which the net charge on the protein is zero. If the pI of α-lactalbumin, β-lactoglobulin, bovine serum albumin, immunoglobulin G and lactose are above the pH of the medium, then the protein will bind to a cation resin. On the other hand, if the protein has a pI below the pH of the medium, it will bind to an anion resin [24].

Soybean β-conglycinin has been described as a functional food ingredient and it is available commercially in all Asian countries. It is a trimeric glycoprotein having complex protein but consists of health promoting properties. A survey of the literature reveals that very few reports have been available in the literature and it is of great interest. Therefore, Miryam Amigo-Benavent and co-workers reported the two-step ion exchange chromatographic method for the separation of α, α' and β subunits of glycoproteins. In this study, chromatographic conditions were optimized to obtain β-conglycinin subunits that are free of contaminating proteins. In the first step, β subunit was separated by means of anionic exchange fast protein liquid chromatography with a retention time of 11 min and α and α' separated together at 17-19 mins and in the second step, α and α' chains were separated at 1.6 and 23.7 min by cationic exchange as shown in Figure 4.

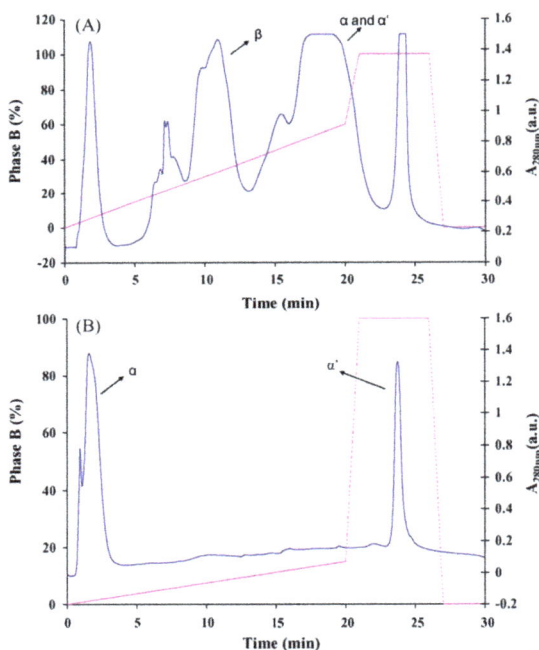

Fig. 4. Ionic chromatographic profiles of β-conglycinin (A) obtained by anionic exchange chromatography and α subunits (B) obtained by cationic exchange chromatography [25].

Two steps based ion exchange chromatography is a sufficient and suitable method to obtain β-conglycinin individual polypeptide chains. This method allowed for obtaining high purity α, α' and β subunits and may be up-scaled for the industrial purpose [25].

Xian-Bing Xu and co-workers developed high-performance anion-exchange chromatography coupled to electrospray ionization mass spectrometry, with a new interface to improve the sensitivity of sugar analysis as shown in Figure 5 [26].

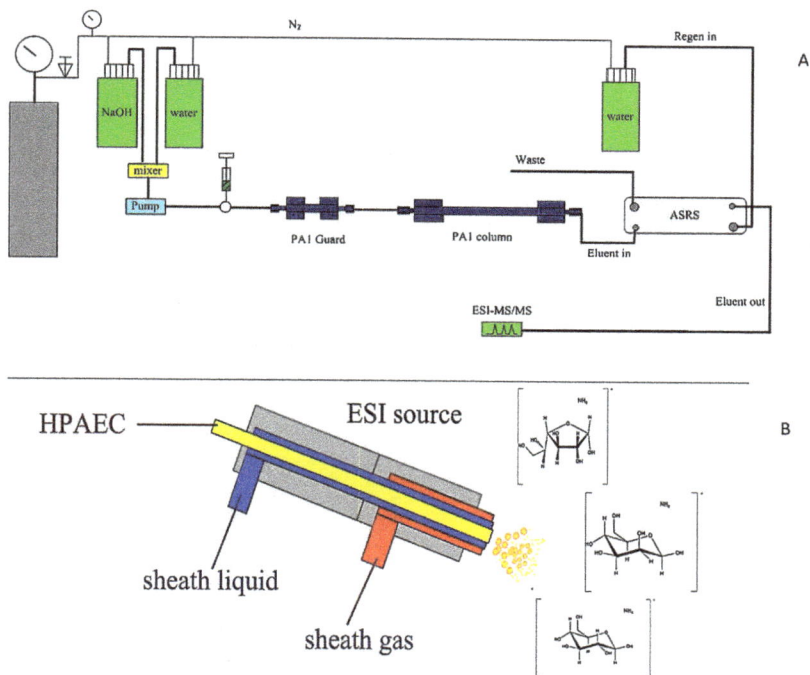

Fig. 5 Schematic diagram of high-performance anion-exchange chromatography coupled to electrospray ionization mass spectrometry for (A) sugars and (B) structure of ESI source with sheath liquid interface [26].

The electrospray ionization contains a sheath liquid with 50 mM ammonium acetate in isopropanol and 0.05% acetic acid, infused at a flow rate of 3 μl min^{-1} at the tip of the ESI probe, requires less ESI source cleaning and promotes efficient ionization of mono- and di-carbohydrates. The obtained results from this investigation suggest that use of a sheath liquid interface rather than a T-joint allows volatile ammonium salts to replace non-volatile metal salts as modifiers for improving sugar ESI signals. The efficient ionization of mono- and di-carbohydrates in the ESI source is affected by the sheath liquid properties such as buffer concentration and type of organic solvent. This method is a powerful tool for the separation, identification, and characterization of co-eluted sugars in pectin samples at trace concentrations [26].

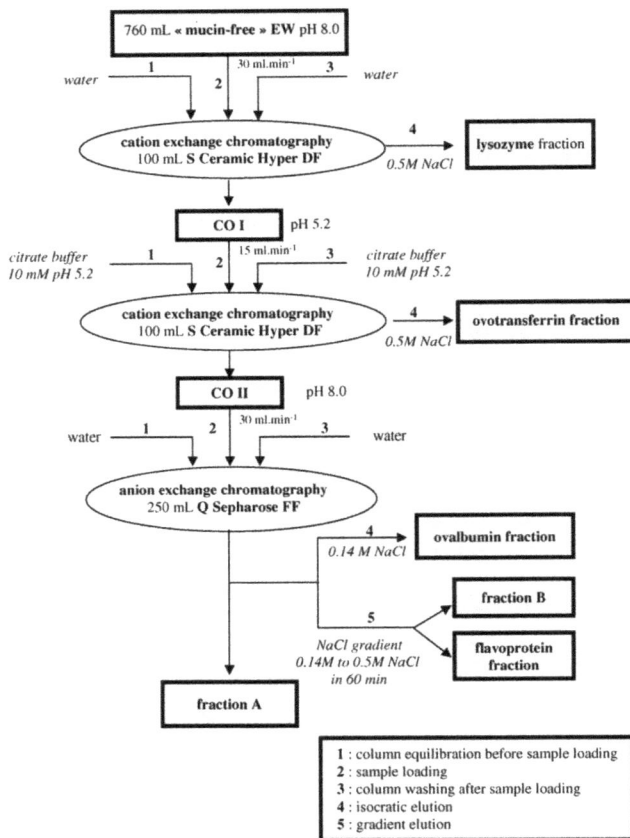

Fig. 6 Hen egg white fractionation process based on ion exchange chromatography. "mucin-free" EW was "mucin-free" egg white protein; COI and COII, were co-product I and II, respectively [27].

Due to the functional biological properties of hen's egg white protein, it is still interesting to enhance the knowledge on the biomolecules and their characterization of egg white portion. Guerin-Dubiard and his co-workers developed a novel protocol to fractionate the mucin free hen egg white using ion exchange chromatography [27]. The Mucin free egg white was fractionated in six steps as shown in Figure 6. The obtained results indicated that the four proteins such as lysozyme, ovotransferrin, ovalbumin and flavoprotein were well characterized with high percentage recoveries of 95, 89, 91 and 100% respectively from the purified fractions, while other two (ovomucoid, ovoglycoprotein) fractions were enriched in recently detected minor proteins in hen egg white due to difficulty in the separation.

Therefore, ion-exchange chromatography is an alternative for purifying ovomucoid, which could be the most convenient for further biological activity studies.

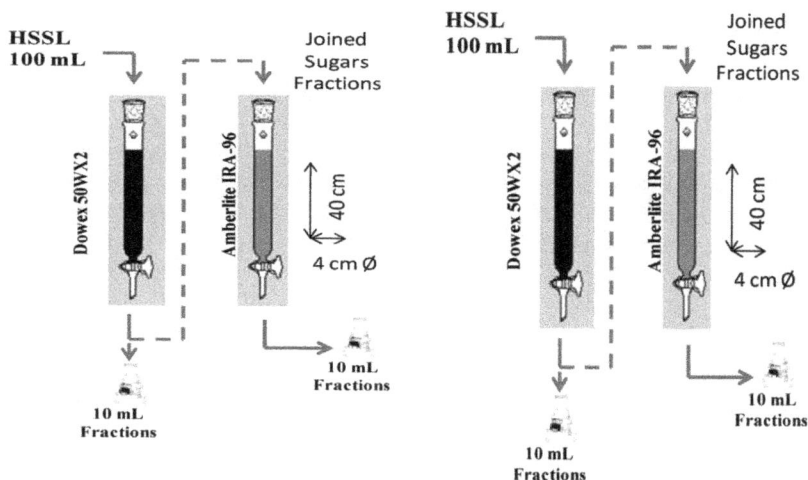

Fig. 7 Schematic representation for the fractionation of sulphite spent liquor from hard wood using cation and anion ion exchange resins connected in series [28].

Sulphite spent liquor is a side product from acidic sulphite pulping of wood, which organic counterpart is composed mainly of lignosulphonates and sugars. The sulphite spent liquor obtained from hardwood is a complex mixture of organic compounds containing lignosulphonates, monosaccharides, oligosaccharides, acetic acid and phenolics and inorganic salts (mainly salts of magnesium and calcium) which were fractionated by using consecutive cation- and anion-exchange resin beds. Keeping this in view, ion exchange chromatographic method was designed in order to fractionate hardwood for the separation of sugars from concomitant inhibitors, such as lignosulphonates, acetic acid, furan derivatives, phenolics, acetic acid and excess of inorganic salts [28]. This experiment has been comprised of two fixed-bed which include cationic and anion ion exchangers connected in series as shown in Figure 7.

The cation exchange column was packed with Dowex 50WX2 resin which eliminates free cations and partially separate sugars from high molecular weight lignosulphonates and furan derivatives. The anion exchange column was packed with Amberlite IRA-96 sorbed remaining lignosulphonates, phenolics and acetic acid. The elution of hardwood sulphite spent liquor through the cation exchange resin bed allowed the fractionation of lignosulphonates, carbohydrates and acetic acid. The high molecular weight fraction of lignosulphonates and a series of sugar-derived furan derivatives were successfully separated from carbohydrate fraction, which still contained monomeric/dimeric lignosulphonates and phenolic compounds. The cations were almost completely retained by cation exchange resin, whereas 5-10 % of Mg^{2+} and Ca^{2+} were jointly eluted with lignosulphonates. This last fact was explained by competition of sulphonic groups of resin and lignosulphonates towards Mg^{2+} and Ca^{2+}. The complete separation of sugars from the remaining lignosulphonates, phenolics and acetic acid was possible after elution through a subsequent anionic resin bed. Such two-step fractionation of hardwood sulphite spent liquor allowed the purification of a sugar solution suitable for subsequent bioprocessing, once the major inhibitors (lignosulphonates, phenolics, acetic acid and high levels of cations) have been eliminated. Therefore, the obtained results showed that the removal of Mg^{2+} (99.99%), Ca^{2+} (99.0%), lignosulphonates (99.6%) and acetic acid (100%), whereas the yield of recovered sugars was at least 72% of their total amount in hardwood sulphite spend liquor [28].

The separation of peptides based on their physicochemical properties offers a selective fractionation not yet used in industrial scale. Cross-flow electro membrane filtration and ion-exchange membrane adsorption chromatography were successfully developed for the fractionation of complex peptide mixtures. In this study, the fractionation steps were generated by tryptic hydrolysis of aqueous solutions of β-lactoglobulin and micellar casein. The fractionation efficiency of both processes, to separate bioactive peptides was determined by analysing the yield of the angiotensin-I-converting enzyme (ACE)-inhibitory peptides in the obtained fractions and by measuring their ACE-inhibitory activities in vitro [29]. While the application of cross-flow electro membrane filtration enables the selective fractionation of a complex peptide mixture into the permeate and retentate fractions, ion-exchange membrane adsorption chromatography offers the opportunity to generate a high amount of fractions by choice of the right elution gradient profile. Because of the high diversity of the inserted hydrolysate in total 13 fractions could be separated. Both techniques were successfully applied for the fractionation of complex hydrolysates and the ACE-inhibitory peptides were enriched in individual fractions, resulting in yields of 52% for β-lactoglobulin (9-14) by ion-exchange membrane adsorption chromatography and 25% for β-casein f(108-113) by cross-flow electro membrane filtration [29].

A novel direct anion chromatographic method coupled to the conductometric detector was reported by Krokidis et al [25] for the determination of EDTA. The anion exchange column was made up of Dionex AS-14 with a dimension of 4 mm × 250 mm. To avoid the interference effect, 10 mM carbonate buffer with pH 11.0 or 10.5. The obtained retention time for EDTA was 5.5 and 9.4 min, respectively. In the case of 120 mM borate buffer with pH 8.5, the retention time was 16.2 min. For 10 mM carbonate buffer with pH 11.0 and isocratic flow rate of 1.0 mL min^{-1}, a linear calibration curve was obtained from 2.7 to 100 μg mL^{-1} ($r > 0.998$), with LOD 0.87 μg mL^{-1} and % RSD 1.5 (5 μg ml^{-1}, $n = 9$). After adopting the optimal conditions, the good resolution was achieved from commonly coexisting anions (chloride, metabisulphite, ascorbate and citrate), and other amino polycarboxylic acids (EGTA, NTA and DTPA). The potential interference of pharmaceutical substances (caffeine, phenytoin, nembutal, tolbutamide, dicumarol, acetylsulphisoxazole and paracetamol) and metal cations (Ca^{2+}, Cu^{2+} and Fe^{3+}) was also examined. The developed ion chromatographic method was successfully applied for the separation and quantification of EDTA different samples such as contact lens care solutions, synthetic injection drug solutions, canned mushrooms and mayonnaise, with good recoveries ranging from 74-108% [30].

A novel direct ion chromatographic method coupled with pulsed amperometric detection of cyanide in drinking waters [26]. In this method, the samples were treated with sodium hydroxide to stabilize cyanide and with a cation-exchange cartridge to remove transition metals. The cyanide was separated by anion-exchange chromatography and detected by pulsed amperometric detection with a waveform optimized for cyanide and used with a disposable silver working electrode. The accuracy and precision of the developed method were evaluated by using five spiked cyanide samples and found that the obtained recoveries were >80%. This method was highly sensitive and has a linear response from 2 to 100 μg L^{-1}, exhibited good recovery, and the detection was selective for cyanide relative to the common anions of drinking water. Dissolved transition metals can interfere with free cyanide determinations in drinking water, but can be removed by treating samples with a cation exchange cartridge prior to analysis [31].

Effective ionic liquid aqueous solution extraction method coupled to ion chromatography was developed for the extraction of three kinds of spices, namely vanillin, ethyl vanillin and ethyl maltol from food products such as biscuit, chocolate and milk powder. In this method, an environmental friendly extracting agent, 1-Octyl-3-methylimidazolium chloride aqueous solution was selected as the extracting medium. A 0.5 g powder of food product was extracted with 5.0 mL of 1-Octyl-3-methylimidazolium chloride aqueous solution (0.3 mol L^{-1}, pH 6.0) under ultrasonication at 50 °C, and then the extract was centrifuged for 10 min at 4000 rpm. The extract was filtered through a syringe filter and injected into ion chromatography system for analysis. The separation of three spices was carried out on an anion exchange column at 280 nm. Under the optimal conditions, good reproducibility of extraction performance was obtained, with the relative standard deviation (RSD) values ranging from 1.9 to 6.3%. The recoveries of spiked samples were found to be 79.8% and 95.8% with a LOD's of vanillin, ethyl vanillin and ethyl maltol were in the range of 20-45 mg kg^{-1}[32].

An indirect ion-pair chromatographic method linked to conductometric detection for the determination of L-carnitine in food supplement formulations was developed and validated by Aikaterini Kakou and co-workers [33]. This method was based on a non-polar (C18) column and an aqueous octane sulfonate (0.64 mM) eluent, acidified with trifluoroacetic acid (5.2 mM). Due to the C18 column, the retention time achieved from this study was found to be 5.4 min and the asymmetry factor of 0.65 which is one of the rapid method available in the literature. A linear calibration curve for L-carnitine ranges from 10 to 1000 μg mL^{-1} (r = 0.99998), with a LOD of 2.7 μg mL^{-1} (25 μL injection volume), a repeatability % RSD of 0.8 (40 μg ml^{-1}, n = 5) and reproducibility % RSD of 2.6 were achieved. The proposed method was applied for the determination of carnitine in oral solutions and capsules with fair recoveries ranging from 97.7 to 99.7% for spiked samples [33].

Sensitive and selective cation exchange chromatographic method, coupled with integrated pulsed amperometric detection was designed for the quantification of biogenic amines (tyramine, putrescine, cadaverine, histamine, agmatine, spermidine and spermine) in fresh and processed meat after extraction with perchloric acid [34]. The method was based on gradient elution of aqueous methanesulfonic acid with post-column addition of a strong base to obtain suitable conditions for amperometric detection with the gold electrode to increase the performance. All analytes were identified in real samples except phenethylamine which is eluted in a zone of the chromatogram rich of interfering peaks with an analysis time of about 68 min [34].

Alliin from garlic powder was extracted and separated by two-step process combining aqueous two-phase extraction (ATPE) with chromatography and the yield values were compared with response surface methodology by optimising the extraction process [35]. The optimal extraction conditions of 19% (w/w) $(NH_4)_2SO_4$, 20% (w/w) 1-propanol, at 30 °C, pH 2.35 with 8.54 % (w/w) NaCl was chosen based on the higher yield. The results obtained by this method was compared with the conventional extraction method in terms of yield values and found to be 20.4 mg g^{-1} versus the original yield of 15.0 mg g^{-1} which was three times higher than the conventional extraction method [35].

Efficient isolation of egg white components is desired due to its potential uses. Existing methods mainly targeted on one specific protein, but this method has made an attempt to co-extract all the valuable egg white components using continuous process by adopting two-step method using anion-exchange chromatography [36]. In this method, Ovomucin was first isolated and the resultant supernatant substance was used as the starting material for ion-exchange chromatography. By using this technique, 100 mM supernatant yielded three other fractions such as ovotransferrin, ovalbumin and flavoproteins by a flow-through fraction. In later stages, the flow-through fraction was also further separated into ovoinhibitor, lysozyme, ovotransferrin and an unidentified fraction which represents 4% of total egg white proteins. Chromatographic separation of 500 mM supernatant resulted in fractions representing lysozyme, ovotransferrin and ovalbumin. This co-extraction protocol represents a global recovery of 71.0 % proteins [36].

Conclusions

In recent years, Ion exchange and adsorber technology have been significantly improved not only with regard to process optimization during the sorption and desorption step but also with regard to the development of highly sophisticated resin materials allowing novel applications. In contrast to many other extraction methods, adsorption and ion exchange technology offers higher selectivity and improved cost efficiency and can be realized with relatively simple technical equipment. However, the prediction of sorption and ion exchange behaviour, when multi compound solute systems are considered, is highly complex and should not be underestimated upon optimization of such processes. Nevertheless, with a steady increase in knowledge of the underlying theory, processes will increasingly be tailored, thus yielding products with well-desired compositions and functional properties, which will certainly expand the application of these technologies to further areas in the future.

Acknowledgement

The authors gratefully acknowledge the financial assistance from Durban University of Technology, South Africa.

References

[1]   P. R. Haddad, P. E. Jackson, Journal of Chromatography Library-Volume 46 Ion Chromatography. Amsterdam: Elsevier Science (1990).

[2]   F. Harm, A. Rumpler, 5th International Congress on Pure and Applied Chem. 59 (1903).

[3]   R. Gans, German Pat No. 197,111 (1906).

[4]   P. De Brunn, German. Pat No. 270, 324 (1914).

[5]   P. N. Engel, German. Pat No. 583, 974 (1933).

[6]   H. Fuertig, F. Wolf, U. Headicke, M. Weber, H. Knoll, German (East), Pat. No. 55, 326, April (1967), Chem. Abstr., 67, 83511z, (1967).

[7]   K. A. Kraus, H. O. Phillips, T. A. Carlson, J. S. Johnson, International Conference on Peaceful use of Atomic Energy 2nd Conference, Geneva, 28, 3 (1958).

[8]   C. B. Amphlett, P. Eaton, L.A. McDonald, A. J. Miller, Synthetic inorganic ion-exchange materials-IV: Equilibrium studies with monovalent cations and zirconium phosphate, J. Inorg. Nucl. Chem., 26(2) (1964) 297-304. https://doi.org/10.1016/0022-1902(64)80073-7

[9]   M. W. Michael U.S. Pat No.2, 543,666, (1951).

[10]   R. F. Boyer, U.S. Pat No.2, 500, 149, (1950).

[11]   D. F. Bradley, A. Rich, The fractionation of ribonucleic acid on ECTEOLA-Cellulose anion exchangers, J. Am. Chem. Soc., 78 (1956) 5898-5902. https://doi.org/10.1021/ja01603a050

[12]   W.C. Bauman, G.B. Heusted, U.S Pat. No.2, 614, 099, (1952).

[13]   E. L. McMaster, Tolksmith, U.S. Pat. No.2, 764, 560,(1956).

[14]   R. Ramaswamy, PhD. Thesis University of Poona, (1970).

[15]   O. Liebknecht, U.S. Pat. Nos. 2,191,060, (1940), 2,206, 007, (1940).

[16]   O. Liebknecht, US. Pat. No.2, 378, 307, (1945).

[17]   Council of Scientific and Industrial Research, Indian Pat. No. 47,446, (1954).

[18]   D. Cozzi, P.G. Desideri, L. Lepri, G. Ciantelli, Alginic acid, a new thin layer material, J. Chromatgr. 35(3) (1968) 396-404. https://doi.org/10.1016/S0021-9673(01)82401-8

[19]   Information on http://www.science.uwaterloo. ca/~cchieh/cact applychem /alsilicate. accessed on 30/6/2016.

[20]   A. S. Behrman, H. Gustafson, Removal of Fluorine from Water A Development in the Use of Tricalcium Phosphate, Ind. Eng. Chem. 30 (1938) 1011-1013. https://doi.org/10.1021/ie50345a016

[21]   Q.-C. Chen, S.-F. Mou, X.-P. Hou, Z.-M. Ni, Simultaneous determination of caffeine, theobromine and theophylline in foods and pharmaceutical preparations by using ion chromatography, Anal. Chim. Acta 371 (1998) 287-296. https://doi.org/10.1016/S0003-2670(98)00301-8

[22]   H. Lee, V. Luiz de MeloSilva, Y. Liu, D. Barile, Quantification of carbohydrates in whey permeate products using high-performance anion-exchange chromatography with pulsed amperometric detection, J. Dairy Sci. 98 (2015) 7644–7649. https://doi.org/10.3168/jds.2015-9882

[23]   Q.-C. Chen, J. Wang, Simultaneous determination of artificial sweeteners, preservatives, caffeine, theobromine and theophylline in food and pharmaceutical preparations by ion chromatography, J. Chromatogr. A 937 (2001) 57-64. https://doi.org/10.1016/S0021-9673(01)01306-1

[24]   S.J. Gerberding, C. H. Byers, Preparative ion-exchange chromatography of proteins from dairy whey, J. Chromatogr. A 808 (1998) 141-151. https://doi.org/10.1016/S0021-9673(98)00103-4

[25]   M. Amigo-Benavent, V. I. Athanasopoulos, M. Dolores del Castillo, Ion exchange chromatographic conditions for obtaining individual subunits of soybean β-conglycinin, J. Chromatogr. B 878 (2010) 2453–2456. https://doi.org/10.1016/j.jchromb.2010.07.013

[26]   X.-B. Xu, D.-B. Liu, X. M. Guo, S.-J. Yu, P. Yu, Improvement of sugar analysis sensitivity using anion-exchange chromatography-electrospray ionisation mass spectrometry with sheath liquid interface, J. Chromatogr. A 1366 (2014) 65-72. https://doi.org/10.1016/j.chroma.2014.09.019

[27]   C. Gu'erin-Dubiard, M. Pasco, A. Hietanen, A. Quiros del Bosque, F. Naua, T. Croguennec, Hen egg white fractionation by ion-exchange chromatography, J. Chromatogr. A 1090 (2005) 58-67. https://doi.org/10.1016/j.chroma.2005.06.083

[28]   D.L.A. Fernandes, C.M. Silva, A.M.R.B. Xavier, D.V. Evtuguin, Fractionation of sulphite spent liquor for biochemical processing using ion exchange resins, J. Biotechnol. 162 (2012) 415-421. https://doi.org/10.1016/j.jbiotec.2012.03.013

[29] E. Leeb, A. Holder, T. Letzel, S. C. Cheison,U. Kulozik, J. Hinrichs, Fractionation of dairy based functional peptides using ion-exchange membrane adsorption chromatography and cross-flow electro membrane filtration, Int. Dairy J. 38 (2014) 116-123. https://doi.org/10.1016/j.idairyj.2013.12.006

[30] A.A. Krokidis, N.C. Megoulas, M.A. Koupparis, EDTA determination in pharmaceutical formulations and canned foods based on ion chromatography with suppressed conductimetric detection, Anal. Chim. Acta 535 (2005) 57-63. https://doi.org/10.1016/j.aca.2004.12.011

[31] T. T. Christison, J. S. Rohrer, Direct determination of free cyanide in drinking water by ion chromatography with pulsed amperometric detection, J. Chromatogr. A 1155 (2007) 31-39. https://doi.org/10.1016/j.chroma.2007.02.083

[32] H.-B. Zhu, Y.-C. Fan, Y.-L. Qian, H.-F. Tang, Z. Ruan, D.-H. Liu, H. Wang, Determination of spices in food samples by ionic liquid aqueous solution extraction and ion chromatography, Chinese Chem. Lett. 25 (2014) 465-468. https://doi.org/10.1016/j.cclet.2013.12.001

[33] A. Kakou, N. C. Megoulas, M. A. Koupparis, Determination of l-carnitine in food supplement formulations using ion-pair chromatography with indirect conductimetric detection, J. Chromatogr. A 1069 (2005) 209-215. https://doi.org/10.1016/j.chroma.2005.02.021

[34] G. Favaro, P. Pastore, G. Saccani, S. Cavalli, Determination of biogenic amines in fresh and processed meatby ion chromatography and integrated pulsed amperometric detection on Au electrode, Food Chem. 105 (2007) 1652-1658. https://doi.org/10.1016/j.foodchem.2007.04.071

[35] X.-M. Jiang, Y.-M. Lu, C.-P. Tan, Y.-L., B. Cui, Combination of aqueous two-phase extraction and cation-exchange chromatography: New strategies for separation and purification of alliin from garlic powder, J. Chromatogr. B 957 (2014) 60-67. https://doi.org/10.1016/j.jchromb.2014.02.037

[36] D. A. Omana, J. Wang, J. Wu, Co-extraction of egg white proteins using ion-exchange chromatography from ovomucin-removed egg whites, J. Chromatogr. B 878 (2010) 1771-1776. https://doi.org/10.1016/j.jchromb.2010.04.037

Keywords

About the Editors

Assistant Professor Inamuddin

Dr. Inamuddin is currently working as Assistant Professor in the Department of Applied Chemistry, Aligarh Muslim University (AMU), Aligarh, India. He obtained his Master of Science degree in Organic Chemistry from Chaudhary Charan Singh (CCS) University, Meerut, India, in 2002. He received his Master of Philosophy and Doctor of Philosophy degrees in Applied Chemistry from AMU in 2004 and 2007, respectively. He has extensive research experience in multidisciplinary fields of Analytical Chemistry, Materials Chemistry, and Electrochemistry and, more specifically, Renewable Energy and Environment. He has worked under different research projects as project fellow and senior research fellow funded by University Grants Commission (UGC), Government of India, and Council of Scientific and Industrial Research (CSIR), Government of India. He has received Fast Track Young Scientist Award from the Department of Science and Technology, India, to work in the area of bending actuators and artificial muscles. He is running one major research projects funded by Council of Science and Technology, Lucknow (Uttar Pradesh), India. He has completed three major research projects sanctioned by University Grant Commission, Department of Science and Technology, and Council of Scientific and Industrial Research, India. He has published 72 research articles in international journals of repute and eight book chapters in knowledge-based book editions published by renowned international publishers. He has published five edited books with Springer, United Kingdom, three by Nova Science Publishers, Inc. U.S.A., one by CRC Press Taylor & Francis Asia Pacific and two by Trans Tech Publications Ltd., Switzerland. He is the member of various editorial boards of journals. He has attended as well as chaired sessions in various international and nation conferences. He has worked as a Postdoctoral Fellow, leading a research team at the Creative Research Initiative Center for Bio-Artificial Muscle, Hanyang University, South Korea, in the field of renewable energy, especially biofuel cells. He has also worked as a Postdoctoral Fellow at the Center of Research Excellence in Renewable Energy, King Fahd University of Petroleum and Minerals, Saudi Arabia, in the field of polymer electrolyte membrane fuel cells and computational fluid dynamics of polymer electrolyte membrane fuel cells. He is a life member of the Journal of the Indian Chemical Society. His research interest includes ion exchange materials, sensor for heavy metal ions, biofuel cells, supercapacitors and bending actuators.

Professor Amir Al-Ahmed

Dr. Amir-Al-Ahmed is working as a Research Scientist-II (Associate Professor) in the Center of Research Excellence in Renewable Energy, at King Fahd University of Petroleum & Minerals (KFUPM), Saudi Arabia. He graduated in chemistry from Aligarh Muslim University (AMU), India. He obtained his M.Phil (2001) and Ph.D. (2003) degree in Applied Chemistry on conducting polymer based composites and its applications, from the Zakir Hussain College of Engineering and Technology, AMU, India. During his postdoctoral research activity, he worked on different multi-disciplinary project in South Africa and Saudi Arabia such as biological and chemical sensor, energy storage and conversion and also on CO_2 reduction. Throughout his academic career, he has gained extensive experience in materials chemistry and electrocatalysis for applications in energy conversion, storage, sensors and membranes. Currently, he is involved in several multidisciplinary research projects, funded by Saudi national research programs (thin film solar cells) and Saudi Aramco. Dr. Amir has on his credit a good number of articles published in international scientific journals, conferences and as book chapters. He has edited six books (publisher: Trans Tech Publication, Switzerland) and he is in the process of editing and writing another two books with Springer and Elsevier. He is also the Editor-in-Chief of an international journal "Nano Hybrids and Composites" along with Professor Y.H. Kim.